CANAL WALKS

OF ENGLAND AND WALES

RAY QUINLAN

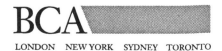

BCA

LONDON NEW YORK SYDNEY TORONTO

First published in three volumes as *Canal Walks: North*, 1993; *Canal Walks: Midlands*, 1992; *Canal Walks: South*, 1992; by Alan Sutton Publishing Limited.

This edition published 1994
by BCA by arrangement with
ALAN SUTTON PUBLISHING LIMITED
Phoenix Mill · Far Thrupp · Stroud · Gloucestershire

CN 5830

Jacket photograph: Clarence Mill and the Bellington Aqueduct on the Macclesfield Canal

Typeset in 10/12 Plantin.
Typesetting and origination by
Alan Sutton Publishing Limited.
Printed in Great Britain by
The Bath Press, Bath, Avon.

CANAL
WALKS

NORTH

> *To Hattie who else?*

ACKNOWLEDGEMENTS

This book would have been impossible without the splendid resources of various libraries: communal ownership in practice. Despite chronic under-funding, the information and help received was substantial.

Help and advice came from many of the local societies and trusts viz.: Dave Edmunds and Pat Jones of the Ripon Motor Boat Club, Richard Willis of the Ripon Canal Society, the Sankey Canal Restoration Society, the Lancaster Canal Trust and the Macclesfield Canal Society. Thanks also to the various employees of British Waterways who, as ever, have been both helpful and co-operative.

Assistance with archive photographs came from Lynn Doylerush of the Boat Museum and Sue Furness of the College of Ripon and York St John. Assistance with, though no responsibility for, the author's pictures came from Mr Ilford and Mr Fuji and two, old and steadily disintegrating, Olympuses fitted with 28 mm and 75–150 mm lenses.

For rest and recuperation during walking trips, thanks go to David and Dorothy in Blackburn and Nigel and Rachel in Mirfield. As before, much appreciation to Taffy for most of the transportation. And, of course, thanks to Mary who continued to provide her wonderful support throughout.

KEY TO MAPS

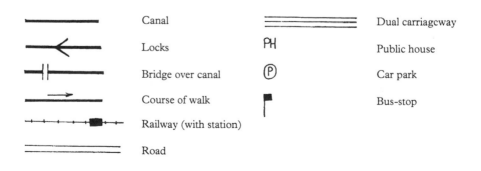

Canal		Dual carriageway	
Locks	PH	Public house	
Bridge over canal	Ⓟ	Car park	
Course of walk		Bus-stop	
Railway (with station)			
Road			

CONTENTS

LOCATION OF WALKS

1. Aire & Calder Navigation at Knottingley
2. Bridgewater Canal at Worsley
3. Chesterfield Canal at Worksop
4. Fossdyke Navigation at Lincoln
5. Huddersfield Canals at Huddersfield
6. Lancaster Canal at Glasson Dock
7. Leeds & Liverpool Canal at Skipton
8. Macclesfield Canal at Macclesfield
9. Peak Forest Canal at Whaley Bridge
10. Ripon Canal at Ripon
11. Rochdale Canal at Hebden Bridge
12. Sankey Brook Navigation at Widnes

INTRODUCTION

In the early planning of this series of three books, I had the feeling that the canals of England were a largely undiscovered resource. I was wrong. It was just that they were a largely unrecognized one.

It's been a year and a half of gestation and nearly 600 miles along forty-six different canals and navigations. Virtually everywhere, I've been amazed at the numbers of people who enjoy our waterways without ever getting into a boat or ever plonking themselves down on the bank armed to the teeth with hook, line and simmering cauldron of maggots. Admittedly there have been times when I've felt completely alone. There are some almost eerie stretches of the Kennet & Avon through Wiltshire, some undeniably deserted sections of the Grantham, and I've never felt as lonely as on the Birmingham Canal in Smethwick. But in other places the canals have become positive tourist attractions. Anyone who has been to the Foxton Locks or to Stoke Bruerne will testify to this. These are the places where day-trippers simply go for a gentle amble along a cultured towpath or for a philosophical stare into the whirling eddies of escaping lock water. Then there are the hiker routes over rougher ground. Here you'll find folk like me with our rucksacks and walking boots. We, identifiable as people who tuck our trousers into our socks, head off into the distance like well-orientated bees even if the driving rain has already seeped into our misnamed water-proofs. The prime sites for us are lines like the three cross-Pennine routes described in this volume.

But those canals are the obvious attractions; the ones where I would have expected to see people. The surprise came in the popularity of the less spectacular and the less well known. These are the waterways which hold no significant attraction, offer no particular 'sight to see', do not present the long-distance challenge but which are still popular for local folk walking the local dog. These are the places for a gentle jog, a brief saunter or an afternoon picnic. You'll see plenty of families out for a Sunday stroll along the Leeds & Liverpool or the Thames & Severn. Seemingly miles from any-where on the Bridgwater & Taunton you'll find an old lady walking the pet mongrel. Under the big sky on Romney Marsh, you'll come across a couple of kids bird watching along the Royal Military. On the towpath of the Sankey Brook Navigation, in the shadow of the massive Fiddlers Ferry power station, you'll pass a pair of mums wheeling a pair of pushchairs with a pair of toddlers admiring the view. I think I said hello to more people in half an hour along the Worcester & Birmingham than in a whole day in the centre of Brum itself. And even by the long-abandoned lengths of canal – such as the Chesterfield above Worksop – you'll find them. Just ambling. Just looking. Just enjoying.

The nation was there before me! The towpaths of the once-derelict and ignored canals are now as popular a place for a walk as anywhere in the country. And yet they are still not recognized as the great national resource that they are. Only a minor proportion of the towpaths of England are designated public rights of way. British Waterways, who let's face it do a pretty good job on our behalf, retain the right to close the non-designated paths at will. Fine now, when the political mood is in the public's favour, but who's to say where and when things may change.

The Inland Waterways Association and the Towpath Action Group have been campaigning for some years that, where possible, towpaths are declared public rights of way and that the genuine rights of access be maintained. We, the people who enjoy these wonderful legacies, should support them in their crusade and, in the mean time, simply keep making the most of it.

The Northern Canals

The northern canals are full of contrast. There's the quiet rural routes of the Chesterfield, the Fossdyke, the Lancaster, vast stretches of the Leeds & Liverpool, the Macclesfield and the Ripon. There's the cross-Pennine journeys up the old mill valleys with the Rochdale and the Huddersfield. There's the busy country around the Peak Forest. Then there's derelict or inner-city sections of canals like the Bridgewater and the Sankey Brook. On top of all that, there's the only 'canal' covered in this series that is still used for the purpose for which it was designed: the Aire & Calder Navigation.

And it's all good walking territory.

Of course, this is just a selection of the canals of the north and although by no means arbitrary, it is strictly a personal choice. Walkers should not ignore the Pocklington or the Market Weighton in Yorkshire. Just south of Manchester there's the Ashton. We never even get near the Sheffield & South Yorkshire or the Stainforth & Keadby. We only scratch the surface of the Calder & Hebble. And then what about the long-lost canals such as the Carlisle, the Barnsley and the Manchester, Bolton & Bury? All these are worthy of further investigation and, with the aid of an Ordnance Survey map and some keen legs, they can be traced both on paper and, often, on the ground.

As with all personal choices, some people's favourite lines or stretches of waterway may have been omitted. There is also the problem of what is a northern canal. The Trent & Mersey Canal certainly goes as far north as the Macclesfield even though the former was included in *Canal Walks: Midlands*. In all these matters, only the author is to blame. I have not tried to be a sage on these matters, merely a stimulus. Walking the canals of the north should be an adventure with plenty to see and to discover for yourselves. And it is quite likely that you will see even more than I did and enjoy them every bit as much.

Walking the Towpaths

The walks in this book are all straightforward and require no special feats of strength or navigation. Towpath walks have two great virtues: they are mostly on the flat and they have a ready-made, unmistakable course to follow. Getting lost should therefore, in theory at least, be relatively hard. The key problem with towpath walks is that if you want to spend most of the day by the canal, circular routes to and from a vehicle or a particular station or bus-stop become difficult. Many of the walks in this volume involve walking one way and returning by public transport. This means that you must check the availability of the bus or train before travelling. Telephone numbers are provided and your local library should have the main British Rail timetable.

Walkers should generally plan for 2 to 2¹/₂ miles an hour so that stops can be made for sightseeing or a break. Head-down speedsters should easily manage three miles an hour on a good track. You should, of course, add a little time for stoppages for refreshment and add a lot of time if you are accompanied by photographers or bird watchers.

No special equipment or provisions are needed to walk the towpaths of England. On a good day and on a good path any comfortable footwear and clothing will do, and you'll be able to leave the laden rucksack at home. However, for longer walks through more remote country you should be more prudent. Even in a drought, towpaths can be extremely muddy and, from experience, it can not only rain virtually any time but usually does. Boots and a raincoat of some sort are therefore advisable. Similarly, although pubs and small shops are often fairly common along the way, it may be useful to carry some kind of snack and drink with you.

This book includes sketch maps that show the route to be taken. However, the local Ordnance Survey map will always be useful and the appropriate map numbers and references are provided in each chapter. Again your local library may well have them for loan.

Finally, the dangers inherent in walking along a waterway are often not fully appreciated. Over the 1990 Christmas holiday, three children died after falling into a lock on the Kennet & Avon at Burghfield. A year later their mother committed suicide having been unable to endure her loss. Locks are deep, often have silt-laden bottoms, and are very difficult to get out of. Everybody, especially children, should be made aware of this. If somebody does fall in, you should not go into the water except as a last resort. You should LIE on the bank and use something like a coat for the person to grab so that you can then pull them out. Better still, keep children away from the edge.

Otherwise, please enjoy.

1
THE AIRE & CALDER NAVIGATION
Knottingley

Introduction

The A&CN is one of the nation's most successful commercial waterways and, unlike most of the network, is thriving and producing a profit even to this day. That novel fact alone should make it worth a visit.

The River Calder has its source in the Pennines north of Todmorden. It becomes the Aire & Calder Navigation at Wakefield from where the line passes via Stanley to Castleford. The source of the River Aire is in Malham Tarn. It becomes the Aire & Calder Navigation at Leeds from where it goes south-east via Woodlesford to Castleford. The combined waters now head east to Knottingley where the navigation splits. A northern branch follows the River Aire to Haddlesey and then the Selby Canal to Selby and the River Ouse. The southern branch, a cut called the Knottingley & Goole Canal, takes traffic to the Ouse at Goole.

For those who like their canals to be narrow, winding, overhung by trees and full of pretty holiday narrow boats, the A&CN will come as a bit of a shock. But it's worth walking just to see what a modern waterway is really like and it is certainly far from uninteresting.

History

From the Middle Ages, the woollen products of Leeds and Wakefield had been exported overland to York and Selby on the River Ouse or to Knottingley on the River Aire for subsequent shipment to Europe. The Aire had been navigable from Airmyn on the Ouse as far as Knottingley since medieval times but the towns to the west needed an improved transport network if they were to compete and grow. As a result, the seventeenth century saw several attempts to extend the waterway along both the Rivers Aire and Calder. A series of bills were presented to Parliament (in 1621, 1625 and 1679) and were defeated, mostly through the objection of the city of York who feared loss of trade.

By 1690 some 2,000 tons of goods p.a. were being moved from the towns of the West Riding by cart at a cost of almost £3 a ton. The merchants of Leeds and Wakefield were thus becoming increasingly convinced of the necessity of a navigable waterway and the scheme was revived. Apart from the export of cloth, the new navigation would facilitate the import of wool and allow greater exploitation of west Yorkshire coal. John Hadley was asked to survey the Aire, and Samuel Shelton to survey the Calder. York Corporation once again objected claiming that it would drain the Ouse. However, Trinity House, in an independent report, disagreed and the A&CN Act received its Royal Assent on 4 May 1699.

The ownership of the navigation was vested in trustees who acted on behalf of the subscribers. These then divided themselves into a Leeds Committee and a Wakefield Committee. John Hadley was engaged as engineer and work started. The plan was to use the natural course of the rivers where possible and to only build new cuts where absolutely necessary. One major exception to this was the Crier Cut near Woodlesford which bypassed a highly convoluted section of the Aire towards Leeds. The work appears to have progressed smoothly and the first boats to navigate from the Ouse reached Leeds on 20 November 1700. The line to Wakefield was completed in 1701. Over the next few years, a series of improvements were also made to the Aire below Knottingley. This included the construction of locks at Beal and Haddlesey in order to raise the water levels along the line. By 1704 the navigation was fully operational. It consisted of sixteen locks (each about 60 ft by 15 ft) and had cost approximately £26,700.

The navigation was worked by contracting the line to lessees who paid a fee to the shareholders, as well as for any outgoings and repairs. In 1704 the annual rent was £800; by 1729 it had risen to £2,600 p.a. at a time when the freight and toll receipts were averaging around £6,000 annually. This meant that the undertakers were already receiving dividends of between 6 and 12 per cent. The traffic at this time was primarily in woollen goods exported from Leeds, Wakefield, Halifax and Bradford. Wool and corn were imported, mainly from Lincolnshire and East Anglia. There was also a healthy trade in coal from Wakefield to the Humber and then along the Ouse to York. Boats took 3–4 days to travel upstream from Airmyn to Leeds or Wakefield and 2–3 days back.

By the 1750s receipts were more than £9,000 p.a., Rawcliffe and Airmyn were developing as navigation centres, and the company was gradually ridding itself of confrontations with the river's millers by simply buying them out. In 1758 Sir Henry Ibbotson and Peter Birt took over the lease which now stood at £6,000 p.a., equivalent to a dividend of 20 per cent. By this time there was a regular coasting service from Airmyn to London and trade continued to rise. About three-quarters of the traffic was in coal, limestone, lime and manure, although the import of wool and the export of cloth remained a significant trade. With the arrival of the canal age, a range of connecting lines was proposed and built. Navigation along the River Calder had been extended towards Halifax and was opened to Sowerby Bridge in 1774. Beyond this, a wholly new cut, the Rochdale Canal, promised a waterway to Manchester. A canal Act passed in May 1770 enabled a new line from the A&CN at Leeds to Liverpool. All three routes would use the A&CN to Hull.

By now the old A&CN was beginning to look rather dated. Many of the lock sills were very shallow and shoals were common during summer months. In an attempt to placate complaints, the undertakers called in John Smeaton to recommend improvements. The objective was to produce a line able to take craft carrying 30–45 tons all year round. His recommendations, published in January 1772, included dredging, rebuilding locks and making a number of new cuts. The suggested cuts included a bypass canal from above Haddlesey Lock on the south side of the river to Gowdall above Snaith with two locks. A second bypass was to run from Brotherton (just below Knottingley). Other new cuts were proposed at Leeds, Knostrop, Woodlesford and Methley. Smeaton also recommended that the undertakers obtain powers to remove obstructions and to build a towpath.

The users of the navigation weren't satisfied with this and demanded more improvements. Such was the demand that rival schemes arose which would have bypassed the A&CN altogether. The Went Canal was to run from Wakefield to the Dutch River at Went Mouth. The Leeds & Selby Canal was to run from Leeds to the Ouse at Selby. However, the A&CN revised its own plans to include new cuts from Ferrybridge to above Beal, a series of bypass cuts from Leeds to Castleford and, perhaps most significantly, a line from Haddlesey to Selby (the Selby Canal). In the event, the rival schemes were defeated by effective lobbying from the A&CN while its own plans were passed. The new cuts (excluding the Woodlesford which wasn't built) were engineered by William Jessop with the Pinkerton brothers as contractors. The line to Selby was opened on 24 April 1778. Apart from improving the route to Hull, it significantly shortened the distance from the Wakefield collieries to York.

The various improvements had cost approximately £60,000 but the prudence of the investment is shown by the fact that toll receipts rose from an average of £22,857 p.a. in 1779–81 to £43,376 in 1791–3. Coal was the single largest cargo although there was a vigorous trade in corn, stone, lime, cloth bales, wool and groceries. The Selby line was such a success that the

A&CN's offices and yards at Airmyn were closed and a new centre opened in Selby. By 1792 trade was so brisk that further improvements could be made. Selby and Castleford Locks were altered and there was a new cut at Cross Channel. The navigation was able to support the promotion of the Barnsley Canal (from near Wakefield to that town) and a line south from the Barnsley to the River Don at Swinton (the Dearne & Dove Canal). There was also a general upgrading of all locks so that boats of 18 ft beam could work the line.

By 1805 prospects must have looked very good with the Rochdale route to Manchester completed and the Huddersfield and Leeds & Liverpool lines due to open. By the time the latter was finished in 1816, the annual toll income had risen to over £70,000. The confidence that this instilled meant that in 1818 the undertakers could propose a wholly new line from the A&CN at Haddlesey to the Dutch River at Newbridge, just six miles from Goole. The new cut would bypass the much convoluted and shoaled lower Aire and thus improve the route to Hull and the River Trent. Rival promotions were in the wind and the A&CN was already feeling the competition from other inland waterways following the opening of the Grand Junction Canal to London. John Rennie, reporting in January 1819, altered the course of the new line to run from Knottingley to Newbridge and then parallel to the Dutch River. He estimated the cost at £137,700. The enabling Act, passed in 1820, also allowed the A&CN to reorganize so that it now had a Board of Directors (although it was still not incorporated).

Several revisions were subsequently made to the plan. The junction with the A&CN was moved upstream to Ferrybridge rather than Knottingley. The line was deepened to take vessels carrying 100 tons. There was also to be a new barge basin and ship dock at Goole so that transhipment would occur there rather than at Hull. Following the death of John Rennie in 1822, George Leather was appointed engineer and to oversee further improvements along both the Aire and the Calder including the widening of more locks to the 18 ft beam. The first commercial vessel to pass along the new Knottingley & Goole Canal did so on 20 July 1826. The new line was 18½ miles long and could be travelled by fly-boat in just three hours. The Goole ship lock was able to handle vessels of 300–400 tons and the dock to which it led measured 600 ft by 200 ft and was 17 ft deep. Alongside these was a barge dock and lock. Ship and barge docks were linked by a short cut as well as a tramway. With the new docks at Goole, the cost had spiralled to £361,484, most of which had to be borrowed. In 1827 Goole was officially designated as a port, and custom facilities were set up.

Meanwhile Leather recommended that the rest of the line be made to a depth of 7 ft to allow the passage of boats carrying 100 tons. Sir John Rennie (junior) recommended a new cut below Wakefield to short cut the meandering line to Castleford. There were also plans to upgrade the River

Aire into what, in effect, was a Ferrybridge to Leeds Canal. But the A&CN hesitated, saying that the Goole Canal had exhausted its funds. This hesitation allowed in further rival plans. One group, headed by Thomas Lee and Shepley Watson, ran for 12 miles along the north side of the Calder from Wakefield to Ferrybridge. Only after the A&CN had pledged to improve the Calder was this plan defeated in Parliament and only then by just three votes. The A&CN employed Thomas Telford to review the state of the Calder above Ferrybridge. He proposed a cut from a point above Kirkthorpe Lock to cross the river via an aqueduct at Stanley Ferry to run to Fairies Hill (south of Methley) where three new cuts would further shorten the route to Castleford. Although Lee and Watson reworked their scheme, it was again defeated in Parliament and the A&CN's own plan received Royal Assent in June 1828. These improvements were estimated to cost £313,570 and an expansion of the Goole Docks added a further £148,850.

The 1830s saw a period of innovation. Steam packets had been used for several years, but in 1831 the company started to employ steam tugs to pull fly-boats on the Leeds to Goole run. The packet-boats between Goole and Castleford could average 6–7 miles per hour. In 1835 there were even two paddle-steamers, each able to carry 100 people, plying the line to Leeds. By April 1835 the continuous line of 7 ft depth from Leeds to Castleford was opened. The Calder Cut didn't open until August 1839 but the new line at 7½ miles was 5 miles shorter than the river route. The cost of the works increased the A&CN's debt in 1835 to £341,100. Repaying it had enforced a reduction in the funds allocated to dividends from £70,000 p.a. in the mid-1820s to, a still extremely healthy, £54,000 throughout the 1830s.

The threat of railway competition was first raised in 1824 when the Leeds & Hull Rail Road Company was proposed. The threat didn't materialize, however, until 22 September 1834 when the Leeds & Selby Railway opened. By now the A&CN was a powerful outfit. Tolls were cut by 40 per cent on 1 October 1834 (goods traffic along the railway was due to begin on 15 December), and by September 1836 it was the railway company that was seeking agreement on keeping the tolls up. The A&CN refused as its income had been only slightly reduced during what could have been a difficult period. In fact, by 1838, toll revenue was higher than it was in 1834. Only when the Selby railway was extended to Hull on 1 July 1840 did the A&CN's revenue begin to fall. From £145,511 in 1840, income was down to £114,654 just five years later. With great rapidity, new railways were opened to provide lines from Leeds and Selby to York, as well as connections to London and the south. In October 1841 the A&CN met with the Calder & Hebble Navigation and the Rochdale Canal with a view to fighting the onslaught. A common pricing policy and a general reduction in tolls was agreed in principle but not in practice. The potential for a concerted effort thus collapsed. In 1845 the Wakefield, Pontefract & Goole Railway (later to merge with the Manchester & Leeds to become the Lancashire & Yorkshire) was authorized, and by 1 April 1848 the line was carrying coal and the A&CN was unable to compete for passenger traffic. While a lot of the business attracted to rail was new, the navigation was still forced to compete by lowering tolls with a consequently reduced income. The main trade along the line by this time was in corn from Lincolnshire being shipped to the West Riding, with coal as a back carriage. There was, however, some measure of agreement between the old navigation and the new railway. At Goole the A&CN spent over £77,000 on new railway docks for which the L&Y paid a rent. In 1855 attempts to produce an agreement with the L&Y and North-Eastern Railways failed and the A&CN found itself in an even fiercer price war. Reductions in tolls had to be made. Receipts which stood at just over £108,854 in 1846 were down to £66,115 in 1856. Agreement was reached during the latter half of 1856 and tolls were raised once again.

In 1855 the Rochdale Canal was leased by four railway companies who promptly raised toll rates as a way of moving traffic onto rail. With increasing competition on its own route, it wasn't too long before the A&CN became receptive to a similar proposal. In 1856 the L&Y and the North-Eastern proposed a 21 year lease in return for £45,000 p.a. as a dividend plus the 4 per cent interest on the outstanding debt of £420,000. The agreement would include an option to apply for an amalgamation Act. However, the shareholders were against this and by 1858 the scheme was dropped.

The A&CN committee now changed philosophy and became determined to keep control of both its own and its neighbouring lines. Feeling that it could better maintain traffic by controlling the Calder & Hebble, it agreed with that navigation to take a 21 year lease with a sum equivalent to an 8 per cent dividend. With this came control of the Barnsley Canal which the C&HN had in turn leased since 1854. The A&CN also continued developing its own line and works. Cargo-carrying tugs came into operation during the 1850s. Pollington Lock, the first up from Goole Docks, was extended by three times its length so that trains of six boats behind a single tug could be used; the 206 ft long by 22 ft wide lock was reopened in October 1860. The locks further upstream – Whitley, Bulhome, Ferrybridge and Castleford – were all similarly treated to enable the passage of tugged trains all the way from Goole to Castleford. The programme of lock extensions was continued to Wakefield (by 1869) and Leeds (by 1873). At the same time, the navigation's engineer, W.H. Bartholomew, introduced compartment boats coupled together into trains towed by a steam tug. These vessels, known as 'Tom Puddings', were to become the workhorses of the A&CN with as many as thirty or forty 'tubs' included in a single train. Although not immediately successful, the use of Tom Puddings gradually restored the A&CN's coal traffic so that by 1897 the Goole Canal carried some 473,061 tons

A loaded train of Tom Puddings makes its way along the Aire & Calder Navigation, probably in the 1950s

National Coal Board

compared with 313,449 tons in 1845. There was also a notable increase in imported timber although the grain-carrying business declined – mostly due to the greater levels of imports from the Prairies into Liverpool.

The end of the century saw the A&CN still developing its line and its port at Goole. New docks were built and in 1884 the A&CN became the conservators of a stretch of the Lower Ouse. They were thus able to improve the stretch to the Trent thereby allowing larger ships to reach Goole. The cost of the work was enabled by the 1884 Act with the creation of £950,000 of debenture stock. By 1902 all the locks were built to a standard 215 ft by 22 ft by 9 ft deep. This enabled Tom Puddings to work the locks in trains of nineteen boats in one movement. The New Junction Canal, which had been suggested by the Stainforth & Keadby Canal Company in 1833, was opened in 1905. The canal, which was built as a joint venture with the Sheffield & South Yorkshire Navigation, runs for 5½ miles from the Goole Canal near Newbridge to Kirk Bramwith on the S&SYN. Although it cost £300,000 to build, trade along the line was slow to develop.

The late nineteenth and early twentieth centuries were otherwise a good period for the A&CN. From an allocation to dividends of just £40,500 in 1880, the figures rose to £60,000 in 1891 and £83,700 in 1909. By 1910 the Tom Pudding compartment boats were carrying 1,297,226 tons of coal p.a.; more than double the figure carried ten years earlier. This rose to 1,563,789 tons out of a total of 2,750,000 tons of coal in total. There were also increased shipments of timber and corn. In 1913 the line moved a total of 3,597,921 tons of cargo.

The First World War caused a slight hiccup in the fortunes of the A&CN; toll revenue of £119,415 in 1913 falling to £33,567 in 1919 before recovering to £71,146 by 1925. But even in the 1930s the committee had sufficient confidence in the future to undertake some further improvements to the Lower Ouse and to construct a new lock (Ocean Lock) from the line into the Ouse at Goole. The new lock could handle 4,000 ton ships. By the Second World War revenue had risen to £193,000 with more than 1³/₄ million tons of coal (increasingly being delivered to waterside power stations) and ³/₄ million tons of other cargo being moved annually.

The next significant phase of the A&CN's history started on 1 January 1948 when it was nationalized and put under the control of the British Transport Commission. The A&CN was probably the jewel in the new

Scene at the coal terminus at Newlands Basin, Stanley Ferry, probably in the early 1900s
The Boat Museum Archive

BTC's crown. Its future as a navigation was never a matter of debate and, unlike most waterways, trade along the line was actively developed. Shipments in 1953 topped 2½ million tons and in 1962 exceeded 3 million tons with revenue at £357,725. At that stage coal still remained the most important trade comprising nearly 71 per cent of the cargo moved. Such activity allowed British Waterways to continue upgrading the facilities along the line. In the late 1960s some minor amendments to the course and capacity of the navigation were made. The line to Leeds, now 10 ft deep, could take 500 ton boats measuring 180 ft by 18 ft 6 in. The line to Castleford was again upgraded to handle craft to 700 tons, and since 1978 the entire line to Leeds has been able to take boats of this size. All the locks on the main line have been electrified. Push tugs were introduced in 1967 and are used to deliver coal to the massive Ferrybridge power station. The tugs push compartment boats each 56 ft long by 17 ft 3 in wide which are able to carry 165 tons. They are pushed three at a time between colliery loading staithes and the power station where they are lifted out of the water and upturned to discharge the load (a process that takes less than nine minutes). Tonnage through the 1970s and '80s still exceeded 2 million tons although toll revenue in 1971 was down to £146,392. The Ferrybridge power station complex alone was receiving 1½ million tons of coal a year in the late 1970s. Even though the last Tom Puddings plied the line in March 1992 (from collieries at Kippax Lock near Mickletown to Ferrybridge B), Ferrybridge C still receives coal via the navigation. Cawood-Hargreaves moves coal using high-sided 'pan' barges that are towed by tugs.

The A&CN is one of the nation's most successful commercial waterways and it is still thriving to this day. Who knows, in an increasingly environmentally conscious world, the navigation may be well positioned as a viable alternative to road transport and all true sceptics can say 'I told you so'.

The Walk

Start and finish:	Knottingley railway station (OS ref: SE 491236)
Distance:	4¼ miles/7 km
Map:	OS Landranger 105 (York)
Car park:	Knottingley railway station
Public transport:	Knottingley is on the Leeds to Goole line (enquiries: (0532) 457676)

From the railway station, walk out to the main road and turn right. Go straight on at the traffic lights and walk on past a row of shops and the Bay

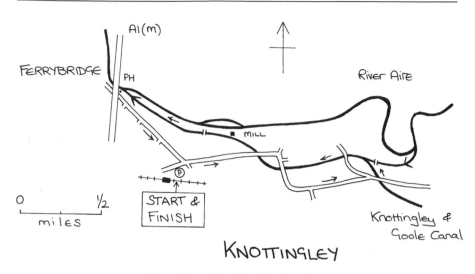

KNOTTINGLEY

The Aire & Calder Navigation

Horse pub. After passing the Working Men's Club, the road crosses the navigation for the first time. Continue around a right bend into Weeland Road. The road crosses the navigation for a second time and bends left to pass Rockware Glass works. At a road junction continue straight on along a road signposted to Goole. This passes the Lamb Inn and John Harker 'shipbuilders and repairers'. After crossing the navigation for the third time, turn left along a small lane signposted to Willow Garth Nature Reserve.

The waterway to your left is the Aire & Calder Navigation or, if you prefer, the Knottingley and Goole Canal. This is the line which took the waterway to the then new port of Goole and thus bypassed the former route along the River Aire to Airmyn or along both the Aire and the Selby Canal to Selby. The new line opened in July 1826 with much celebration. A procession of four steam packets and fifty sailing vessels made their way from Ferrybridge to Goole. They were accompanied by bands playing music and crowds waving flags. Starting at 10 a.m., the cavalcade reached Goole at 4 p.m. to be welcomed by a twenty-one gun salute and, for the VIPs at least, a slap-up meal at the Banks Arms Hotel.

Towpathers who venture along the A&CN today will feast in a slightly different way and that is in the sight of working vessels. Nowhere else in any of the three *Canal Walks* books in this series (which cover forty-six different canals) will you be able to see a line being used for the purpose for which it was built. There are broadly two types of vessel and you are almost certain to meet one or other and most likely both. The first group are the tankers. One line is run by Whitfleet Limited (part of the Whitaker Group) and includes vessels which sail under the names of *Humber Energy* (which passed me as I made my way down the lane), *Humber Navigator, Humber Jubilee* and *Humber Progress*. These boats ship loads of petroleum products from the Humberside refineries to various storage depots along the A&CN. The tankers can move up to 650 tons per load. The second type of tanker is the effluent tanker such as the Dean & Dyball Shipping *Trentaire*. *Trentaire* is presently on contract to Yorkshire Water to ship effluent sludge (some 450 tons at a time) from Knostrop (near Leeds) to Goole where the contents are transhipped for subsequent dumping at sea. *Trentaire* has a sister ship, *Trentcal*. Both were built in Le Havre in 1957 and were originally designed to carry wine. This may be considered to be a bit of a come-down but, at just over 180 ft long by 18 ft wide, they are the largest vessels that regularly ply the A&CN.

The lane leads to a small bridge, Trundles Bridge, over a junction canal to the River Aire. Turn right here briefly to follow the line round to see Bank Dole Lock, the first on this, the old, route to Selby. This short link between the Knottingley & Goole Canal and the River Aire was built at the same time as the K&GC to allow traffic to pass on to Selby and thence to York. Just beyond the lock, the junction between the cut and the river can be seen. Return to the bridge and walk on (following a signpost for the 'Knottingley Canal Walk') along a grassy area with Harker's boatyard on the far bank. Harker's, established in 1929, once built and maintained barges and coasters here although when I passed, the site appeared quiet and declining. At one stage Harker's operated their own large fleet of bulk fuel tankers.

The towpath goes under Shepherd's Bridge and then on under Caslane Bridge. Just after this, on the opposite bank, is the Steam Packet pub followed by the site of Rockware Glass, manufacturers of a wide variety of bottles and jars. At one time the factory had its supplies of silica sand delivered by barge; some of this being specially shipped in from Belgium.

We continue under Jackson's Bridge before the canal swings right to go under Gagg's Bridge and on past Kings Mill. Today the mill hides behind high fences and trade is continued with the aid of numerous, noisy lorries. In former times the mill received its grain supplies from the navigation. And before that it used the River Aire, which is just beyond the mill buildings to the right, as a source of power for its water-wheels.

After going under another road bridge, Mill Bridge, we walk along an increasingly narrow stretch between the A&CN to the left and the River Aire to the right. The land around here has been treated to an environmental improvement scheme that is being run by the Wakefield Groundwork Trust. Despite the proximity of industry and the ever-present rumble from the nearby A1(M), it's a pleasant airy spot with the Ferrybridge power stations acting as a clear book-end on the horizon. A weir on the Aire brings the level of the river up to that of the canal and at one point we cross sluice-gates between the canal and the river. When

The 460 ft long
Ferrybridge Flood Lock at
Kellingley

offered the choice, take the left-hand path which will bring you round to
Ferrybridge Flood Lock. It is an enormous structure at 460 ft long. About
two-thirds of the way along is a lock-keeper's lookout box. Cross the canal
via the footbridge at the far end of the lock from where there are splendid
views back downstream along the length of the lock and upstream towards
the A1(M) flyover and the power stations. If you're lucky, you may well be
able to watch the second type of commonly seen commercial boats as they
manoeuvre through the lock: the coal boats that deliver fuel to the
Ferrybridge power stations.

There are nine Cawood-Hargreaves coal barge trains altogether and they
are numbered CH101–109. Each train can carry 500 tons of coal from the
Kellingley deep mine to the Ferrybridge C station in three containers. When
empty, the vessels float high on the water but when full they wallow along,
only barely above the surface. On arrival at Ferrybridge, each compartment
is lifted out of the water and poured into a large hopper from where the coal
is moved along a conveyor to the station. Ferrybridge C, I'm assured, can

A Cawood-Hargreaves coal barge train passes through Kellingley on its way to the Ferrybridge power stations that are in the distance

burn 1,000 tons of coal an hour. About 40 per cent of this is supplied by these boats which are the natural successors to William Bartholomew's Tom Puddings which went out of service as recently as March 1992.

On the far side of the footbridge, turn right to walk along the front to the Golden Lion pub. Here the River Aire and the Knottingley and Goole Canal form a junction. Before 1826 traffic took the Aire route to the River Ouse. To your left, as you look at the navigation, is a fine view as far as the eighteenth-century road bridge which is now limited to pedestrian traffic and probably grateful for it.

To return to the railway station, turn left and walk up the hill to the traffic lights. Turn right to reach the station.

Further Explorations

The A&CN is not a haven for walkers. A lot of the more easterly stretches pass through rather bland, flat landscape while much of the more interesting westerly end is sadly not accessible. Indeed, I was told by British Waterways that it was dangerous! Hopefully one day it will be possible to walk from Wakefield to Knottingley but at present this is not feasible.

Despite this, the stretch of canal at Stanley Ferry is worth a visit and a mooch. Stanley Ferry can be found on OS Landranger 104 (Leeds & Bradford) at ref: SE 355231. There is a car park at the Ferry Boat Inn. From there, steps lead up to the pub and to the towpath. Just across the waterway are a line of small terrace houses which were built by the A&CN company in the 1880s to house the banksmen and their families.

Originally the A&CN between Castleford and Wakefield used the River Calder and it was opened for traffic in 1701. There had been a crossing over the river since Roman times and a ferry operational since the seventeenth century at the latest. Stanley Ferry had thus grown into a small community which included a hostelry for passing traders. By the mid-1820s traffic along the A&CN was burgeoning and it was decided to upgrade this section of the waterway. Thomas Telford was engaged to improve the route and he did so by producing a characteristically straight course which reduced the distance between Castleford and Wakefield by 4 miles. As part of those improvements Telford took his new line over the meandering Calder at Stanley Ferry.

Before taking a closer look at the crossing, it is well worth having a stroll along Telford's oh-so-straight waterway. To do so, turn right and walk along the line to the footbridge, which in one book on my shelf is called Ramsden's Swing Bridge. Once across, turn right and walk on to pass two bridges to Broadreach Lock. The stretch of the canal here has been extensively reclaimed and was the site of the 1992 National Waterways Festival. It was formerly the site of various buildings and yards associated with Parkhills colliery. Navigation-side loading staithes were used to charge waiting barges with coal for shipment to Hull, Grimsby, Goole, and even London. The staithes were still operational in 1975 when coal was shipped out by the Leeds Co-op. The colliery closed in December 1982. Just beyond the lock, the canal and the River Calder rejoin en route for Wakefield which makes itself evident to the right.

Once at the end there's nothing for it but to return along the towpath. Don't cross the footbridge initially but continue past the cottages and on to see where the A&CN crosses the Calder. There are actually two aqueducts here. The original, built as part of Telford's scheme, was designed by George Leather. It has two bow-shaped, cast-iron arcs that support the trough on either side while the ends are supported on stone abutments. The rather fine structure carries a trough that is 165 ft long, 24 ft wide, 8 ft 6 in deep and which holds 940 tons of water. Work on the aqueduct started in November 1834 with iron cast by William Graham & Co. of Milton Ironworks, Elsecar, on foundations constructed by another contractor, Hugh MacIntosh. The aqueduct was opened on 9 August 1839 at a cost of nearly £50,000. It is now scheduled as an ancient monument.

By the start of the 1970s the old aqueduct was suffering, both as a result of subsidence and due to being repeatedly rammed by passing vessels. As part of the upgrading of the Wakefield section to a 700 ton standard, a new aqueduct was opened in November 1981. Although decidedly more utilitarian than its neighbour, the new structure has a deeper draught and allows larger vessels to make their way upstream. It was designed by Husband & Co. and is 237 ft long, 29^1/$_2$ ft wide and a little over 14 ft deep.

As you return back to the footbridge, you will note the basin on the pub side of the navigation. This is Stanley Ferry Basin (or the Lofthouse Basin). It was originally built in 1840 and was served by a narrow gauge railway which brought coal from Lofthouse down to the canal. The railway was closed in 1924 when lorries started to do the work, and traffic as a whole came to a stop in the 1940s. Since then the basin has been used for various purposes including an oil distribution depot and a wheat wharf.

Stanley Ferry is also the site of a British Waterways yard. This too has a history that dates back to the earliest days of the line. In 1833, when the Calder Cut was being built, the land purchase included the area between the waterway and the access road to the ferry. The land was leased for a while to a firm of boat builders who constructed a dry-dock and various buildings. In 1873 the operation was taken over by the A&CN as its principal repair yard and, with certain additions, it remains much the same today.

Recross the canal at the footbridge to visit the pub and to return to the car.

Further Information

While there is no A&CN society as such, the Inland Waterways Association has its West Riding branch based in Leeds. The address of the head office in London is given in Appendix B.

For historical information on both the navigation and the vessels that use it, the standard work is:

Smith, P.L., *The Aire & Calder Navigation.* Wakefield Historical Publications, 1987.

2
THE BRIDGEWATER CANAL
Worsley to Barton

Introduction

The Bridgewater Canal has the accolade of being the first in England to be built under the powers of a canal Act. Perhaps more importantly, it was THE model for those that came after. Once the success of the Bridgewater was evident then the rest were almost bound to follow. And it was a success. It was talk of the profits being earned on the Bridgewater that drove otherwise well-adjusted minds into canal mania.

The original line of the Bridgewater runs from Worsley in north-west Manchester through Eccles (where it crosses the Manchester Ship Canal on the Barton Aqueduct) via Trafford Park into the centre of the city at Castlefield. Later the canal was extended west from Worsley to Leigh where it now forms a junction with the Leeds & Liverpool Canal. At Trafford Park, a second, more southerly, line leaves a junction with the original to head through Sale to Altrincham and Lymm. It then passes through the outskirts of Warrington before meeting the Trent & Mersey Canal at Preston Brook. The route then turns abruptly west to its modern-day terminus in central Runcorn. In a more glorious age, two lines of locks ran down the hill to the docks and the River Mersey.

The town canals aren't the natural habitat for the dedicated walker but yard for yard they are usually the more interesting ones. The Bridgewater is no exception. Besides, going to Worsley is almost a pilgrimage.

History

In 1736, with the Mersey & Irwell Navigation nearly complete, the M&IN company started to consider the possibility of building a waterway to the coalfields around Wigan. The original plan had been that the Douglas Navigation would serve the area but the construction work had been slow and the M&IN saw coal deliveries to Manchester as a potentially important trade. The M&IN engineer, Thomas Steers, was asked to carry out a survey of Worsley Brook with a view to making it navigable to Worsley with a new cut to Booth's Bank (towards Wigan). An Act was passed in 1737 but with construction of the Douglas Navigation once again under way, the project went no further.

By the time the third Duke of Bridgewater succeeded to the title in 1748, exploitation of the Wigan coalfield was increasing and many new turnpike roads were being built. The idea of a canal from the coalfield into Manchester was raised by a group of businessmen from the city in 1753. A line was proposed to run from Wigan via Leigh and Worsley to Eccles and Salford. William Taylor undertook a survey and a bill was put to Parliament. Although lost, the idea stayed with the young duke's agent, John Gilbert, and his guardian, Earl Gower. Both had industrial interests in Worsley and the earl had already employed James Brindley to survey a line that would become the Trent & Mersey Canal. The duke must have been impressed by the idea of a canal. In 1758 he visited his Worsley estates in order to assess the feasibility of a waterway to carry his coal directly into central Manchester, and in 1759 he purchased a tract of land in Salford with the view of making it the southern terminus of the planned canal.

The duke's scheme was for a level canal from Worsley to Salford with, if possible, a connection with the Irwell. When preparing the bill, it is widely recognized that the duke had learnt from the failure of the 1753–4 Bill and addressed many of the issues that were raised in opposition at that time. He promised that canal-delivered coal would be cheaper than that brought by road and promised to detail the price (not more than 4d. a hundredweight) before the bill was submitted. Lime and manure were to be carried free. The water supply would come entirely from Worsley Brook (there had previously been objections to extraction from the Irwell). The cost of the entire enterprise would be met from his own pocket. Interestingly, the duke's bill not only included details of a line from Salford to Worsley but also a second canal from Worsley to Hollins Ferry on the M&IN.

The Act, with no opposition, was passed on 23 March 1759 and the construction work began almost immediately. James Brindley was appointed engineer with John Gilbert overseeing the operation on behalf of the duke. By the beginning of 1760, there were already two miles of waterway from Worsley to Patricroft and two miles of the line towards Hollins Ferry which reached to Botany Bay Wood just north of Irlam.

Work had also begun within the colliery at Worsley where an underground canal system was to take the specially constructed boats, later to be known as 'starvationers', virtually to the coal face. The boats were typically

Starvationers at Delph Mine, Worsley, *c.* 1910

Ware/The Boat Museum Archive

47 ft long by 4½ ft wide and had prominent 'ribs'. When the system was fully operational, a train of five boats was pulled by a horse or a pair of mules into Manchester. The underground tunnels were 10 ft wide with a headroom of about 8 ft. The tunnels were built with loading bays and passing spaces and there were two other canals at different depths. These didn't connect directly with the main line but were able to convey coal to it via vertical shafts through which coal was delivered to waiting starvationers. The boats were propelled from the mine by opening a sluice and washing them out. A third underground canal, from Walkden, was connected to the main line via a 151 yd long underground inclined plane. Eventually there were to be more than 30 miles of underground tunnel at Worsley.

By this time, there had been a modification to the original line into Manchester. The new route was to pass from Patricroft over the Irwell via an aqueduct, over Trafford Moss and then on to a terminus which was eventually settled at Castlefield. The Hollins Ferry branch was dropped and never completed. This change of plan was enabled by an Act in March 1760. The aqueduct over the Irwell was opened on 17 July 1761 and boats, much to the apparent disbelief of the onlookers, sailed over the river with 'not a single drop of water' oozing through the masonry to the river below. By December 1761, the line had reached Stretford and, on 1 August 1765, the first wharf at Castlefield was operational.

While the main line was being completed, the duke was hatching his scheme for an extension of the Bridgewater Canal that would compete directly with the M&IN to Liverpool. Brindley had already surveyed the route by January 1762 when the duke published a pamphlet on the scheme in which he described the M&IN as 'imperfect, expensive and precarious'. The duke's line, 9 miles shorter than the M&IN route passed west from Longford Bridge, over the River Mersey and Sale Moor to Altrincham. The line then crossed the River Bollin to Lymm. The last stretch continued west-wards, descending by a flight of ten locks to the Mersey at Hempstones, a point on the Mersey to the east of Runcorn. The new line was to be sup-plied from the same Worsley springs that kept the main line in water. The scheme was opposed vigorously, and unsuccessfully, by the M&IN. The western end of the line was altered by an Act of 1766 so that it joined the Trent & Mersey Canal at Preston Brook before making its way to the Mersey. With this, the locks were moved to Runcorn.

On 21 March 1776 the line to Runcorn was open for traffic and the duke had a canal that connected Manchester with the Midlands via the T&M as well as a route to the profitable coal market of Liverpool. Hadfield & Biddle estimate the cost of the canal to this point at approximately £80,000. The duke, who by this time had been forced to borrow to complete his scheme, was soon more than well repaid. His income on freight carriage between Liverpool and Manchester alone in the period from 1770 to 1779 was £21,472. The two years 1780 and 1781 earnt him £7,381. And these fig-ures do not include the sale of the duke's own coal in central Manchester; a trade that grew dramatically.

To support the trade in Liverpool, the duke purchased land on the banks of the Mersey and had his own dock and warehouses built. There was also a healthy passenger traffic between Warrington and Manchester. For a fare of 10d. basic or 2s. 6d. first class, patrons could spend five hours being liber-ally supplied with tea and cakes while journeying between the two towns. The service was a popular one with annual receipts in the late 1780s approaching £3,000. By this time, the canal was carrying about 265,000 tons of freight p.a. of which 64,000 tons was coal; over 30 per cent of the traffic was from Liverpool. Annual receipts for this period averaged over £60,000.

By the turn of the nineteenth century, the English canal network was beginning to take shape. In 1804 the Bridgewater had a junction with the Rochdale Canal at Castlefield; a link which opened the way for Worsley coal to reach markets to the east as well as providing a route for limestone and merchandise coming west. Castlefield was now frequented by no fewer than twelve different carrying companies including the duke's own, Pickfords and Hugh Henshall & Co. (which was owned by the T&M). By 1821 there were twenty-one companies and it was possible to catch a fly-boat all the way to London. An additional line north was forged when the Leigh branch of the Leeds & Liverpool Canal joined the Bridgewater at Leigh. This link had started life when the duke obtained an Act in 1795 enabling the extension of the line from Worsley to Leigh. The route opened in 1799 and was joined with the L&L in 1820 when that company built a branch south. The trade from Manchester to Runcorn, however, was one that was bitterly fought for by the M&IN who, by 1804, had a new cut all the way into Runcorn. To work this, the M&IN introduced a steam packet which could whisk passengers from Manchester to Runcorn in just eight hours (the Bridgewater boat took nine) for immediate transfer to the Liverpool boat. This competition seems to have halted the otherwise ever increasing growth in receipts along the duke's canal. The duke responded by ending an agreement on rates and by raising tolls sharply. There was then a sudden and dramatic rise in profits from an average of £24,441 p.a. for 1797–9 to £50,736 p.a. for 1803–5.

The great canal duke died on 8 March 1803 and his canal property passed to the Marquess of Stafford. The duke willed that a group of trustees be established to further the interests of the collieries and the canal line which were to be managed by Robert Bradshaw. The change had little affect on the business during a period of relative calm. For the years 1806–24, receipts and profit levels remained static at approximately £120,000 and £45,000 respectively. The Bridgewater and the M&IN had a toll agreement for the prosperous Manchester–Liverpool route and it was accepted that if one company altered their rates then the other would follow suit. The two companies also followed each other in terms of new facilities. After the M&IN had built itself a substantial new basin at Runcorn, the Bridgewater built a second line of locks at its terminus in the town. The tidal lock was then extended and a new slate wharf added.

The first suggestion of a railway between Liverpool and Manchester was seen in a newspaper on 2 October 1822. The Bridgewater and the M&IN were united in their opposition to the scheme which both recognized as potentially damaging to their interests. The opposition was successful and the Liverpool & Manchester Railway Bill of 1825 was defeated. The Marquess of Stafford, however, was already coming down in favour of the idea of railed ways as the answer to the transport problems of the nineteenth century. He decided to support the new railway and to invest £100,000 (20 per cent of the capital) in it. As if to salve his own conscience, he also invested £40,000 in improving facilities along the Bridgewater. The M&IN was obviously displeased at this change of heart and said so but by then the battle was lost. The Act that enabled the historic Liverpool & Manchester

Railway to be built was passed on 1 May 1826. It was opened on 15 September 1830.

The waterway companies responded to the new competition by reducing their tolls, albeit without panic. Bradshaw must have found himself in a difficult position and was unable to form an alliance with the M&IN to control rates. It appears, however, that he remained a firm supporter of canal transport despite the sentiments of his employer. Bradshaw reduced tolls in order to maintain trade, but with a consequent slight reduction in receipts (from £175,997 in 1830 to £142,251 in 1833) and a substantial fall of profits (from £47,650 to £17,473).

When Lord Stafford died in 1833, ownership of the collieries and the canal passed to Lord Francis Egerton, the Marquess' second son, and James Sothern became manager. The two new incumbents were to often differ over policy. Egerton was keen to produce agreements with the railway and the M&IN over toll rates; Sothern wasn't. It wasn't until 1837 that Egerton was able to buy Sothern out and replace him with James Loch (as superintendent). The three companies then stopped a damaging series of rate-cut rounds and a kind of harmony was installed. In 1838 a junction was formed between the canal and the M&IN at Castlefield (via the Hulme Locks). This was presumably a boon to both lines but the construction of the Manchester & Salford Junction Canal (across central Manchester) to link the M&IN with the Rochdale, bypassed the Bridgewater junction at Castlefield and must thus have been a blow to Bridgewater prospects. In the event the M&SJC (which was bought by the M&IN) was not a great success and had little effect on the Bridgewater.

The agreement lasted until 1840 when the M&IN saw brighter prospects outside the alliance than in. It had forged a link with the Manchester & Leeds Railway, had plans for turning its line into a ship canal, and felt that it had a strong competitive base. The other two partners retaliated and the M&IN was forced to rejoin the pact. However, by December 1841, it was the Liverpool & Manchester Railway who saw its share of the business falling and who unilaterally reduced its cotton rates. This was followed in 1842 by an agreement between the LMR and the Manchester & Leeds for through rail traffic in direct competition with the waterways. By this time the average annual tonnages for each company's trade between Manchester and Liverpool were: Bridgewater 200,856; M&IN 141,813; and LMR 164,625. Although another rates agreement was made in 1843, vigorous unseemly competition between the three companies soon returned. The only solution, as James Loch realized, was for the Bridgewater to buy the M&IN and they duly did so on 17 January 1846 following an Act of 1845. The price was £400,000 although Egerton (or Lord Ellesmere as he had become) also took on a debt of £110,000. Funds were raised in bonds and by borrowing on mortgage but the Bridgewater was still doing well enough to afford the purchase. Profit in 1844 amounted to more than £85,000 with nearly 1,300,000 tons of cargo being carried along the line. About a quarter of this trade was carried by the Bridgewater's own carrying fleet.

Despite this success, James Loch was already reviewing the possibility of selling both the trustees' waterways to a consortia interested in converting them into railways, ship canals or both. Railway competition had been ever more serious during the late 1840s and by January 1850, a new agreement with the London & North Western Railway (into which the LMR had been absorbed in 1846) forced the trustees to accept 50 per cent of the Liverpool–Manchester traffic as against 66:33 in previous agreements. The canals going over the Pennines were by now all railway owned; a fact which further weighed against the Bridgewater trustees.

The one area of the canal's trade that was burgeoning was its carrying business which was now operating into the Midlands. It was thus opportune for James Loch to start discussions with the Great Western Railway who was looking to establish a carrying link into the north-west. But agreement on a lease or sale of the line to the GWR was not forthcoming, and when Loch died in 1855 no agreement was impending.

With Loch's death, Lord Ellesmere's third son, the Hon. Algernon Egerton became superintendent. This change of management brought about a change in attitude towards the railways. The canal would no longer be looking for railway associations or take-overs but would seek to strengthen its links with other waterways and build a viable long-distance trade. The 1 1/4 mile long Runcorn & Weston Canal, which ran from the town to a basin at Weston Point a little further along the coast towards the Weaver Navigation, was opened in 1859. Improvements were made at Runcorn Docks with a new large tidal basin completed in 1860. Similar improvements were made to the trustees' docks at Liverpool.

In 1870 the Bridgewater carried 2,490,715 tons of cargo with about 44 per cent being handled by the trustees' carrying companies or associated companies. Despite this evident high level of trade, the trustees continued to seek a railway association. In 1871 the chairmen of two railway companies, Sir Edward Watkin and W.P. Price of the Manchester, Sheffield & Lincolnshire Railway and the Midland Railway respectively, formed a new company, the Bridgewater Navigation Company, and purchased all of the canal interests of the trustees for £1,115,000. The deal was completed on 3 July 1872 with Edward Leader Williams appointed as general manager.

The new canal company set about bringing its possession into the steam age. Leader Williams introduced steam tugs, which could each tow three barges on the lockless section of the line from Runcorn. Perhaps more significantly, the company looked to improving the now declining M&IN. Accommodation at Runcorn was improved including the opening of the Fenton Dock in July 1875. By this time the Runcorn complex was handling

500,000 tons of cargo p.a. including exports of salt, coal, pitch and earthenware, and imports of flint, china clay, iron, grain and sand. The dock complex covered 16 acres of water and 37 acres of quay.

Business along the BNC's waterways in 1884 was still brisk. Some 2,815,018 tons of cargo were moved along the lines about half of which was shipped into the estuary, and 800,000 tons of cargo, mostly coal, crossed the Barton Aqueduct towards Manchester with little or no business taking the return route. By now just 30 per cent of the traffic was carried by carriers associated with the BNC. Traffic in and out of Runcorn was still totalling about 500,000 tons p.a. The BNC was able to pay 8 per cent dividend throughout the 1880s with profits hovering around £60,000 p.a.

Despite the profits, a key threat to the BNC arose during the course of the 1880s. The Act for the Manchester Ship Canal was passed on 6 August 1885. The Act not only enabled the construction of the new waterway but provided powers for the new company to purchase its potential rival, the BNC, for £1,710,000. The purpose behind the move was that the MSC was to be built mostly along the course of the old M&IN. The ship canal was open to traffic on 1 January 1894 and officially declared open by Queen Victoria in the following May. As part of the new MSC, Brindley's old

Brindley's original Barton Aqueduct which was demolished to make way for the present swinging aqueduct in 1894

Ware/The Boat Museum Archive

Bridgewater Canal Barton Aqueduct was demolished and replaced by the present swing aqueduct which was designed by Leader Williams.

Although the MSC developed a significant level of new business of its own, it was inevitable that there would be a drop in traffic along the Manchester to Runcorn line of the Bridgewater. However, this was more than adequately compensated by the business created for the Bridgewater at the Manchester terminus where the old canal formed a valuable link with the Rochdale and the line to the Leigh branch and Wigan. The continued high levels of carriage along the route to the Leeds & Liverpool meant that, by the time of the First World War, carriage was still nearly 2,000,000 tons p.a. But thereafter, the business declined. By the time of the Second World War, tonnage was down to just 815,391. In the 1960s, by which time the canal no longer reached down its western locks to the Runcorn Docks, traffic was down to a quarter of a million tons. Trafford Park power station continued to receive its coal deliveries by barge until 1972 and by 1974 all the freight traffic had gone. The Bridgewater Canal was not nationalized in 1948 and is still owned by the Manchester Ship Canal Company. Today it is a popular holiday route and part of the well-known Cheshire Ring.

The Walk

Start and finish:	Worsley Green (OS ref: SD 747004)
Distance:	5¹/₂ miles/9 km
Map:	OS Landranger 109 (Manchester)
Car park:	Near A572 roundabout
Public transport:	British Rail Patricroft

From the A572 roundabout (near junction 13 of the M62), take the B5211 road to Eccles and park in the free car park which is about 100 yd after the turning. Leave the car park and turn right to go over the canal bridge, Worsley Bridge, in Barton Road. Before going down to the canal, cross the road to admire the view. Here is the Bridgewater, glowing yellow-orange with the iron hydroxide (ochre) that has leached from the old coal mines.

Before starting the main walk, pass down the left-hand side of the canal to walk in front of the Packet House with the canal to the right. From here in the summer months, visitors can take canal boat trips. In fact, this has been the embarkation point for travellers along the canal since the Duke of Bridgewater's time. Packet-boats operated to destinations as far as Runcorn, Warrington and Manchester. In 1841 a packet-boat left Worsley for central Manchester at ten in the morning and six in the evening. The trip took just

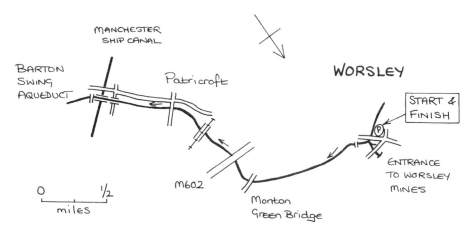

The Bridgewater Canal

two hours and cost 9d. first class or 6d. second. The fine, and much photographed, Packet House, by the way, was built in the late eighteenth century. The Tudoresque frontage is strictly Victorian imitation, being added in the 1850s under the guidance of the Earl of Ellesmere.

Continue on and across the small footbridge (called the ABC Bridge because it has twenty-six planks) to an island with the oldest building in the district, built in 1725 – and hence before the canal. In the 1760s the building housed a nailmaker's workshop and from that it takes its name to this day. Worsley Road Bridge is to the right. Underneath the bridge are entrances to some underground rooms reputed to be stables for the horses used in the mine's horse-gins. Walk on with the Nailmaker's Workshop to the right up to the road. Cross here and bear left to School Brow. Here on the left is The Delph and the entrance to the Duke of Bridgewater's mines. The Delph itself was once a sandstone quarry and much of the stone was used to build many of the canal's bridges.

There are two entrances to the 30 miles of tunnels that form the Worsley coalfields. The tunnels were probably started when the canal was first built and were still being extended into the 1840s. In their heyday, the underground canals became quite a tourist attraction and many good and famous folk came to take a subterranean trip. As mentioned earlier, the underground canals were on three levels, connected partly by an inclined plane. The tunnels are mostly brick-lined and still in relatively good condition. The starvationer boats used in the tunnels were moved primarily by opening a sluice which then washed them out, although some legging or simple pushing must have been needed in places. Once outside, the full boats were towed into Manchester by horses and empty ones legged into the tunnels along the now-still waters. The starvationers

A canal arm to The
Delph off the
Bridgewater Canal at
Worsley

were last used in 1889 by which time a steam railway had been built. The
Worsley field was last mined in the 1960s.

To start the main section of the walk, return to Worsley Road Bridge and
cross the road. Take the steps to the left of the bridge near Bank Cottage.
These lead down to the canal. Turn right to go under the bridge with the
canal to your left. Here on the left is the canal entrance to The Delph as well
as the Packet House.

The towpath leads round right and on with a fine-looking boat house on
the opposite bank. This was built by Lord Ellesmere, the Duke of
Bridgewater's great nephew, to house a royal barge built for Queen
Victoria's visit in 1851. An information notice back at the Packet House tells
the story of how the rain-soaked crowds who attended the occasion were so
noisy that they scared one of the horses pulling the royal barge, causing it to
leap into the canal. History does not recall whether the queen was amused.

Go under the footbridge (which goes to the Worsley Green) with the
Bridgewater Hotel to the right. The footbridge was built at the behest of the

Earl of Ellesmere in 1901. Just opposite Brinks Boats is an information notice that explains the history of the area. On the left are some dry-docks and on the right a converted granary. The granary was originally a forge with a water-wheel worked by Worsley Brook. The dry-docks date from the early 1760s and are reputed to be the earliest on the entire inland waterway system. Most of the barges that worked on the canal and in the mines were built here. Just beyond the granary on the right are the barely recognizable remains of some old limekilns.

The path goes through a gate and turns right along a very straight long embankment. The new houses on the left bank are on a site formerly occupied by colliery basins. The canal goes under Monton Green (road) Bridge, changes colour to the more normal canal green, and bends right to go past The Bargee Inn and Restaurant. On the right is a new residential development called 'The Waterside' (built on the site of an old textile mill building), and on the left a housing estate. The canal then goes under the M602. After this, on the right, is a still extant mill – Eccles Spinning and Manufacturing Co. Ltd – now used by the mail-order group Great Universal. The canal goes under a railway bridge (which carries the old Liverpool & Manchester Railway), past the Wellington pub and beyond, with the Barton Road close to the right. After Boat Building Services, continue under Patricroft Bridge with Bridgewater Mill looming from the left bank of the canal. Here is the diminutive Packethouse Pub. The route continues with a basin to the left now occupied by the Worsley Cruising Club. One of the basin warehouses has an overhanging loading shelter. Shortly the towpath ends and walkers are forced onto the pavement. Continue on to traffic lights and then straight over the crossroads to reach the metal girder bridge which crosses the Manchester Ship Canal. Here on the left is the Barton Swing Aqueduct.

The Manchester Ship Canal was built on a scale larger than anything else attempted in England. It was opened on 1 January 1894 after seven years of construction and an expenditure of over £14 million. To make way for the new waterway, five railways were diverted and a whole series of new bridges were built to accommodate the required headroom of 70 ft. The canal was built to a depth of 26 ft (later deepened to 28 ft) and is capable of taking vessels of 10,000 tons. There are five sets of locks: at Mode Wheel (Salford), Barton, Irlam, Latchford and a sea lock at Eastham. Each set contains a large lock which measures 600 ft long by 65 ft wide together with a smaller version for tugs and other small vessels. The line was a phenomenal success, carrying millions of tons of cargo p.a. and turning Manchester into a major industrial city.

The swing aqueduct that takes the Bridgewater over the MSC is widely regarded as one of the wonders of Victorian engineering. Originally, Brindley had built a three-arched stone aqueduct here at Barton. Altogether (including the approaches) it was 600 ft long. The central arch spanned

57 ft and the two side arches were 32 ft each. It was 36 ft wide: 18 ft of waterway, 6 ft of puddled lining and the remainder packed with earth. With the advent of the MSC, the old aqueduct had to go and this new structure, designed by Edward Leader Williams, replaced it in 1894. The new aqueduct comprises a 235 ft long wrought-iron tank, 18 ft wide and 7 ft deep. To allow shipping to pass along the MSC, the aqueduct swings on a central pivot by means of sixty-four cast-iron rollers. This manoeuvre is undertaken with a full caisson of water and there are hydraulically operated lock-like gates at each end to seal the tank. The small gap between it and the canal side is bridged by a rubber-shod iron wedge which is moved into place by hydraulic dams. All this is worked from the control tower on the central pier. A close-up view of the aqueduct can be obtained by continuing over the bridge and turning left up a driveway and then left onto a path which leads to some steps and a viewing bridge.

The control tower of the Barton Swing Aqueduct that carries the Bridgewater over the Manchester Ship Canal near Eccles

It is possible to continue along the Bridgewater to Trafford Park and central Manchester; however, to return to Worsley, you can either catch one of the many buses that cross the MSC here or return by foot along the canal.

Further Explorations

There's a pleasant afternoon stroll of about 7 miles along the Bridgewater from Altrincham to the canal village of Lymm on the Greater Manchester/Cheshire borderlands (OS Landranger 109, ref: SJ 682873). I chose to park in the well-signposted free car park near the library in Lymm and then took the North Western bus no. 37 (enquiries: (061) 228 7811) to the bus station in Altrincham. From there, continue along New Road into Barrington Road. At a T-junction, turn right along Manchester Road to cross the canal just after Navigation Road. After passing over the canal (on Broadheath Bridge), walk into the car park of Halford's superstore and bear right around the Altrincham, Sale & District Sea Cadets' building to reach the canal. Turn right so that the canal is on the left.

The line from Manchester to Runcorn was the second phase of the construction of the Bridgewater Canal and was enabled by an Act of 1762. The canal was open to Altrincham and the first tolls were taken on 6 June 1766. Within the year, coal prices in Altrincham had halved. The line between here and the end of the walk, at Lymm, took three years to build and it wasn't until 1776 that the route to Runcorn was fully operational.

The towpath immediately passes under the Manchester Road and continues on with some industrial buildings to the left (including an old cotton warehouse with a covered loading bay) and some new DIY stores to right. Although for some distance the path is accompanied by industry and a new housing estate, after going under Seamon's Moss Bridge (at Oldfield Brow), the canal reaches countryside. After meandering around some open farmland, the line reaches Dunham Town at Dunham School Bridge (turn left here for the Axe and Cleaver pub) and then Dunham Town Bridge.

The canal now crosses the Bollin embankment; a high and surprisingly 'modern' feature for this, one of our oldest canals. It offers towpathers a fine airy position and good views over the surrounding countryside. En route we pass over three small aqueducts. Dunham Underbridge, the River Bollin and Bollington Underbridge. The embankment has not been without its problems. In 1971 the southern side near the Bollin Aqueduct collapsed. Although there was some debate as to whether the line would ever reopen, the company eventually built a 300 yd long concrete channel which was rewatered in 1973. The embankment also offers views left to the National Trust's Dunham Massey Hall.

At the end, the canal bends right to Agden Bridge. A straight section now leads past a series of mooring spaces and Lymm and Hesford Marinas. After crossing the B5159 via Burford Lane Underbridge (aqueduct), we pass Grantham's Bridge and then Lloyd Bridge on the outskirts of Oughtrington. It is now just a short distance to the A6144 bridge in Lymm which you'll recognize as it has a separate footbridge ably protected by lime-green railings.

The history of Lymm is closely linked with that of the canal. In the early nineteenth century, the town was an important stopping place for the Manchester to Runcorn (and hence Liverpool) packet-boat. The vessel stopped here for 15 minutes to allow passengers to stretch their legs and gain refreshment from a local hostelry before proceeding onwards. Reports suggest that more often than not all 120 seats were full for the journey which, in 1784, took 3½ hours from Castlefield or 6½ from Runcorn.

If you cross the bridge and continue along the right-hand bank, you will return to the library car park. In doing so, you pass over an aqueduct over Bradley Brook. Otherwise go left across the bridge into the village centre.

If you wish to visit the Manchester terminus of the canal at Castlefield Junction, see Further Explorations in the Rochdale Canal chapter.

Further Information

The canal was not privatized and hence is not managed by British Waterways. The owners are:

The Manchester Ship Canal Company,
Admin Building,
Queen Elizabeth II Dock,
Eastham,
Wirral L62 0BB.

There is no canal society as such but the Inland Waterways Association has a Manchester branch. For the address of the branch secretary contact IWA head office in London (see Appendix B).

For more details of the history of the Bridgewater:

Hadfield, C. and Biddle, G., *The Canals of North West England,* Vols. I and II. David & Charles, 1970.

3
THE CHESTERFIELD CANAL
Norwood Tunnel to Worksop

Introduction

The Chesterfield is, perhaps surprisingly, one of the earliest of the nation's canals. Surveyed originally by James Brindley, it is of similar vintage to the Oxford Canal. If you look closely, its age shows. You certainly wouldn't have caught Thomas Telford building the Shropshire Union along such a meandering line, and would he have shied clear of the straight course from Retford to Gainsborough? Of course, we can only guess at the answer and while we do, we can enjoy a magnificent waterway.

The Chesterfield Canal starts its life at West Stockwith where it forms a junction with the River Trent a few miles downstream from Gainsborough. From there the line heads south-west to pass to the west of Gringley on the Hill before turning south-east to Clayworth. After running south to the town of Retford, the canal heads west to Worksop where the current head of navigation is reached just short of Morse Lock. The now partially, or sometimes wholly, derelict line runs up to the summit pound and through the Norwood Tunnel. From the western portal of the tunnel, the line formerly descended through Killamarsh before heading south to Staveley and then west to its terminus at Chesterfield.

Here is a fine country walk along a canal on the verge of restoration. For that reason, it's worth walking now so that you can bore the grandchildren later with tales of what things used to be like in the bad old days.

History

For generations, the trade of Chesterfield together with the lead and stone of the eastern Derbyshire hills had been carried by road to Bawtry where it was loaded on to a boat, shipped down the River Idle to West Stockwith and then on to the River Trent for export to the rest of the country and, via Hull, the world. It was a cumbersome route and, at times, the River Idle was unnavigable through drought or flood.

With the opening of the Bridgewater Canal in 1763, the first flush of canal-building swept the country, and the businessmen of Chesterfield were soon discussing the potential for a canal to form a direct link between the town and the Trent. James Brindley, the canal engineer of the age, was approached in late 1768 and asked to undertake a survey. Brindley, who was building both the Trent & Mersey and the Staffordshire & Worcestershire at the time, delegated the initial inspection to one of his assistants, John Varley. The promoters of the new line initially asked Varley to survey a line from Chesterfield to Bawtry (which Brindley promptly costed at £100,000) but, with the growing interest of the people of Retford, a second route was surveyed by Varley in June 1769. The new line would run from Chesterfield to Worksop and Retford and then on to West Stockwith and the Trent; thus bypassing the River Idle altogether.

On 24 August 1769 the first public meeting was held at The Red Lion (now the Golden Ball) at Worksop. Brindley reported that a canal along the new line was practicable and estimated the cost as £95,000 or £105,000 if the eastern terminus were taken further up the Trent to Gainsborough. The meeting agreed on this latter option. However, by the following January, Brindley had decided that the West Stockwith Junction was to be preferred for reasons of cost, a faster completion date, the need for fewer locks and the avoidance of a tunnel through Castle Hill (between Retford and Gainsborough). The promoters of the new line suggested that it would be an important carrier of a range of different cargoes: coal, earthenware, lead, timber, millstones, limestone, roof tiles and gravel for mending roads. Of these, coal was always considered to be the most important. Brindley thought that with the opening of the canal, Derbyshire coal would be cheaper than the Yorkshire coal coming via the Don Navigation to Gainsborough. The canal would also import goods not regularly available in the land-locked towns of Chesterfield, Worksop and Retford: fine wool, rice, oils, wines, sugar, tobacco and fresh groceries.

Despite the opposition of John Lister, the owner of the navigation rights on the River Idle, a bill was presented on 23 January 1771 and the Act gained Royal Assent on 29 March. Powers were given to raise £100,000 in £100 shares with a further £50,000 should it be needed. Tolls, on what was to be a narrow boat canal, were set at $1\frac{1}{2}$d. per ton on coal, lead, timber and stone and 1d. on lime. The project was promoted in London as well as locally and the issue was fully subscribed by July 1771. Brindley was appointed chief engineer at £300 p.a. but, as he was rather too over stretched to be regularly on-site, John Varley was made resident engineer with an annual salary of £100.

A three-rise lock on the Norwood flight in *c.* 1900. The building on the right is a water-powered sawmill

Dudley PL

The construction work started in the early autumn of 1771 with 300 men working on the two ends of the biggest engineering project along the line: the 2,850 yd long Norwood Tunnel. Work was also underway on the reservoir dam further up the hill at Pebley. Most of the construction work was undertaken by contractors who were employed for specific sections of digging or the building of particular bridges or locks. Brindley's plan was to complete the line section by section so that the canal company could open its waters for trade. This had the benefit of earning revenue as well as enabling the movement of some of the heavy construction materials that his contractors needed. Brindley's plans for the summer of 1772 included the amassing of the materials (primarily timber and stone) that were needed to build the flights of locks at each end of the summit pound: the Norwood flight westwards to Chesterfield and the Rother valley, and the Thorpe and Turnerwood Locks eastwards to Retford and the Ryton valley.

Sadly, in September 1772, James Brindley died while surveying the Caldon Canal. Initially John Varley was allowed to continue as acting chief engineer but Hugh Henshall, Brindley's brother-in-law, was appointed to the position at £250 p.a. With a less esteemed, if not necessarily less able, engineer, the company felt in a stronger bargaining position and insisted that its new appointee spend at least fifty-six days a year on site. It is perhaps a good job that it did. In the summer of 1773 Henshall found that some of the work at the Norwood Tunnel, a section contracted to John Varley's father and two brothers, wasn't up to standard. More examples of malpractice and lax management became apparent, further reflecting badly on the Varley family. It appears that John Varley was not directly blamed and the new regime under Henshall was able to make good the faults, rationalize the management and accounting systems and make excellent progress throughout 1774.

The earlier proposal that the Chesterfield should be a wide canal had been defeated, but now Retford Corporation and various shareholders from the town agreed to bear the expense of widening the line from Stockwith to Retford so that it could take the wide beam boats that worked the River Trent. This move was agreed by the canal company in May 1775 and the work got under way. Meanwhile, during the same month, Norwood Tunnel was officially opened. Three vessels sailed through the hill carrying about 300 people and a band who played appropriate airs. The journey took just over an hour from end to end. The new tunnel was 2,850 yd long, 9 ft 3 in wide and 12 ft high. It's said that the tunnel was so straight that it was possible to see the light at the other end. In that same month of May, work was proceeding to Chesterfield as well as on the Retford section of the line. By the beginning of April 1776 the canal was open from Killamarsh to the new basin at Stockwith. By August 1776 the canal was complete from the turnpike road at Norbriggs to Stockwith.

Following some difficulties in buying the necessary land at Stockwith, the lock into the Trent from the Chesterfield Canal was finally officially opened on 4 July 1777. In all, the line was 46 miles long with sixty-five locks. The day was celebrated with the arrival in Chesterfield of a boat from West Stockwith which, having been duly welcomed by a group of shareholders, was unloaded on to wagons. The shareholders, together with a band playing celebratory tunes, then led a procession into town. However, such festivities were short-lived. As early as September 1777 there were problems with Norwood Tunnel. Various individuals had dug coal from the ground over the tunnel and had damaged the roof by lessening the weight that compressed the arch. Mining subsidence was also to be a continuing problem over the years and was eventually to lead to the closure of the tunnel.

With the completion of the line, John Varley left the company and Richard Dixon, formerly the company's bookkeeper, took responsibility for engineering. Probably the first thing he had to consider was the construction of the Lady Lee arm for which permission had been given in March 1778. The arm was built for a Mr Gainsforth to link his stone quarry near

Worksop with the main line. Dixon also had to sort out the inadequate water supply on the Norwood Tunnel summit pound. In July 1779 penalties were imposed on empty or lightly laden boats travelling under 12 miles when using the locks. This unsatisfactory regulation was rescinded when Woodhall and Killamarsh Reservoirs were finished around 1790. Further capacity was added at Harthill by 1806.

The canal company also found its financial reservoirs equally inadequate. An attempt to raise additional funds by calls on shares was unsuccessful and £53,000 was borrowed on mortgage in order to complete the canal. The consequence of this was that a high proportion of the early trading profits were needed simply to service the loan. Furthermore, although traffic on the partially completed canal in 1774 had reached 42,693 tons, by 1778 levels had actually dropped to just 34,077 tons as the canal experienced the effects of a recession brought on by the American War of Independence. Thus the hoped-for revenue did not materialize at once and was very slow to increase.

Despite this shakey start, trade did improve enough for the first dividend of 1 per cent to be paid in 1789. The gross income for that year was £8,230 and net profit was £2,780. The total tonnage carried was 74,312. Of this some 42,379 tons (57 per cent) was coal. The rest consisted of stone (7,569 tons), corn (4,366 tons), lime (3,955 tons), lead (3,862 tons), timber (3,444 tons) and iron (1,544 tons). Other traffic included pottery and beer. The opening of various linking tramways resulted in a steady increase in traffic so that by 1795, the company was able to pay a dividend of 6 per cent. In 1826, 103,000 tons were carried yielding £13,582 in tolls. Among the loads shipped in 1840 was a cargo of Anston stone destined for the new Houses of Parliament.

This modest prosperity continued into the middle of the nineteenth century but the Chesterfield was never to emulate the huge success of some of the canals built at the same time. Perhaps the key drawback was the fact that the line only had one junction with the outside world: at West Stockwith. Over the years there had been numerous suggestions for links with other canals or simple extensions. As early as 1771 Brindley had suggested a canal from Swarkstone on the Trent & Mersey Canal to Chesterfield. Another proposal had a line connecting with the Sheffield Canal that was eventually proposed as part of the Grand Commercial Canal project. This scheme, which would have joined the Peak Forest, Sheffield, Chesterfield and Cromford Canals, was proposed in 1824 at a cost of £574,130. Also during the 1820s a westward extension to Barlow and Calver was suggested. In 1852 the Sheffield and Chesterfield Junction Canal was proposed. None of these plans came to fruition, and by the 1840s it was all too late anyhow.

The Sheffield and Lincolnshire Junction Railway (from Sheffield to

Gainsborough) was proposed in 1844. In response, the Chesterfield Canal Company formed the Manchester and Lincoln Union Railway in 1845 with a view to converting parts of the canal into railway. Such was the threat that the two companies formed a joint committee and agreed to future amalgamation. Under an Act of 7 August 1846 the M&LUR was authorized to construct a railway from Staveley to the canal at Worksop. The Act also enabled the amalgamation of the railway and the canal as the Manchester and Lincoln Union Railway and Chesterfield and Gainsborough Canal Company, and then to further amalgamate with the S&LJR (by now called the Manchester, Sheffield and Lincolnshire Railway). The canal lease was valued at £147,912 and the Chesterfield company was dissolved. The Manchester and Lincoln Union Railway amalgamated with MS&LR on 9 July 1847. The Act stated that the new company was to keep the canal in good order and to maintain reasonable tolls.

During the course of all the manoeuvring, the canal had become neglected and, in places, difficult to navigate because of subsidence. As a consequence, in 1848 the new company set about a programme of maintenance. This lead to an increase in revenue and even the railway company began to carry goods on the canal. But the scheme to turn part of the canal west of Norwood into a railway was not forgotten. Plans were considered throughout the 1870s and early 1880s but were not progressed. In 1888 the canal still carried 62,075 tons but toll receipts had declined to just £2,793. By 1905 the tonnage had fallen to 45,177 (of which coal comprised 15,408 tons and bricks 11,070 tons) and there were only forty working boats still operating. By now, the length between Staveley and Chesterfield had become unnavigable due to subsidence.

Between 1871 and 1906 some £21,000 had been spent repairing Norwood Tunnel. In 1904 the minimum headroom was down to 4 ft 10 in and although there was an official timetable for boat movement, the relative infrequency of traffic meant that boatmen used the tunnel as they pleased. The final *coup de grâce* for the tunnel came in 1908 when a roof collapse under the Harthill to Kiveton Park road closed the through route. In fact this closure meant that no commercial traffic travelled beyond Shireoaks.

The new century saw a series of amalgamations and absorptions with ownership of the canal passing initially to the Great Central Railway and then, in 1923, to the London & North-Eastern Railway. The new owners continued to maintain what was left of the waterway. The tidal lock at West Stockwith, for example, was enlarged and repaired in 1923–5 and the line was kept free of weed. The Second World War saw a mini-revival in trade with the carriage of munitions and a restored coal-carrying business from Shireoaks colliery basin. After the war this coal trade stopped and the carrying of bricks from a factory at Walkeringham ceased in 1955. In 1961 it was proposed to close the line and applications to use it for pleasure boating

Tomlinson's yard at Stockwith Basin in the early years of the twentieth century
Ware/The Boat Museum Archive

were refused. The last commercial traffic moved warp (a type of silt from the mouth of the Idle that was used for metal polishing in the cutlery trade) from West Stockwith to Walkeringham in 1962.

The 1950s were, therefore, a rather bleak decade for the Chesterfield Canal, and its future was not bright. The Inland Waterways Protection Society carried out a survey of the line in 1958–9 and rallies were held at Chesterfield and Worksop in order to publicize the plight of the canal. On 24 May 1960 a public meeting was held in Chesterfield to consider the future of the waterway. It was proposed that the line from West Stockwith to Worksop be retained for pleasure boating and that the section up to Kiveton, and from Spinkhill to Chesterfield, be kept as a water supply channel. However, the stretch from Kiveton to Spinkhill was to be infilled and sold. The recommendations were, therefore, somewhat of a curate's egg.

The 1968 Transport Act secured the future of the canal from Worksop to the Trent as a cruiseway but the whole section between Worksop and Chesterfield was allowed to decline and sections have been infilled and even built on. The campaign of the Inland Waterways Association to retain the rest of the waterway was supported by the setting up of the Retford and Worksop (Chesterfield Canal) Boat Club in 1962. The Boat Club and (since 1976) the Chesterfield Canal Society have continued to lobby for the restoration of the line beyond Worksop. Volunteers have made a start and the first lock (No. 1 at Lockoford Lane) was restored in April 1990. The key problems for the restorers will be the reopening of the Norwood Tunnel and re-routeing of the line around Killamarsh where houses have been built on the original course. The hope is that the Chesterfield will once again be fully navigable within the early years of the new century.

The Walk

Start:	Norwood Tunnel – eastern portal (OS ref: SK 500825)
Finish:	Worksop railway station (OS ref: SK 585798)
Distance:	7 miles/11 km
Map:	OS Landranger 120 (Mansfield & The Dukeries)
Outward:	British Rail Worksop to Kiverton Park (enquiries: (0522) 539502)
Car park:	At Worksop station
Public transport:	British Rail Worksop

Leave Kiveton Park station and walk towards the signal box. Turn right to reach the canal. Don't go immediately to the towpath but follow the advice of a public footpath sign and walk along a dirt track which leads to an entrance to a factory (right). Go straight on following some overhead power cables. The path goes over a canal feeder and on through grassland, initially to the right and then to the left of the power cables. Eventually the path

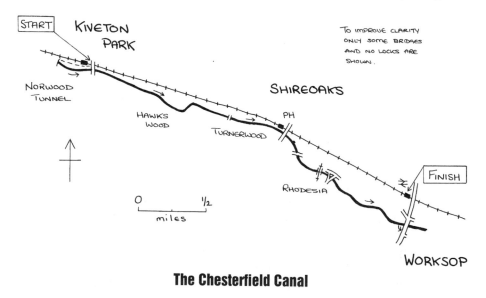

The Chesterfield Canal

splits and a left-hand branch goes down and over the entrance to Norwood Tunnel so that you are now facing back to the station and walking on the right-hand bank of the canal.

Norwood Tunnel is 2,850 yd long and took just four years to build. It was officially opened on 9 May 1775 and was closed in 1908 following a roof collapse about 1/4 mile from this, the eastern, portal. It is actually a relatively shallow tunnel and many authors have suggested that later engineers, such as Thomas Telford, may well have built most of it as a cutting. The big problem for the canal society is, of course, how to restore it. Although very firmly bricked up, I'm told that this end is regularly inspected and is in good condition. The western end however has been plugged at a point where the M1 goes over the top. Much of the ground in between consists of coal reserves and it has been proposed that restoration of the tunnel could be undertaken in conjunction with the mining industry. Recent developments in that business may, however, rule that out for some time.

The towpath along the first half-mile of the canal was being cleared when I made my visit here and it was a hive of activity. While the weeds and brushwood were being cleared, classes of biology students were studying the burgeoning wildlife. Half-way along this first stretch back to Kiveton Park station, the feeder crossed earlier enters the canal. The water comes from the Harthill and Pebley Pond reservoirs which are 2 miles to the south-west near Harthill. The feeder canal, the Broad Bridge Dike, is marked on OS Landranger 120. These reservoirs were built in 1806, nearly thirty years after the canal had opened, in order to overcome what was becoming an embarrassing shortage of water in this, the summit, pound.

At the first bridge, Dog Kennel Bridge, the towpath changes sides and we continue under two bridges with Hawkswood now dominating the scene to the right. After a further bridge (Pudding Dike Bridge), the canal swings left and then right to reach a bridge and the start of the Thorpe flight of locks. There are fifteen of them altogether although many are presently only barely discernible and part of the excitement of the next half a mile or so is to see if your lock tally matches the official number. The trick is to recognize the two triple staircase locks and the two double staircase locks. For those unfamiliar with a staircase lock, it is one in which the top gate of one lock acts as the bottom gate of the next. Easy if the lock-gates are in position, harder when the locks are as dilapidated as this. The key is to look for the recesses for the lock-gates along the sides of the lock chamber. As a clue, the sequence goes: a triple, three singles, a triple, a single, a double, a single and a double. While attempting to recognize the locks, the canal passes through some wonderfully evocative woodland. The expanded lagoons between the locks add to the mystery of the place.

The Thorpe Locks end at Turnerwood where there is a small hamlet and an old quarry basin. After this we continue past the Turnerwood flight of seven locks. After the fourth, the Brancliffe feeder from the River Ryton enters the canal. After the seventh, we go under a bridge and cross an aqueduct (over the River Ryton which marks the border between South Yorkshire and Nottinghamshire) before arriving at Shireoaks. Cricket fans can sit here awhile to watch the game in the canalside Steetley Sports Ground (one batsman hit a rather fine straight six on my visit). Those who aren't fans can walk up to the road bridge to visit The Station pub (Shireoaks station is to the left) or the small shop to the right. The road bridge has been culverted and will need rebuilding before the canal can be restored above this point.

Continue by crossing the road and taking the steps on the other side back down to the towpath. After a short distance the path goes over a roving bridge which marks the entrance to the old colliery basin. Coal traffic continued here until the Second World War and it is hoped that it may be busy again some day although this time as a marina. Our route now continues past a lock and a much rebuilt lock-keeper's cottage and on past two more locks to the outskirts of Rhodesia. After a road bridge, a lock and a rail bridge, we enter the centre of the village where the canal survives going under the new A57 flyover only to be culverted under two road bridges. At the first of the two, the towpath changes sides to the right-hand bank. We are now heading into the outskirts of Worksop and we pass Deep Lock, a bridge and then Stret Lock and a road bridge. This is shortly followed by what remains of the Lady Lee arm on the southern (towpath) side. An otherwise incomprehensibly positioned roving bridge still crosses the old channel. This branch formerly led to the Lady Lee Quarries half a mile to the south-west.

From here it is only a short distance to the last of the derelict locks, Morse Lock, and the head of the navigation. After passing a winding hole that marks the site of the Inland Waterways Association's 1988 Worksop Water Festival, the canal and towpath suddenly have a much more well-kempt appearance. The large area of water to the left here is Sandhill Lake; a flooded quarry that is now used for fishing and watersports. The centre of Worksop is now quite close and after passing our second canalside cricket ground (this one is used periodically by Nottinghamshire), we arrive at the former site of Worksop wharf – now a car park. In its heyday it was lined with warehouses and maltings and boats carried coal in and malt out. On the other bank was the Shireoaks Colliery Company coal wharf. The walk ends at Town Lock. Here there is a commemorative stone erected in 1977 to celebrate the canal's bicentenary.

It isn't possible to walk on past the lock. The land on the other side of the bridge is used by British Waterways as a depot and it is they who have blocked the towpath from here on. To complete the walk, therefore, we have to return to the car park and walk around the perimeter of a DIY shop to the road and turn left. If you cross the road and look on to the BW depot, you will see a rather fine old warehouse that bestrides the canal. This was formerly owned by Pickfords who at one time was an important canal carrying

Worksop Town Lock is wedged between the shops in the centre of town

company. They only later turned to road haulage. The canal society's booklet reports that on the undersurface of the warehouse arch are trapdoors through which goods were lifted or lowered from and to waiting boats. To complete the walk, continue along Carlton Road where the railway station can be found a few hundred yards along on the left.

Further Explorations

The whole of the extant Chesterfield Canal is open to walkers and those armed with an Ordnance Survey map should be able to navigate themselves along the entire 32 miles from Norwood Tunnel. With readily available public transport, the 11 mile section between Worksop and Retford makes for a quiet (apart from a length that runs parallel with the A1 near Ranby) afternoon stroll. Starting at the Retford end, take the A638 road to Newark and cross the canal at the Carolgate Bridge. Turn right to walk along the right-hand bank. The Chequers Inn at Ranby (roughly 6 miles) makes a good resting point. The towpath changes sides just before Osberton Lock (the first out of Ranby) and then changes back again two bridges further on at Manton Turnover Bridge. Should you prefer to divide the walk, East Midlands bus no. 42 (enquiries: (0909) 47577) stops at the Chequers Inn.

A shorter walk (about 2¹/₂ miles) that is chock-full of interest starts at West Stockwith (OS Landranger 112, ref: SK 790947) on the banks of the River Trent. For those without a car, Retford & District bus no. 96 runs to the village from Gainsborough. For those with a car, park in the small car park opposite the White Hart Inn. Walk back to the road and turn left to cross the River Idle. Even before the Chesterfield Canal, West Stockwith was an important port for Chesterfield and east Derbyshire traffic. Prior to the canal, goods were shipped overland to Bawtry, then on to the River Idle and along to Stockwith where it was transhipped on to the vessels that plied the Trent. The village has thus been a busy and relatively prosperous place since the fourteenth century, as the fine houses on the eastern bank suggest.

Walk up the track to the left of both the river and its massive vertically lifting floodgate. Our way goes up on to the flood banks and on to the second barrier where there is a drainage pumping station for the surrounding low-lying agricultural land. The path follows the river around a series of bends and then goes through a gate near the old pumping station at Misterton Soss. This rather fine old building (it dates from the 1830s) used to pump water up from the Mother Drain, a collecting channel for the many criss-crossing drains that service the surrounding countryside.

Just past the pumphouse, take the road left which goes past a kennels, bends around to go under a railway bridge and bends left to reach Misterton and a bridge over the Chesterfield Canal. Here on the right is Misterton Bottom Lock and, beyond Station Road Bridge, Misterton Top Lock. Return to the start of the bridge and walk along the left-hand bank of the canal and past (or via) the Packet Inn. The straight section of canal runs under the railway and on for half a mile. When you reach a bridge, go up to the road (near the Water Boat Inn) and cross both it and the canal to go down some steps to the right-hand side of West Stockwith Basin.

Today the basin is a popular mooring place for pleasure boats but it was once the loading and unloading site for the vessels from the canal and the barges and keels that worked the Trent. West Stockwith Lock, the route to the River Trent, can be seen on the eastern side of the basin. The building to the left of the lock-houses is the lock-keeper's office and on the right is an

West Stockwith Basin in 1992

old warehouse (the date 1789 is enscribed on the wall facing the river). At one time there was also a stables and a blacksmith's shop.

Walk past the warehouse building and cross the Trent-side gate of the lock. Just on the right, in front of the lock-keeper's house, is a winch which was used to pull boats on the river into the lock. Continue walking along the embankment of the Trent with the basin now to the left. This path soon reaches the road, the bridge over the Idle and the car park.

Further Information

The Chesterfield Canal Society says that 'nothing less than full restoration of navigation to Chesterfield' will satisfy it. It'll be a big job but a worthy one and those who wish to help should contact:

The Chesterfield Canal Society,
30 Park View,
Kiveton Park,
Sheffield S31 8SE.

The society publishes two very useful booklets which will expand on the information already provided in this chapter. The first is a brief history entitled *The Chesterfield Canal*; the second a towpath guide entitled *The Chesterfield Canal (West Stockwith to Rhodesia)*.

A more substantial volume on the Chesterfield is:

Roffey, James, *The Chesterfield Canal*. Barracuda Books, 1969.

4
THE FOSSDYKE NAVIGATION
Saxilby to Lincoln

Introduction

Although we are all familiar with the long, straight Roman roads that criss-cross the country, it is perhaps less well known that the Romans also made much use of water transport. In Lincolnshire it is actually possible to see and walk along a Roman canal! The Fossdyke can arguably be called the oldest artificial line of inland navigation in England; built fifteen centuries before the Exeter Canal and sixteen before the Sankey Brook. And those with a sense of rightness in these things will be glad to know that it's pretty straight.

At its western end, the Fossdyke Navigation forms a junction with the River Trent at Torksey, about 16 miles downstream of Newark. From there the line heads east, through Torksey Lock to Drinsey Nook where it meets the A57. After passing through Saxilby, the Fossdyke heads on under the shadow of Lincoln Cathedral until it reaches Brayford Pool and the junction with the River Witham, 11 miles from the Trent.

The broad, flat landscape of Lincolnshire isn't to every walker's taste – one boating guide I've read describes it as boring – but the fine prospect of the cathedral and the delightful river front in central Lincoln should be sufficient reward for anyone venturing forth.

History

The town of Lindum was established in AD 48 by the Roman IXth legion at the junction of Ermine Street (which went north) and the Fosse Way (which went south-west to Exeter). It was an important strategic position and the town grew rapidly, encouraged by the surrounding rich agricultural land. Although the roads were clearly of great importance to the Roman occupants for the movement of troops, the inward shipment of bulk materials was always going to be more of a problem. In addition, it would clearly have been an advantage to have used the productive Lincolnshire soils as a source of provisions for the garrisons further north. Thus the Roman mind turned to the potential of water transport.

There are two artificial canals in Lincolnshire which are generally accepted as being of Roman origin. They form just part of an extensive Roman waterway system which, although probably originally built for drainage purposes, provides through communication from the Cambridge area as far north as York. The Car Dyke is 56 miles long and runs from the River Nene east of Peterborough to the River Witham, a few miles below Lincoln. The 11 mile long Fossdyke (or Foss Dyke or Fossdike or Fosdig) connects the Witham with the Trent at Torksey. It is generally accepted that both canals were built as a way of connecting the farming area of the fens with the military garrisons of the north. But, although the Cambridgeshire Car Dyke has been shown to date from Roman times, there is no absolute proof that these artificial waterways in Lincolnshire are Roman. All that can be demonstrated is that they are pre-Norman and from then on archaeologists have had to conject and surmise. Car Dyke has been tentatively dated at AD 60–120 and as such it has been suggested that it formed part of Emperor Hadrian's period of imperial planning. Although the dating of Fossdyke has proved more difficult, a bronze statuette of Mars was found at Torksey during dredging in 1774 and an inscribed sepulchral tablet was found at Saxilby. These finds have lent weight to the theory.

Both dykes were probably built by simply digging out the earth from a relatively easy terrain and throwing it up to form banks on either side. There is certainly no suggestion that the construction of the navigation would have presented any great feat of engineering even though both lines pass through low-lying land liable to flooding. When the canal engineer John Rennie saw Car Dyke he said that 'a more judicious and well-laid out work I have never seen'. No doubt part of the route used the course of rivers. On the Fossdyke, for example, the builders were able to use (albeit with some straightening) the line of the River Till for the first 4 miles west from Lincoln. The overall width of both dykes was approximately 100 ft although the navigable line was probably only about 30 ft. Archaeologists have suggested that the line was plied by barges similar to those that are seen on relief sculptures of the time.

The position of Lincoln on this system of internal waterways was almost certainly responsible for the prosperity of the town during Roman times and it became a kind of inland port. The remains of a Roman quay, for example, were found during the construction work on the telephone exchange at the corner of Broadgate and Rumbold Street. All the way along, the town's river front must have been furnished with quays for the large volume of trade that was handled. Lincoln certainly became an important transhipment port for grain going north. The dyke would also have been used for local traffic. Excavations at Littleborough have revealed 'colour-cooked wares' from the Nene valley, near Peterborough, that were almost certainly brought by water. Roman pottery kilns are known to have been built along the Fossdyke, for example at the racecourse in Lincoln and at Little London near Torksey. The proximity of the waterway may suggest that finished pots were shipped to market using the waterway network.

What happened to the dykes during the Dark Ages remains unclear. It is, however, generally considered that when the Danes invaded in the 870s, they overran Lincolnshire by the simple expedient of sailing down the Humber, the Trent and then along the Fossdyke into Lincoln itself. From then it isn't until 1121 that the navigation again appears in the history books. Symeon of Durham records that the Fossdyke was reopened for traffic following dredging work carried out for Henry I. How long the line had been closed is unknown although it was possible that it had become unnavigable at the time of the Norman invasion. The line must certainly have been navigable in the reign of Henry II as a small Cistercian nunnery was built on its banks near Torksey at that time. The house became an object of benevolence for the trading community, a point perhaps enhanced by the fact that the nunnery was dedicated to St Nicholas, the patron saint of sailors.

Between the twelfth and fourteenth centuries the Fossdyke carried the wool of the Midland counties to Lincoln for export to Flanders. In later days it carried corn, wool and malt to the West Riding, Yorkshire, returning with linseed cake, manure, coal and manufactured goods. In 1319 a trip from Lincoln to York took two days. All this activity spurred the growth of the town of Torksey which was able to exploit its position at the junction with the Trent. It is said that some of the inhabitants sought to exploit the situation rather unfairly. In the thirteenth century Robert of Dunham (a village on the Trent above Torksey), who was a bailiff for William of Valence, was levying a toll of a half penny for every ship that passed from Lincoln along the Fossdyke to Dunham. This action was probably illegal but obviously highly profitable. The records suggest that 160 ships a year did the voyage.

By the middle of the fourteenth century the line was beginning to decline and wool was being carried by road. The dyke was becoming seriously silted and in July 1335 a petition of complaint was sent to the king. He responded by setting up a commission to survey the dyke and to clear it. It appears that contractors were being paid to maintain the waterway but they weren't doing so. Although the situation was remedied, problems continued off and on for nearly two centuries. In 1518 the Bishop of Lincoln, William Atwater, was appointed commissioner and he stimulated interest and managed to start a fund-raising scheme. But on his death in 1521 the plan faltered and was quickly forgotten. The dyke languished in ill-repair for over a century. The situation was so bad that the state of the navigation was recorded in poetry by John Taylor (1580–1653). The poem, 'A very merrie wherry-ferry voyage' describes the Fossdyke as a 'ditch of weeds and mud' and records that a single stretch of 8 miles took 'nine long hours'.

In the early years of the eighteenth century, Lincoln council was struggling to maintain the navigation which had been vested to them under an Act of 1671. Revenue was enough to pay for minor works but it was clear that a major dredging effort was needed which was estimated to cost £3,000. Although a one-third share on the line had been sold to James Humberton in 1672, he was unable to assist and the council sought external help. This was found in the shape of Richard Ellison (of Thorne in the East Riding, Yorkshire). It was agreed that Ellison would maintain the channel to a depth of 3 ft 6 inches in return for a rent of £75 p.a. (£50 to the council, £25 to Humberton). The deal was sealed on 18 September 1740 and the revamped line was opened in 1744.

The reopening of the waterway helped restore Lincoln from the period of decay into which it had deteriorated since the sixteenth century. This revival is mirrored by the steadily increasing toll receipts along the Fossdyke: from just £700 in 1751 to £1,499 in 1774 to £5,908 in 1814. By 1805 there was a horse-drawn packet-boat that plied the line to York. By now the dyke had become a major conduit for wool en route via the Trent, the Humber, the Ouse and the Aire & Calder Navigation to the towns of the West Riding. The line also carried away corn, ale and pit props with the town receiving coal in return.

The lease on the dyke passed down the Ellison family over the years and, it has to be said, that they tended to enjoy the income without any particular inclination to maintain the line. In 1836 Richard Ellison IV became the owner of the lease at a time when income had risen dramatically, from £5,000 in 1813 to £8,000 in 1825; a situation no doubt aided by the gradual development of the rest of the country's canal system. The development of the trade at this time is no better illustrated than by the fact that some 40,000 tons of corn, harvested p.a. in Lincolnshire, was being shipped via the Calder & Hebble Navigation and the Rochdale Canal to feed the cotton workers of Lancashire. However, with this growth in traffic was a concomitant growth in complaints about the condition of the navigation. Again it was said that the Ellison family were creaming off the profits with no view to improve the line.

The Fossdyke meanwhile was silting up and legal battles followed in which many complainants sought to wrestle control of the waterway back into public control. The Lincolnshire farmers were producing and the merchants handling a steadily increasing volume of goods for despatch to

Yorkshire and Lancashire and it's no wonder they were concerned about their main route to market. The trade was largely in the hands of merchants whose yards and warehouses flanked the Brayford Pool. By 1836 it was said that the condition of the Fossdyke was so bad that a trip from Hull to Lincoln took as long as one from England to the United States, i.e. three weeks. The bridge at Torksey was a major impediment as vessels could not get under it when the Trent was high. The canal channel was reported to be too shallow and shoals near Saxilby meant that vessels with half a cargo had to discharge into lighters. Linseed cake and coal were becoming scarce in Lincoln. It is perhaps understandable that things were getting heated. A compromise bill in which it was proposed to make the navigation capable of drawing 5–6 ft of water was drawn up. The plan was to spend £40–50,000, shared equally between Ellison and public funds, to improve the line. However, the scheme was lost in a wave of ill-feeling. In 1839 Ellison, perhaps eager to rid himself of a headache, offered to sell his lease to the council for £16,000 but they resolved that the price was too high particularly as a railway to Gainsborough could be built for less.

Thus Richard Ellison IV set about improvements under his own steam. The whole line was dredged and the stretch from Drinsey Nook to Torksey widened. By this time 40,000 tons of coal p.a. were being delivered from the Nottingham coalfield. In return, the quays of Brayford Pool exported wheat, barley and flour to Manchester, cattle to Rotherham and Manchester, and wool to the West Riding. In return came hardware, cotton goods, earthenware, linen, woollens, timber, linseed cake and oil.

The first rumblings of railway competition were heard in 1825 when a line from London to Cambridge with extensions to Lincoln and York was proposed. It wasn't forthcoming immediately and not until 1846 did the Lincoln & Nottingham Railway provide a route to the Midlands. This was followed in 1848 by the Manchester, Sheffield & Lincoln Railway which connected the city with Lancashire and the West Riding. Ellison clearly can't have thought too long about ridding himself of his heirloom. On 21 December 1846 he granted a 894 year sublease to the Great Northern Railway at a rent equal to his net revenue plus 5 per cent, namely £9,570 p.a. Although the council and businessmen of Lincoln were alert to the dangers of monopoly, the railway soon made its market by capturing both the passenger and the freight traffic. By 1852 the packet-boats which regularly plied the waterway were gone. Freight traffic between 1848 and 1868 dropped by 70 per cent. Although the Great Northern was officially obliged to maintain the waterway, Fossdyke was gradually being more and more neglected and the channel steadily silted up. In 1854, for example, it took one boat owner 6 hours to drag his boat from the Pyewipe Inn into Brayford Pool (a distance of a little over a mile). Although some minor dredging was carried out, the situation did not improve very much. Despite this, water

Brayford Pool in the early 1900s

The Boat Museum Archive

transport still had its protagonists and barges still served the Brayford Pool. At the time a boat from Lincoln to Hull took 24 hours compared with the train which took 30 hours and cost more.

By the turn of the twentieth century Fossdyke was still carrying more than 75,000 tons of cargo p.a., including timber, agricultural produce and general merchandise. Perhaps the key factor in the continuing use of the line was the fact that most of the works in Lincoln still had river frontage. The main problem for the carrying companies was that there was not a constant water depth of 5 ft; an issue that didn't concern the railway company particularly.

Although by the First World War the navigation was declining, the use of the channel as a drain was still important and there was some traffic. In 1914 William Franklin Rawnsley, in *Highways and Byways in Lincolnshire*, states that the view of Lincoln Minster from Saxilby with 'the sails of barges in the foreground as they slowly make their way to the wharves . . . is most picturesque'. The barges in use at this time were 'low, round-nosed . . . with widespread canvas'. Even in 1949 Arthur Mee in *King's England* describes the barges making their way along the Fossdyke. Today the commercial traffic has gone but full use is made of the waterway by leisure cruisers who follow the route once taken by the Roman legions.

Keel under sail at Saxilby in 1930

W.E.R. Hallgarth/Ware/The Boat Museum Archive

The Walk

Start:	Saxilby railway station (OS ref: SK 892752)
Finish:	Lincoln railway station (OS ref: SK 975709)
Distance:	6¹/₂ miles/10 km
Map:	OS Landranger 121 (Lincoln)
Return:	British Rail Lincoln to Saxilby (enquiries: (0522) 539502)
Car park:	At Saxilby station
Public transport:	British Rail Lincoln

This is a pleasant stroll but there is a length of road walking (albeit on a pavement) at the beginning that may deter some. One possible way to avoid this is to take the Roadcar bus no. 352 (enquiries: (0522) 532424) which runs from Lincoln to Gainsborough. Some drivers may be prepared to stop at the Broxholme turning thereby avoiding the stretch along the road.

From Saxilby station, walk out to the High Street and turn right to go past the post office. Continue into Bridge Street where the Fossdyke appears on the right. Walk past The Ship and The Sun Inn. Just before a road bridge, bear right up a slope to the A57 Gainsborough Road. Continue over a junction where the A57 now becomes the Lincoln Road. Continue along the

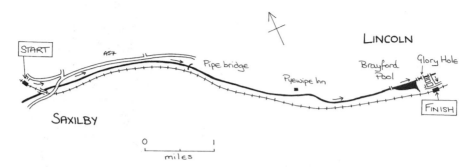

The Fossdyke Navigation

pavement until you pass the turn off for Broxholme. Cross to the opposite side of the road and go over the River Till which enters the Fossdyke from the left. Just after the crash barrier, a footpath sign points along the left-hand embankment of the dyke.

After the traumas of the road, the solitude of the dyke gradually takes over and it gets quieter and quieter until you feel that you're about to come to a dead stop. After rounding a bend and passing a high pipe bridge, the lofty position of the embankment allows fine views of Lincoln Cathedral in the distance. By now there's only the herons and the fishermen, both of whom sit down by the dyke staring mournfully at the waters. The banks here appear solid but they are apparently made of sand, occasionally quicksand, and in former times were a constant headache to lessee Richard Ellison who found that they often collapsed. This was particularly the case after a vessel had run against them.

The land to either side of the dyke here was always susceptible to flooding and before the nineteenth century was inevitably under water. In 1804 John Rennie was called in to advise and, under a scheme known as the Lincoln West Drainage, an area of 4,000 acres was drained. By 1806 the plan was complete and one local correspondent is said to have reported that 'fat beef was growing on Swanpool where fishes lately swam'. The scheme was undertaken partly by the construction of an embankment and partly by the insertion of catchwater drains to intercept upland water.

Two miles from the road, the line bears left and then goes under the new flyover that forms part of the Lincoln ring road. About a hundred yards further on we pass the canalside Pyewipe Inn (Chef & Brewer). Charles Hadfield suggests that the name Pyewipe derives from peewit – that is the lapwing. The dyke passes over an aqueduct (which crosses a drainage dyke). The Fossdyke then, unusually, meanders slightly before reaching the outskirts of Lincoln. The first signs of this are the golf course to the left and then the British Waterways depot (the East Midlands Navigation Lincoln office) to the right. By now the towpath has broadened to a dirt track.

The path reaches a road (Fossbank). Continue close to the dyke along a pedestrian alleyway to reach an electric lift bridge which marks the entrance to Brayford Pool. Continue along a cobbled alley to a road. Bear right to go past the home of the Lincoln Boat Club and Bunnys Wine Barge. This leads to a pedestrianized area from where there are good views across the pool to the river disappearing over in the far right corner. In medieval times, the bulk of what is now Brayford Pool (and then known as Bradeford) was simply a marsh with a ford across it. The area was probably first used as a quay in Roman times. But until the eighteenth century Brayford had been a pool of mud and sand without wharves on its banks, surrounded by gardens and orchards. With the passage of the Horncastle Act in 1792 (see Further Explorations), trade was improved. Gardens were turned into coal-yards or covered with warehouses and the pool quickly became a port. By 1810 the pool could admit vessels of 50 tons alongside the quay and was no longer dry in summer. Today it is a popular spot for cruising boats who are warned that, although wide, it is rather shallow away from the dredged through channel that runs along this side.

The walk now passes the Royal William IV pub; a fine place to sit outside on a warm summer's lunch-time. If not stopping, continue past the stop gates and under the road bridge to walk by the side of the River Witham. Until a bridge was built by Richard Ellison here in 1801, the only way

Brayford Pool in 1992

pedestrians could cross was via a series of stepping stones and then only in summer. This path leads round the backs of some big high-street stores and winds around a solicitors building along North Witham Bank to reach the most famous waterway sight in the area: the Lincoln High Bridge and the Glory Hole.

The Glory Hole is a single, vaulted arch that is 22 ft wide. It was originally built in the twelfth century although the buildings that sit atop it date from the sixteenth century. When the Horncastle & Tattershall Canal Act was passed in 1792, the possibility for boats going from the Brayford Basin through the Glory Hole and on to the River Witham was increased. However, at that time, the channel was too shallow to allow such traffic to pass. Goods were off-loaded on one side of the bridge, transported by land to the other side and reloaded. The H&TC Act therefore enabled the new company, together with the Witham and Sleaford Navigations, to deepen

The Glory Hole in central Lincoln

the channel under the bridge to 3 ft 6 in. This was done in 1797 and the cost shared between the interested parties.

The path now goes up some steps to the left to reach the pedestrianized shopping centre. Continue down the other side and walk on along Waterside North; a pleasant city waterway which is surely a model for others. Cross the River Witham at a footbridge and walk straight on (i.e. turn right from the towpath course) and along Sincil Street to reach the railway station.

Further Explorations

The Horncastle & Tattershall Canal runs for 11 miles from the small market town of Horncastle south to a junction with the River Witham. The Tattershall Canal was originally a separate entity. It was built in 1786 by John Gibson and John Dyson and ran for about a mile from the River Witham at Tattershall Ferry to the village. The new scheme for a line to Horncastle owed a considerable amount to a local dignitary, the botanist Sir Joseph Banks, who saw the benefits that would accrue to the town if it had a route to export its agricultural produce and for the import of salt, coal and other heavy goods. The canalization of the River Bain was enabled by the Horncastle Navigation Company Act passed on 11 June 1792. This Act enabled the new company to buy the Tattershall Canal as well as to deepen the channel under the Lincoln High Bridge (see above).

William Jessop surveyed the line but did not become involved in its construction. This was a pity as the building of the canal was dogged by a string of incompetent engineers. By the time the line was opened to Dalderby (2½ miles south of Horncastle) in 1797, the company had run out of money and this remained the terminus for the next five years. Further funds were raised following an Act in 1800. This allowed John Rennie to plot a new, straighter course to Horncastle and the line was fully opened on 17 September 1802. It had cost £45,000, four times Jessop's original estimate. In all there were twelve locks (measuring between 71 and 75 ft long and 15 ft wide) and a fall of 84 ft to the Witham. The completion of the canal made Horncastle into a mini boom town. The traditional agricultural businesses (such as wool merchants) expanded greatly and many new ones, such as the manufacture of farm implements, traders in artificial manures and coal merchants, were started. The Horncastle Gas Light and Coke Co. was also able to open in 1833. From 1826 a steam packet passenger service was operating from Horncastle to Lincoln and Boston.

Up to 1812 profits from the line were used to pay off the not inconsiderable debt, but by 1813 trading showed a credit of nearly £1,000 and a dividend of 5 per cent was issued. Dividends varied between 5 and 7 per cent from then until 1856. In the early 1850s the canal was moving nearly 10,000 tons of coal a year into Horncastle and taking just over 5,000 tons of corn, wool and other goods back on to the Witham.

In 1848 the Great Northern Railway's line from Lincoln to Boston was opened and a branch to Horncastle was completed in 1854. The canal company responded by reducing its tolls so that receipts stayed at roughly £850 p.a. during the first decade of railway competition. But trade gradually declined and the last dividend of 1½ per cent was paid in 1873. An attempt to spruce up the line during the late 1870s was not very successful and in 1889 the local council proposed to close it. This was done on 23 September 1889. Despite this, coal was still being shipped in to Coningsby in 1910. Norman Clarke reports that the line could easily be restored for use by pleasure boats if only the spirit were willing.

It is possible to walk along various stretches of the canal and there is a fascinating short stroll of about 2½ miles from the centre of Horncastle. The town is on the Lincoln to Skegness A158 on OS Landranger 122 (Skegness) at ref: TF 258695. There are a number of small car parks near the Market Square. Alternatively Roadcar bus no. 6 goes to Horncastle from Lincoln (enquiries: (0522) 532424).

From Market Square turn left to walk along High Street and then right along Bull Ring. Before this road reaches the A158, turn right along Wharf Road with the canalized River Waring to the left. This is the site of the old Horncastle Canal South Basin. Continue along Wharf Road until the canal goes under the, comparatively new, road bridge. Cross the main road at the pedestrian lights and then turn right to walk along the grassy left-hand bank of the canal. The path goes over the South Ings Drain. The drain was built to straighten the rather convoluted course of the Old River Bain which it joins about a mile to the south. The sluice (or staunch as it is called) was used to control the level of water in the canal. The path passes in front of the swimming pool and then on to a footbridge that goes over the course of the canal which bends left here to start its line to the River Witham. The swimming pool was originally a dry-dock that was owned by the navigation company. It was sold and converted into a pool in 1875. Near here is a comparatively recent weir. For the full walk, turn left to walk down the left-hand bank. This route takes you past the site of Horncastle Lock. At the first bridge (near Thornton Lodge Farm), cross and return on the opposite bank (following the course of the old railway line and now the Viking Way). This returns you to the other side of the footbridge where you will meet those who didn't do the full walk and who have merely crossed the bridge and turned right.

The path now follows the course of the River Bain and heads north to the North Basin. We cannot follow it but have to walk out to a road. Turn right to pass the Maypole House School to reach the Lincoln road. Cross this and go along West Street. Bear right into Bridge Street where the road crosses the River Bain. The road leads back to Market Square. Before jumping in the car, turn left past the post office to go along St Lawrence Street. This leads to the site of the old North Basin opposite Watermill Road. The wharf is now a car park. Apart from the watermill, the basin here was surrounded by two windmills, and malt kilns. There was also a tannery and currier's workshop, a boot and shoe factory, a saddler, a basket maker's, a basket maker's, coal merchants, cutlers, gasfitters, ironmongers and steel merchants, a chandler, a nail maker, a blacksmith, a coachbuilder, a pipe maker, a glover, and a plumber. Perhaps the most vigorous business down here, however, was the beerhouse and brothel run by a Mr Daft.

Further Information

There is no Fossdyke society as such but the Inland Waterways Association has a Lincoln & South Humberside branch. For further information contact the London office (see Appendix B).

The history of the Fossdyke is described in:

Boyes, J. and Russell, R., *The Canals of Eastern England.* David & Charles, 1977.

For information on the Horncastle & Tattershall Canal:

J. Norman Clarke, *The Horncastle & Tattershall Canal.* The Oakwood Press, 1990.

5
THE HUDDERSFIELD CANALS
Mirfield to Marsden

Introduction

While the Sir John Ramsden's Canal to Huddersfield was a minor and highly sensible arm of the Calder & Hebble Navigation, the Huddersfield Narrow Canal was a bold and almost reckless endeavour. It attempted to scale the Pennines head-on: rising 436 ft from Huddersfield via forty-two locks in just under 8 miles and burrowing through the Pennine Hills along the longest canal tunnel ever built in Britain. This is the kind of stuff that great adventures, if not great profits, are made of.

Sir John Ramsden's Canal forms a junction with the Calder & Hebble Navigation at Cooper's Bridge between Mirfield and Brighouse. From there, it runs for 3³/₄ miles to Aspley Basin in Huddersfield. The SJRC and the Huddersfield Narrow Canal form a junction, in theory at least, near Huddersfield University. The HNC then starts its 19³/₄ mile route over the Pennines to Ashton-under-Lyne. En route it passes through Milnsbridge and Slaithwaite before reaching Marsden and the Standedge Tunnel. Exactly 3 miles 418 yd further south-west, the line re-emerges from the hill at Diggle and starts its downward trek through Greenfield and Mossley to reach Stalybridge where it has been infilled and built over. The final stretch begins on the other side of town and continues for just a mile and a half to Ashton-under-Lyne where the canal meets the Peak Forest and Ashton Canals at Portland Basin.

The Tame Valley is highly industrialized but is always pleasant and/or interesting and is often both. It also has the carrot of leading on to the wide-open spaces of the central Pennines. This means that, even without boats, it's one of the finest towpath walks around.

History

The Act which enabled construction of the Calder & Hebble Navigation received its Royal Assent in June 1758. By 1774, 21½ miles of river from the Aire & Calder Navigation at Wakefield to Sowerby Bridge were open for traffic via twenty-seven locks each measuring 57 ft 6 in by 14 ft 3 in. With this important waterway from the east coast into the mill town of Halifax in place, it wasn't too long before the prospects for a navigable line to Huddersfield came under consideration.

The first mention of a branch canal to Huddersfield was made in the C&HN Act of 1758. In 1766 Robert Whitworth surveyed a line of 3¾ miles from Cooper's Bridge (about 1½ miles west of Mirfield) to King's Mill, Huddersfield and several petitions to Parliament followed without success. By 1773 the major local landowners, the Ramsden family headed by a still under-age Sir John Ramsden, sponsored another survey, this time by Luke Holt and John Atkinson. In the bill, promoted by the trustees of Sir John in 1774, it was claimed that the canal would benefit Huddersfield through the cheaper import of raw materials. But, as Charles Hadfield points out, the Ramsdens owned virtually the whole of Huddersfield at that time and the family must have seen the venture as a safe, and potentially highly lucrative, investment. The Act, for what was to become known as Sir John Ramsden's Canal (but which is also known as the Huddersfield Broad Canal), received its Royal Assent on 9 March 1774. The Act proposed a line of nine locks built to take Yorkshire keels measuring 57 ft 6 in long by 14 ft 2 in wide. The cost was estimated at £8,000. Tolls were set at 8d. a ton for coal, lime and stone, and 1s. 6d. a ton for other goods.

The canal, which opened in the autumn of 1776 at a cost of £11,974, was a modest success from the off. Huddersfield was becoming an important wool-spinning town and was growing rapidly. The SJRC thus became an important conduit for the export of its products as well as for the import of raw materials. The canal carried textiles, coal, lime, stone, timber, wool, corn, glass and other goods. To cope with this, new docks, wharves, and warehouses were built at Aspley Basin in Huddersfield where there were also a number of hostels built for the canal workers.

Meanwhile, on the western side of the Pennines, the Ashton Canal had been built from a basin at Ducie Street, Manchester to Ashton-under-Lyne. This project had been authorized by Parliament in June 1792 and it wasn't too long before it was realized that if the SJRC and the Ashton Canal were linked, the result would be the shortest water route between the east and west coasts. A group of Ashton Canal shareholders began to consider the idea in May 1793 and, following a survey by Nicholas Brown, Benjamin

Outram proposed a line at a cost of £178,748. On 4 April 1794 the Act was passed to enable the construction of a narrow canal (to save both water and costs) of 19³/₄ miles by seventy-four locks from the Ashton Canal via Marsden to the SJRC at Huddersfield. The most significant engineering work on the line was to be a tunnel through Standedge – a distance of 3 miles 176 yd (later extended). The Act authorized share capital of £184,000 and provided powers to raise a further £90,000 if it was needed.

Construction work started in July with Benjamin Outram as engineer. By November 1796 the line was open from Ashton to Stalybridge and from Huddersfield to Slaithwaite and, presumably, some toll income was being received. However, the difficulties of building the Standedge Tunnel were already becoming evident and by 1798 it was becoming increasingly hard to find a contractor willing to take the job on. The situation was not helped by tenuous finances and the fact that some sections were so poorly constructed that they had to be rebuilt. When Robert Whitworth acted as deputy to Outram during the latter's illness, he described the masonry and earthworks as the worst of any he had seen.

By 1799 the canal was virtually complete with the northern length from the tunnel to Huddersfield, and the line from Ashton to Saddleworth, finished. But only 20 per cent of the tunnel was complete. With finances exhausted, the company tried to raise £20,000 by mortgage but only succeeded in raising £8,000. Further actions from the company brought in some additional funds but it was still short. Serious floods in 1799 didn't help and another £7,000 was needed to undertake repairs. The situation was remedied by an Act in 1800 which enabled the company to raise an additional £20 on each share already issued. The tunnel was still a major headache. At one stage it was thought that a railway should be built across Standedge to link the two ends of the canal, but this idea was rejected as impracticable. By June 1801 only 1,000 yd of the tunnel were cut with another 1,000 yd only partially bored. The company's financial problems had been partly solved when it raised £48,190 by the issue of new shares but the tunnel remained unopened.

Meanwhile one John Rooth had started a small carrying business on the western side of the tunnel and by April 1801 he had been made superintendent for the canal. By October, with Outram having resigned, he was given the additional responsibility for the tunnelling work which was proceeding, if slowly, from each end. The work was paid for by the creation of further new shares in 1804 and then by calls of £16 on all shares in 1806. These actions raised a further £125,914. In 1806 Thomas Telford was called in to survey the work to date and to make recommendations on how to proceed. With his plan and the additional funding, the company was able to complete the line. Although the company's bankers and the Diggle reservoir were both broke during the course of 1810, December of that year saw the tunnel

Workmen repairing a breach on the Huddersfield Canal at Golcar, probably at the turn of the nineteenth century

Ware/The Boat Museum Archive

finally finished. It was (and is) the longest canal tunnel in Britain and cost a total of £123,804 against Outram's estimate of £54,187.

The grand opening of the Huddersfield Narrow Canal (and hence the Huddersfield trans-Pennine route from Manchester to Hull) occurred on 4 April 1811. The completed canal was just under 20 miles long and comprised some seventy-three locks built to the narrow boat gauge of 72 ft by 7 ft. The canal had five aqueducts: Stalybridge, Royal George (Greenfield), Saddleworth, Scarbottom and Paddock. It had ten reservoirs, four of which held more than fifty million gallons: Haigh, Slaithwaite, Red Brook and Swellands. The total cost was £273,463.

Unfortunately trade did not come to the HNC either as freely or as fast as anticipated. In 1816 the canal was described as 'a bad speculation for the subscribers and . . . a warning to others how they engage in such tremendous works'. The key problem was that by 1811 the Rochdale Canal had already established itself and was cheaper to use. In 1813, for example, traffic along the HNC totalled 40,460 tons compared with the Rochdale's 290,508 tons. This competition forced the combined HNC, SJRC and Ashton Canal to reduce their tolls. In addition, a plan to charge 1s. 6d. to pass through Standedge Tunnel was dropped. The situation was not helped by the fact that the HNC had been built to Brindley's narrow boat gauge as used on the Ashton and Peak Forest Canals to the south. The classic narrow boat was, however, too long for either the SJRC or the Aire & Calder Navigation both of which took barges (Yorkshire keels) measuring 57 ft 6 in by 14 ft 2 in. As a consequence cargo had to be transhipped at Huddersfield. To overcome this problem a number of short narrow boats were constructed, able to work both lines as well as the waterways in the Calder Valley. Their success, however, was always going to be limited given their reduced cargo-carrying capacity. Attempts by the HNC to convince the Aire & Calder to make its locks longer were never successful. The HNC were similarly unsuccessful in convincing the Rochdale to join with it to fix prices. This predicament was not helped by the problems with the water supply (the canal was often closed because of drought) and a new reservoir capacity had to be built. On top of all this, maintenance of the HNC was expensive and there was criticism of bad workmanship on the locks. This forced the company into a further round of building and repair.

Despite the difficulties, trade into the 1820s grew gradually so that by 1822 receipts were £12,284. Cargo included coal, corn, lime and limestone as well as general merchandise. The towns along the route, like Marsden, Slaithwaite and Milnsbridge benefited from the canal as did the many mills along the valley. In the following years the company appeared to have much

Lock at Holme Mill, Slaithwaite, in the early years of the twentieth century

Ware/The Boat Museum Archive

to do but still the money did not come in to pay off the capital invested, and it wasn't until 1824 that the shareholders received their first dividend of £1 per share. Even so this did not mean an end to the company's money problems and authority was given to borrow up to £10,000 to promote trade. Five unfruitful years followed during which a new warehouse in Manchester was built and reservoir capacity expanded. As a consequence no dividends were paid until 1831 which again saw £1 per share.

Trade continued to increase slowly and in 1833 a dividend of £1 10s. was paid, rising to £2 in 1837–9. But in 1835 both the HNC and the SJRC attended a meeting with the Aire & Calder, the Calder & Hebble, the Rochdale Canal Company and the Ashton Canal Company to discuss the coming of railway competition. Some tolls were cut but in many ways this was a phoney war. By 1838 the company was forced to reduce its tolls in the face of direct competition from the Manchester & Leeds Railway and the battle began in earnest. Although the committee was able to pay a dividend of £2 in 1841, by 1842 it was unable to pay anything. Tolls of £2,435 in 1841 had dropped to £1,720 in 1842. Further toll cuts were implemented in 1843 followed by economies of £2,000 a year but this still failed to raise enough trade.

In 1844 the Huddersfield & Manchester Railway was promoted for a line from Stalybridge to Cooper's Bridge, a route that ran almost parallel to the combined HNC and SJRC. The HNC committee who had been asked to recommend a new plan for the company was quick to take the hint. It soon reached agreement with the H&MR for amalgamation. Following the passage of the necessary Act on 21 July 1845, the new Huddersfield & Manchester Railway & Canal Company paid £183,730 for the HNC and £46,560 for the SJRC. The canal was a valuable asset to the railway company. Standedge Tunnel, for example, was used extensively during the building of a new rail tunnel to move spoil as well as for ventilation and drainage.

Unlike many canals which operated in direct competition with a parallel railway, the HNC/SJRC worked (at least in theory) in co-ordination with its younger neighbour. However, in 1847 both the railway and the canal were leased in perpetuity to the London & North Western Railway. The LNWR's enthusiasm for the canal was somewhat less overt. The year after it was absorbed by the LNWR, the SJRC had carried 120,207 tons, but from then on through traffic declined. There was a brief revival in 1856 when the Aire & Calder tried to rebuild a carrying trade to Manchester via Huddersfield but, generally, cross-Pennine trade, and hence that on the HNC, evaporated. Local traffic on the SJRC held up comparatively well with eighteen to twenty-five boats a day working the waterway in 1863. In addition, the LNWR began to sell canal water. In 1875 the extraction rate was some 200,000 gallons a day.

In the early 1890s, 75,000 tons of cargo a year were being exchanged with the C&HN; three-quarters of which was to and from Huddersfield itself. The cross-Pennine route, however, was flagging. When the canal tunnel closed for repairs during 1892 and 1893, the LNWR shipped the canal cargo by rail and when the tunnel reopened that trade stayed with the railway. The canal to either side, however, remained active. Remarkably enough the HNC carried 161,899 tons in 1898 compared with 169,487 fifty years earlier.

By 1905 the transfer from the C&HN amounted to 53,850 tons to and from Huddersfield with 11,020 tons going on along the HNC. Little of it was reaching the tunnel and traffic over the summit ceased in 1905. Trade on the lower sections of the HNC continued after the First World War but was all but gone by the Second. In 1944 an Act of Parliament was obtained by the London Midland & Scottish Railway (into which the LNWR had been absorbed in 1923) to abandon the line. On 1 January 1945 the C&HN bought the SJRC together with a half-mile length of the HNC for £4,000. The rest was then abandoned. At this time eighty barges a month were carrying power-station coal. As it was now part of the C&HN, the SJRC was nationalized on 1 January 1948.

The last recorded through passage from Ashton to Huddersfield took place in 1948 when the *Ailsa Craig* took Tom Rolt, Robert Aickman and others on what sounds like an eventful journey (the trip is described by Rolt in *Landscape with Canals* (Alan Sutton, 1986)). This famous excursion, which occurred at the very beginning of the canal restoration movement, was only possible by the application of brute force from twelve men supplied by the British Transport Commission who hauled the boat over the shallows. During the course of the trip *Ailsa Craig* sank at least once and became wedged in the then still barely open Standedge Tunnel. The lock-gates on the HNC were removed in the early 1950s and, in October 1953, the coal trade along the SJRC ceased. Although the HNC was considered long gone by the British Transport Commissions' Board of Survey in 1955 and by the Transport Act of 1968, the SJRC was listed as a cruiseway under the 1968 Act. Luckily the channel of the HNC was maintained as a reservoir for local industry and with the formation of the Huddersfield Canal Society in 1974, restoration began. Since then a phenomenal amount of work has been carried out with the objective of seeing cross-Pennine boat traffic by 4 April 2001 at the latest. Perhaps the most encouraging evidence of the success of the venture occurred in 1988, when an Act of Parliament was passed to officially allow boats to navigate the restored lengths of the HNC.

The Walk

Start:	Mirfield railway station (OS ref: SE 203195)
Finish:	Tunnel End, Marsden (OS ref: SE 039119)
Distance:	14^1/$_2$ miles/23 km (or shorter)
Map:	OS Landranger 110 (Sheffield & Huddersfield) plus small portion on 104 (Leeds, Bradford & Harrogate)
Return:	British Rail Marsden (or Huddersfield) to Mirfield. Hourly weekdays, every two hours on Sunday(enquiries: (0484) 545444)
Car park:	Mirfield station (southern side)
Public transport:	British Rail serves Marsden, Huddersfield and Mirfield

With the proximity of the railway for the entire length, this walk can be conveniently divided into two shorter lengths: one along the Sir John Ramsden's Canal from Mirfield to Huddersfield of 6^1/$_2$ miles (10 km) and one along the Narrow Canal of 8 miles (13 km).

Mirfield to Huddersfield

From the car park at Mirfield station, go down the steps and turn right to go under the railway bridge. Within a short distance the road crosses the Calder & Hebble Navigation. Turn left to walk along the right-hand bank.

The C&HN runs from a junction with the Aire & Calder Navigation at Wakefield, for 21^1/$_2$ miles along the Calder Valley to Sowerby Bridge near Halifax. For part of its route it uses the River Calder although some sections consist of artificial cuts which shorten or simplify the line. Built to provide the woollen merchants of Halifax with cheap water transport, the C&HN received its Royal Assent in June 1758 and was open to vessels (measuring 57 ft 6 in by 14 ft) as far as Sowerby Bridge in 1774 at a cost of £75,000. The line was a prosperous one and even when the railways came to the Calder Valley in the 1840s, it continued to be busy. Part of its success was due to the fact that, with the Rochdale Canal at Sowerby and the SJRC at Cooper's Bridge, it formed a component of two of the three cross-Pennine canal routes. A wide range of cargo was carried including coal, wool, stone and foodstuffs. Competition began to have an effect in the 1860s but it was only after the First World War that traffic levels dropped dramatically. Despite this, trade continued until after the Second World War when many

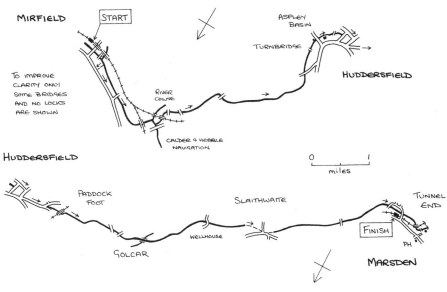

The Huddersfield Canals

barge owners saw a threat from nationalization and either sold up or ceased trading. The last cargo vessel to navigate the full length of the C&HN reached Sowerby Bridge in 1955 while the last into Brighouse ran in the early 1960s. Coal traffic from Wakefield to Thornhill power station, Dewsbury, continued for some time after this. The C&HN was always viewed relatively favourably by the British Transport Commission and was classified as a cruising waterway in the 1968 Transport Act.

The walk starts near a maltsters on the left bank and then soon passes a boatyard, complete with a covered dry-dock. Many hundreds of wooden-hulled Yorkshire keels were built at the Ledgard Bridge yard in years gone by and motor cruisers are now repaired here. The canal engineer William Jessop and contractor John Pinkerton had a dry-dock for many years at Mirfield made out of an old cut they had replaced in 1776.

We leave the Mirfield cut via Ledgard Bridge Flood Lock and Ledgard Bridge itself to meet the River Calder for the first time. After crossing a redundant railway bridge, the towpath ends and we are forced to follow the path round to the A644 Brighouse to Dewsbury road. Turn left and walk on for approximately half a mile. Just after The Pear Tree Inn, turn left into Wood Lane. The first bridge crosses the River Calder and the second goes over the Battyeford cut. Turn right along Waterside Walk and back on to the left-hand bank of the navigation.

This short, quiet section ends at another footbridge and stop lock where

Entrance lock and lock-house at the start of the Sir John Ramsden's Canal

the C&HN and the Calder once again merge. Go straight on to cross the river and turn left to continue along the right-hand bank. This leads to a further artificial cut which appears as a right turn near the first of the Cooper's Bridge Locks. Walk along the cut to reach the next road bridge which is Cooper's Bridge (carrying the A62 Huddersfield–Leeds road). Go up to the road and turn left to cross the cut. A second bridge crosses the river. From this to the left, one can see the start of the SJRC. Take a path on the far side of the bridge that goes left down to the canal. Cross the small bridge and turn right to walk along the left-hand bank of the canal and alongside lock 1 (Cooper's Bridge Lock), the entrance lock from the River Calder. The lock is typical of the SJRC; being a broad lock measuring 57 ft 6 in by 14 ft 2 in. The lock-keeper's cottage still stands on the right.

From here the canal swings left to run parallel with the River Colne (to the left). The route from here to Huddersfield isn't aesthetically pleasing but the curious mixtures of organic solvent smells will keep you constantly at wonder as to what precisely the factories along the way are up to. After bending right, the line goes under the Huddersfield to Leeds railway and on to lock 2 (Colnebridge Lock) and Colnebridge Road Bridge. The view is now dominated by the brick viaduct built in 1910 to carry the Midland Railway extension to Huddersfield. It's a splendidly built structure with some delicate features that wouldn't shame a garden wall. Sadly it hasn't seen active service since the track was lifted in 1937. After the viaduct a short stretch leads to Ladgrave Lock (No. 3) and a dye-works. The Huddersfield Canal Society's booklet points out that the proximity of the bridge at lock 3 limits the length of the balance beams and makes it impossible for the whole length of the lock to be used.

After Longland's Lock (No. 4) the canal goes under a footbridge and then another old railway viaduct; once part of the Huddersfield to Kirkburton line. The canal now bends right under Leeds Road Bridge (A62). Just beyond it is an old mill in which the loading doors have been converted into windows. There are now extensive playing fields opening up the view to the left and the first sight of Huddersfield ahead. Turnpike Road Lock (No. 5) is quickly followed by Riding (or Riddings) Lock (No. 6) and Fieldhouse Green Lock (No. 7) which has Fieldhouse Lane across its lock tail bridge. After passing the 2 mile post, another short pound brings us to Falls Lock (No. 8) and then Red Doles Lock (No. 9) with its lock tail bridge, lock cottage and rather dilapidated outbuildings.

We have now reached the summit pound on the SJRC. The line goes under a concrete bridge and past the council incinerator. We then go under the girder bridge which carries Hillhouse Lane and walk on to the next bridge: Turnbridge or Locomotive Bridge. This extraordinary structure, built in 1865, is a lifting bridge. By operating the wheels and chains on the far bank, the bridge deck rises parallel with the waterway. It is called Turnbridge because of the swing bridge that formerly occupied the site.

It is now a short walk into Aspley Basin. The basin was originally the terminus of the SJRC and was later the transhipment basin for cargo undertaking the full cross-Pennine route; the narrow boats of the HNC being too long for the locks of the SJRC and the Yorkshire keels being too wide for the locks of the HNC. The basin was equipped with numerous wharves, cranes, mills and warehouses and must have been a busy spot. Indeed, the basin was still receiving coal into the 1950s. Many of the original buildings have been flattened; the most notable clearances occurring in 1963 when the Wakefield Road was widened. There are some (barely) surviving original buildings and a basin. There is also, for those contemplating the full walk to Marsden, Ramsden's Landing Restaurant and PJ's Café Wine Bar.

The line of the canal now continues under the busy Wakefield Road ahead. To reach it, turn left to walk past the café/wine bar and on to St Andrew's Road. Turn right and then right again along Wakefield Road. After 100 yd, turn right to go down some steps and under the road. The path now leads to the environs of Huddersfield University. Almost immediately on the right is a former transhipment warehouse. It was built in 1778 for Sir John Ramsden and used for wool storage. Next to the warehouse is a crane which consists of a cast-iron base with a wooden jib. There is another crane sited further towards the main university building. This was once positioned on a small canal arm which looped around to the right.

Walk on along the path in between the buildings and over the Shore Mill Goit which feeds the SJRC with water from the River Colne. Shortly the line bends sharp left to a culverted bridge which, now, separates the SJRC from its narrow sister and, apparently, houses the university's sewer. Beyond this, the SJRC originally went straight on across Firth Street to join the Colne near Kings Mill. The HNC, however, bends right to the first narrow lock: Stanley Dawson Lock (1E – where E stands for east of the Standedge Tunnel). The lock and the following stretch of canal are enclosed by the surrounding mills to form a kind of dank chasm through which the HNC attempts to find a path. In fact these mills provided the final traffic for the canal with the last load of coal being delivered to the wharf above the lock in the early 1950s. The wharf can be picked out by the stone setts and what remains of the mooring rings. After Coal Wharf Lock (2E) is another coal wharf serving the adjacent Priestroyd Mill and then Queen Street South Bridge which still bears the Ramsden coat of arms. From here the canal is

The first lock on the Huddersfield Narrow Canal in Huddersfield

completely blocked by a 'new' building. In restoring the HNC, the canal society hopes to burrow underneath. To do this the line will have to be lowered by shifting lock 2 to the other side of the tunnel.

To continue the walk or to complete this stretch to Huddersfield, turn left and then right along Colne Road. Turn right into Chapel Hill. Go up the hill, over the line of the canal and on to a major road junction. If continuing the walk, turn left to go along the Manchester Road – signposted A62 to Oldham. If completing the walk here, go straight on into the pedestrianized town centre (New Street). Walk on for about 1/4 mile and turn left into Westgate and then second right into Railway Street. Huddersfield railway station is on the left.

Huddersfield to Tunnel End, Marsden

If starting from Huddersfield railway station, walk on to Railway Street and turn right to Westgate. Turn left and then right along New Street. This pedestrianized street soon arrives at a major road. Bear half right along the Manchester Road – signposted A62 to Oldham – here you will meet those who are continuing from Mirfield.

The second half of the walk starts along the A62 from where the canal can be seen wending its way through the factories in the valley to the left. Eventually the road crosses the canal at Longroyd Bridge. Longroyd Bridge Lock (4E) can be seen on the right-hand side. To reach the towpath, walk around to the left of some buildings into a car-spares garage and then right to reach the canal. On the right is the lock. Turn left to walk along the left bank. Continue round a bend and under a railway viaduct (the Huddersfield to Penistone line). Beyond the viaduct the canal turns sharply to Paddock Foot Lock (5E) and then, almost immediately, Paddock Foot Aqueduct. The positioning here is most unusual with the lock by-wash weir starting in the middle of the aqueduct span. The buildings beyond the aqueduct are part of an iron foundry while across the canal is a boilermakers. This is closely followed by Mark Bottom Bridge.

Stoney Battery Road Bridge has been replaced by an embankment and the canal culverted. Shortly we pass Mark Bottom Lock (6E) and arrive at Fountain Lock (7E), the first of the flight of five that proceeds through the mills of Milnsbridge. Contrarily a notice on the wall of one factory overlooking the canal states that bathing is prohibited. As this option is thus excluded, we carry on along the next stretch and past locks 8E and 9E (Roller and Isis Locks). Market Street/Whitely Street Bridge is then followed by Spring Garden (10E) and Library (11E) Locks. After Morely Lane Bridge we pass lock 12E (Rough Holme Lock) with its small lock hut, and the canal crosses the River Colne via Golcar Aqueduct. Shortly thereafter is a rare sighting for the HNC of Golcar (or Holme Mills) Swing Bridge.

By the time we reach locks 13E/14E (the Ramsden Locks), we feel that we are entering into true Pennine scenery and finally beginning to leave the old mill buildings behind. Above 14E the towpath forms a causeway between the canal and Ramsden Mills mill-pond. There are sluices into the dam where the canal swings right near the river. Appleyard Bridge is followed by Golcar Brook Lock (15E) and the Westwood Locks (16E and 17E). In 1894 Lowestwood Lane Bridge was widened with iron. After 17E there is a brick lock hut and the remains of a wharf. We pass locks 18E (Can) and 19E (Holme) then Lees Bridge which was originally a swing bridge.

Locks 20E (Spot) and 21E (Waterside) are followed by the village of Slaithwaite where the canal comes to an abrupt, culverted, end. This section of the canal was infilled in 1956 when there was little hope for the line. Luckily nothing has been built on top and the course of the old waterway can still be followed as a path through an area of well-kempt grass. Eventually the path reaches a car park; the boundary of which marks the former site of Pickle Lock (22E). Continue on to the road where there are a small selection of shops. The canal line continues along a grassy border and across a road and a small area of parkland (opposite the Shoulder of Mutton pub) to the tail bridge of lock 23E (Dartmouth Lock). The lock itself has been infilled and now sits underneath a picnic site but just a little further on water has been restored.

The scenery now becomes more wooded. Slaithwaite reservoir, one of ten that served the canal and one of the biggest at 68,200,000 gallons, lies up the hill to the right. After the infilled lock 24E (Shuttle Lock), a road crosses. We pass locks 25E and 26E (Shaker Wood and Skew Bridge Locks) and start our ascent to Marsden. Locks 27E–30E (Mill Pond, Waring Bottom, White Hill – beyond which there is a new road bridge – and Bank Nook) follow. At Booth Lock (31E) a former winding hole can be seen with a canal cottage dating from 1858. The canal then bends to the right and we pass Cellars Clough Mill and go under a footbridge which crosses to the nearby Sandhill cottages. To the right of the next lock (32E – Pig Tail), is Sparth reservoir (8,150,000 gallons) which is, along with Slaithwaite reservoir, one of only two that still supplies the canal with water. Locks 33E–38E (Sparth, Cellars, Moorvale, White Syke, Smudgers and Colne) follow in short order. Ahead we obtain a clear view of the lumpy heaps on the side of Standedge; spoil that was dumped during the excavation of one of the four tunnels that pass through the hill.

The canal now goes under Warehouse Hill Bridge and on to Warehouse Hill Lock (39E). The canal then swings right past an old wharf and some cottages and on through a cutting that is, in fact, made up of a succession of heaps of tunnel spoil. We pass locks 40E and 41E (Hopper and Dirker Locks), go under two bridges and arrive at Marsden railway station and the appropriately named Railway Lock (42E). With this the canal reaches the summit pound which, at 645 ft, is the highest point of any artificial water-way in Britain. There is now just half a mile to Tunnel End. The line goes under a road bridge and on through a narrowed section which was formerly the site of a bridge. We go under the railway to reach an old canal ware-house that is now used as a British Waterways maintenance depot. The warehouse was originally served by its own canal arm and was used when this was the terminus for the eastern length of the canal. Cargo was tran-shipped onto pack-horses and carts for road transport over Standedge. In the warehouse wall can be seen the top of the arch of a former covered boat dock. From the towpath, at the head of the winding hole, can be seen the two single and the double railway tunnels with the stepped outfall of Tunnel End reservoir (22,650,000 gallons).

Follow the path round and across the final bridge to reach Tunnel End and the Tunnel End cottages. These were built in the 1840s to house the tunnel-keepers who guarded the comings and goings through the hillside. The cot-tages now house the Canal and Countryside Centre which has an exhibition on the canal and sells books and leaflets. Beyond the cottages is the entrance to Standedge Tunnel which at 5,456 yd long (extended to 5,698 yd after the rail-way was built) (4,950 and 5,209 metres respectively) is the longest canal tunnel ever built in Britain. It must also be one of the deepest as in places it is over 600 ft below the surface of Standedge. Only portions of the tunnel are lined and, in those parts that are not, the rock surface is very rugged and rough with the dimensions varying considerably: sometimes just 7 ft by 7 ft and at others opening out into a sizeable cavern. The tunnel took sixteen years to build and cost £123,804 – roughly 40 per cent of the total cost of the HNC. It was finally opened on 4 April 1811 when 'upwards of five hundred people' passed through singing 'Rule Britannia' and other patriotic songs.

There is no towpath. Boats were 'legged' through by men who lay on their backs and walked the vessel along. Given the wide variations in the bore of the tunnel, it must have been a difficult job and it isn't surprising that the passage of a fully laden boat took four hours. Boats were admitted according to a strict schedule. They were allowed to enter the west end between 6 and 8 a.m. and 5 and 8 p.m., and the eastern between 12 and 2, day and night. Boatmen could measure their passage using cast-iron dis-tance plaques set in the roof at every 50 yd. In 1816, in order to speed things up a bit, the company considered using a steam tug. After much thought, and some experimentation, a tug service was introduced in 1824. This was, however, abandoned in 1833 in favour of more efficiently orga-nized leggers. Subsequently, the engineer John Raistrick suggested the cre-ation of an artificial current through the tunnel although this was never tried. The tunnel was closed in 1944 due to rock falls and subsidence but the HC Society are confident that the channel can be restored albeit after the expenditure of several million pounds.

Although the canal tunnel was the first through Standedge, it wasn't the last as there are no fewer than three railway tunnels. The first (a single-line tunnel) was opened in 1849 to house the Huddersfield & Manchester Railway & Canal Company line. It took just three years to build; a job made much easier by the fact that the canal tunnel was already in place. A second single-line tunnel was opened in 1871. The final tunnel is the double-track tunnel used today. This was completed in 1894.

To return to Huddersfield or Mirfield, retrace your steps to the railway station next to lock 42E.

Further Explorations

The Ashton end of the HNC is open for walking from Stalybridge to the western end of Standedge Tunnel. For a walk of 4^1/$_2$ miles, park the car near Greenfield railway station – which can be found on the A669 Holmfirth–Oldham road (OS Landranger 109, ref: SD 992047) – and take the train to Stalybridge (enquiries: (0484) 545444).

Leave Stalybridge station and turn right to go under the railway. Bear right along Market Street to pass the Staveleigh Medical Centre and an open bus station. Continue past The Talbot and turn right into Melbourne Street. This road crosses the River Tame and becomes pedestrianized. Some 50 yd after The Friendship pub, the road goes over a bridge without apparently crossing anything. This marks the spot where the HNC once went through central Stalybridge on its way to a junction with the Ashton Canal at Ashton-under-Lyne (to the right). Turn left here to walk into Armentieres Square car park. The canal originally crossed the square to go to the right of the parish church ahead. Our route takes us along Corporation Street to the left of the church. The canal originally took a course through the Delta Crompton Cables Factory on the right. At the end of Corporation Street turn right into Mottram Road. About 200 yd along, go past a left turn and walk on for a further 100 yd to some greenery and a noticeboard that announces 'The Tame Valley'. Turn left. It is unlikely that the section of the HNC through Stalybridge will ever be restored. As an alternative the canal society have proposed to build a new navigable line along the River Tame which follows a similar course to the canal albeit slightly to the north of the original route. The eastern end of this diversion will rejoin the original route near the site of a disused railway viaduct.

The first lock along the now extant, if a little overgrown, canal is the infilled 7W. From here the line of the waterway winds its way through the outskirts of Stalybridge to a point where the aforementioned old railway bridge crosses the waterway. The new line of the canal, the Stalybridge bypass, will leave from here. We, meanwhile, continue onwards to another infilled section with an electricity substation on the left. The route wends its way underneath an electricity pylon and then, a couple of hundred yards further on, passes over the site of lock 8W (Bywith Lock), indicated by a slight rise in the ground level. A short distance on and the canal is once again in water although the next bridge (a former swing bridge) has been culverted. Just beyond the bridge was the site of Hartshead power station of which only the landscape scars and a tall, partly demolished, conveyor bridge remain.

Keep walking and you will finally reach the restored section at Black Rock Lock (9W). This is shortly followed by Avenue Lock, Terrace Lock (10W and 11W) and the 220 yd Scout Tunnel. A towpath goes through the tunnel but those who suffer from claustrophobia or who have small children may prefer to take the path that passes over the top. The tunnel is brick-lined for the first few yards and then cuts through solid rock. Whitehead's Lock (12W) will cause some problems in restoration as there has been some ground movement locally which has resulted in the whole structure becoming twisted. Above the lock, the canal becomes increasingly more of a refuse tip but improves after a road bridge and Wharf Cottage Lock (13W) where the canal is dominated by the large red edifice of Mittan Mill and the stone-built Woodend Mill. This is Mossley. Here the line swings right to Woodend Lock (14W) and the towpath changes sides to the left-hand bank.

The canal bends left to Roaches Lock (15W) near a pub of the same name. After a road bridge we reach Gas Works Lock (16W) where we recross to the right bank via the tail bridge. After Division Lock (17W) the canal goes under Division Bridge (formerly the Lancashire and Yorkshire boundary) and crosses the River Tame via the stone-arched Royal George Aqueduct. The Royal George is the largest of the seven aqueducts on the canal and has been concrete-lined to help stop leakages. We have now reached the outskirts of Greenfield. The canal goes through Keith Jackson Lock (18W), under a road bridge and on to Royal George Lock (19W). The path then rises to a road at the curiously named Well-i-Hole Bridge, under which the canal is culverted. Turn left and go up the hill. Greenfield station is at the top on the left.

Further Information

The Huddersfield Canal Society looks after the interests of both the narrow canal and Sir John Ramsden's. Together with BW and the local authorities, the society hopes to reopen the line in 2001. It can be contacted at:

Huddersfield Canal Society,
239 Mossley Road,
Ashton-under-Lyne,
Lancashire OL6 6LN.

The society publishes a towpath guide and the Kirklees Metropolitan Council publishes an excellent Information Pack about the canals. Both are available at the Tunnel End Canal and Countryside Centre.

For more historical detail the SJRC is covered in:

Hadfield, Charles, *The Canals of Yorkshire and North East England*, Vols. I and II. David & Charles, 1972.

The HNC meanwhile is covered in:

Hadfield, C. and Biddle, G., *The Canals of North West England*, Vol. II. David & Charles, 1970.

6
THE LANCASTER CANAL
Lancaster to Glasson

Introduction

Although a waterway link to the south was planned, and a section (now part of the Leeds & Liverpool) actually built, the Lancaster Canal was never physically joined to the rest of the canal system. Instead the company opted for a 5 mile long railed way from Preston to (what is now) the L&L at Walton Summit (near Whittle-le-Woods). Despite this isolation, the line was so successful that it actually took a lease on its rival railway company. It's worth visiting for that piece of effrontery alone.

The present course of the Lancaster starts in Preston and heads west to Salwick before turning north to Garstang. At Galgate the Glasson arm runs west to Glasson Dock and the Irish Sea. The main line, meanwhile, continues north to Lancaster. On the northern outskirts of the town, the canal crosses the River Lune via one of the most famous aqueducts in the land. The route now runs parallel with the coast to go through Hest Bank to Carnforth. The small village of Tewitfield marks the northernmost limit of current navigation. The closed section north of here plays cat and mouse with the M6 for a while before heading west to go through Hincaster Tunnel. Although dry, the canal can still be traced from Hincaster all the way into Kendal.

The Lancaster Canal stands gloriously alone. Isolated from the rest of the English canal system, it has a kind of proud aloofness and seems none the worse for it.

History

Plans for a navigable line to link the Leeds & Liverpool Canal with the towns of Lancaster and Kendal had been in the wind since the 1760s.

Indeed, in November 1771, a meeting at Lancaster Town Hall had called for a survey of a suitable line. Robert Whitworth carried out two, of what turned out to be several, surveys but none seem to satisfy the would-be promoters and the scheme was dropped. Further plans for a waterway were aired during the 1780s, all of which evaporated in waves of indecision and faltering enthusiasm.

By 1791 the nation was entering into a period which was to become known as canal mania and the people of the north-west were not able to escape the craze. On 4 June 1791 a group of businessmen from Lancaster petitioned for a canal to link the town with the rest of the canal network. A meeting on 8 June resolved to promote a canal and by October, John Rennie was carrying out a survey. Rennie's plan, described to a meeting on 7 February 1792, was for a 75½ mile long broad canal which started at Westhoughton (in the middle of the Wigan and Bolton coalfields). From there the line would go north to Clayton Green and then pass down thirty-two locks to an aqueduct over the River Ribble to Preston. After Garstang the line would cross the River Wyre via an aqueduct to Lancaster, Tewitfield and Kendal. There were to be branches to Duxbury and Warton Crag (near Tewitfield). At this time the southernmost junction with the rest of the canal network was still a matter of debate; the choice being to join either the Bridgewater Canal or the Manchester, Bolton & Bury. With this issue still in limbo, the meeting resolved unanimously to proceed and nearly a quarter of a million pounds was raised on the spot. The Lancaster Canal Act was passed on 25 June 1792. The company was authorized to raise £414,000 with an extra £200,000 if needed. John Rennie was appointed as chief engineer with Archibald Millar as resident. A further branch to a sea dock at Glasson was enabled by an Act of 10 May 1793.

With the construction work underway in early 1793 (John Pinkerton, among others, taking one of the contracts), John Rennie set about finding himself a southern terminus. An early agreement with the Duke of Bridgewater to form a junction with his canal at Worsley fell through. This was then followed by an agreement with the Leeds & Liverpool Canal Company in which it was resolved to form a junction between the two lines at Heapey. Meanwhile the construction of the canal proceeded apace, albeit with recriminations for poor workmanship by both Pinkerton's men and those of a local contractor John Murray. By 1795 the company took control of the undertaking itself with Archibald Miller supervising the works more closely. In order to do this, Henry Eastburn was made resident engineer for the section south of the Ribble.

By 1795 a stretch of 4 miles, from Bark Hill to Adlington (now part of the L&L near Chorley), was opened and the first coal traffic was moving in July 1796. By 2 November 1797 the length from Preston to Tewitfield was opened with a modest celebration which included the firing of guns in front

The barge *Zion* SS taking a party along the Lancaster Canal to Sedgwick
Ware/The Boat Museum Archive

of Lancaster town hall. South of Preston, a continuous line from Bark Hill to Johnson's Hillock (near Whittle-le-Woods) was completed in 1799. The problem that now faced the company was how to join the two ends of its line over the River Ribble. The cost of building a line with locks descending to and ascending from an aqueduct over the river was estimated by the new resident engineer, William Cartwright, at £180,945. A tramroad way, however, was estimated at £60,000 and it was this that was built. Thus, the northern section of the Lancaster was never connected to the southern and has hence always remained separated from the rest of the canal network.

An Act of 1800 enabled the raising of a further £200,000 which paid off the company's outstanding debt and funded the new tramroad. Although there was talk of an alternative route along an extended Douglas Navigation (see Chapter 7), the tramroad scheme was supported by John Rennie and William Jessop and work started. The route involved building a canal/rail interchange basin at the Walton summit (near Whittle-le-Woods) and another in Preston, together with a tunnel through Whittle Hill. The tramroad, which was powered by stationary steam engines, was opened on 19 January 1804.

The company was now able to operate over the majority of its new line and its financial prospects improved enormously. Revenue of just £4,853 in 1803 (for which a 1/2 per cent dividend was issued) was nearly doubled in

1804. In 1805 a dividend of 1 per cent was paid. At that time the northern end of the navigation remained at Tewitfield and the company still had to complete its route into Kendal. There had been a suggestion for a tramroad but, despite the cheaper cost, this was rejected. The company opted for a canal line and the work proceeded. Meanwhile, even by 1810, the southern end of the canal was still not connected to the rest of the canal network. However, the Leeds & Liverpool were by then pushing to complete their cross-Pennine route between Blackburn and Wigan and an agreement between the two companies was forged. The L&L would use the Lancaster between Johnson's Hillock and Bark Hill from where the L&L would build a short connecting line to Wigan.

With constant delay, due to the impact of the Napoleonic Wars and a repeated lack of funds, the full route, including the Hincaster Tunnel, wasn't open until 18 June 1819. On what must have been a memorable occasion, an audience of 10,000 celebrated on Castle Hill, Kendal. The town's bells were rung and 120 dignitaries consumed a splendid dinner in the town hall. The Lancaster Canal opened twenty-seven years after it was started at an approximate cost of £600,000.

Even before the line was open it was doing relatively well. In 1817 income was regularly in excess of £19,000 p.a. although the dividend was pegged at 1 per cent. Once the full route was open, however, the line really began to flourish. The small town of Kendal, for example, went through a mini-boom period. This increased prosperity was shared by the canal whose income grew by about a third almost immediately. Revenue in the year 1823 was £28,874, with income coming roughly equally from the two ends of the line.

A panoramic view of Kendal, thought to be at the end of the nineteenth century, shows the canal terminus and docks

Ware/The Boat Museum Archive

The volume of traffic carried, however, was more disparate. Of the total cargo of 459,000 tons in 1825, 303,000 tons were moved on the southern line (including the tramroad) and 156,000 tons on the northern. Most of the traffic on the southern line, some 71 per cent, was coal. Despite the growing level of trade, dividends were pegged at 1 per cent until 1825. In 1826, with the Glasson branch opened for traffic, the company was able to raise its dividend to 1½ per cent. With the failure of its bankers and increased lending rates on its outstanding loans, however, the company reverted to a more prudent 1 per cent from 1828 onwards.

This, albeit comparatively modest, success of the Lancaster was not to last for long. In 1831 Parliament passed an Act which authorized the construction of the Wigan & Preston Railway. The new line was to follow the course of the Lancaster Canal for virtually the entire way. The canal company realized its vulnerability, particularly in respect of the now outdated tramroad section that linked its two watery halves. A number of schemes were proposed varying from abandoning the entire line south of Preston to amalgamating with the new competitor. One option, that of converting the tramroad into a locomotive-powered railway, was considered in more depth. George Stephenson was asked to look into the possibility and returned with an estimate of £11,895. However, the company was sceptical about the practicality of the scheme.

Meanwhile the Wigan & Preston (now known as the North Union Railway) opened for business in 1834 and another line, that from Bolton to Preston, followed shortly thereafter. With this, the Lancaster reached an agreement in January 1837 in which the railway company would lease the canal's tramroad for £8,000 p.a. and build a new canal/rail interchange basin in Preston. The deal was a bad one for the railway, who, just a year and a half later, obtained an Act which allowed it to use the North Union's line into Preston. Thus the Bolton & Preston Railway found itself paying a lease for a line it didn't need. Although a reduced lease rate of £7,000 was later agreed, the canal company appears to have been unenthusiastic about cancelling the agreement.

By 1836 the canal's revenue had risen to £33,000 p.a. with 550,000 tons of cargo being moved along the line. As a result, the dividend in 1837 was increased to 1¼ per cent. In 1840 trade was still improving, despite the railway, and income stood at £34,200. But shortly thereafter it was agreed that the tramroad should be closed and the canal company abandoned trade south of Preston. Instead it concentrated on expanding the coal business along the northern waterway, supplied by the railway companies to the Preston terminus and exported via Glasson Dock.

In 1837 competition for the northern traffic arrived with a bill for the Lancaster & Preston Junction Railway. The new line would share a station with the North Union in Preston and would follow the canal north. By June

1840 the railway was open to Lancaster and it may be supposed that the good days for the Lancaster Canal were over. The canal company, however, came out fighting. It halved its passenger fares; an act which enabled it to maintain the numbers who used the route. In addition, the new railway fell out with the North Union and a delay in the building of the Bolton & Preston Railway meant that trade was less brisk than predicted. Thus by 1842 it was the railway company that was in dire financial straits and, contrary to most of the railway-canal competitions, it was the canal company that took a lease on the railway. The deal was one in which the Lancaster was to pay £13,300 p.a. (equivalent to a 4 per cent dividend) lease plus the interest payments on the railway company's outstanding loan. The lease was to last twenty-one years. This arrangement was enabled by an Act of 3 April 1843. Very quickly it appears that the canal company began to exploit the situation. Cargo-carrying became a monopoly and the company was able to cream off relatively high profit levels. Passenger train fares were increased sharply and, at one point, all the seats in third class were removed so that more passengers could be fitted in. Meanwhile the Lancaster & Carlisle Railway was going ahead. An Act was passed in June 1844. The L&PJR, viewing the possibility of a better deal, extracted itself from the Lancaster lease and, instead, leased itself to the new L&CR; an agreement that came into effect on 1 September 1846. The canal company did not surrender its legal control of the railway willingly. Although requested to hand over the line on 1 July 1846, it refused and sought legal action to support its case. The L&CR ignored this and ran its trains to Preston as if it were the lessee and refused to pay any tolls for doing so. Although negotiations were started, the canal company refused to pay its rent in 1848 and each company commenced legal action against the other. At one stage it looked as if the argument may subside when the canal company came close to selling itself to L&CR but this fell through.

After a fatal accident on 21 August 1848, the companies were instructed to sort the matter out by the Railway Commissioners and this they duly did. On 13 November it was resolved that the canal company was to be paid £55,551 (a sum decided following arbitration) for the unexpired portion of the lease. The canal company then gave up control of the line on 1 August 1849. Overall, the deal meant that the Lancaster had made a profit of nearly £70,000 from its railway lease; a sum which allowed it to pay off its debts, pay a dividend bonus of nearly 2 per cent and still have funds for contingencies. This must have been welcome to the shareholders of a company who, even at its peak in 1846, was still only able to pay a 2¹/₂ per cent dividend.

In 1850 the new cordiality between the canal company and the railways was cemented when the canal agreed with the L&CR to share business. The canal company was to take the coal and heavy goods traffic (plus that to Glasson Docks) in return for giving up its passenger and general merchandise trade. Although the company had offered to sell its southern end to the Leeds & Liverpool in 1845, the deal that was eventually reached in 1850 meant that the L&L leased the merchandise tolls at £4,335 p.a. for a period of twenty-one years.

The 1850s saw the canal company starting its own coastal trade with five steamers. It also leased its own quay in Belfast. In 1857, perhaps more significantly for the long term future of the canal, the London & North Western Railway was expanding its interests in the north-west and took control of the L&CR. All traffic agreements appear to have come to a close almost immediately, and trade along the Lancaster went into terminal decline. By 1860 the company was seeking to sell itself to the LNWR and in 1864 it succeeded. The Lancaster Canal Transfer Act, passed on 29 July 1864, enabled the LNWR to lease the northern end of the canal in perpetuity for £12,665 p.a. The southern end was leased to the Leeds & Liverpool Canal for £7,075 p.a. By this time the northern half of the tramroad had declined and was now closed. The southern half (Walton Summit to Bamber Bridge), which still had some remaining traffic, was eventually closed in 1879.

In 1885 the LNWR offered to buy the canal company at a price equivalent to £43 15s. per share. The shareholders agreed and the canal was sold on 1 July 1885. The Lancaster Canal Company was finally dissolved on 1 January 1886. In the early years of the LNWR's ownership, the line continued to carry coal along the main line, and various goods, such as grain, minerals and timber, from Glasson. In 1889 the line carried 165,005 tons of cargo and had an income of £18,728. By the turn of the century traffic was dwindling. Cargo loads were down to 130,396 tons and income to £13,984.

In 1944 the London, Midland & Scottish Railway (into which the LNWR had been incorporated in 1923) proposed to close the Lancaster; a move that was defeated by a consortium of local interests. However, by then the trade north of Lancaster was restricted to coal deliveries to Kendal gasworks, a total of 7,500 tons p.a. This business was moved to road transport in the autumn of 1944. The final cargo from Preston to Lancaster was moved in 1947 to Storey's Mill, White Cross, Lancaster. The canal was nationalized in 1948 and, like so many of the nation's waterways, its future looked bleak. The 1955 Board of Survey report for the British Transport Commission placed the canal in category 3: waterways having insufficient commercial prospects to justify their retention. While this was not put into immediate effect, the line from Stainton (near Hincaster Tunnel) was closed and the top 2 miles into Kendal infilled. In 1964 a section in Preston was also filled and a number of other sections were flattened or allowed to decay. In 1968, as a final indignity, the M6 was built across a culverted canal in three places. The northern section above Tewitfield was then sealed by the construction of the A6070.

In 1963, when the future of the canal looked at its bleakest, a group of enthusiasts formed the Lancaster Canal Trust (as the Association for the Restoration of the Lancaster Canal). Since then they have vigorously fought to keep the canal in water and to reopen the closed sections. The situation was certainly improved (if not resolved) when the Transport Act of 1968 listed most of the line as a cruiseway. The year 1990 saw the formation of the Northern Reaches Restoration Group and a reinvigoration of the scheme to re-establish the waterway in full. The isolation of the northern stretch above Tewitfield will not be ended until either the motorway is raised or the canal lowered. A feasibility study by a group of consultant engineers has suggested that this could cost £17.5 million, so it may be some time before Kendal sees boat traffic. But where there's life . . .

The Walk

Start:	Lancaster (OS ref: SD 473617)
Finish:	Glasson Dock (OS ref: SD 445562)
Distance:	9 miles/14 km
Maps:	OS Landranger 97 (Kendal & Morecambe) and 102 (Preston & Blackpool)
Return:	Lancaster City Transport or Ribble Buses nos. 86, 88 and 89 operate from Glasson Dock to Lancaster (enquiries: (05240) 841109)
Car park:	Signposted in Lancaster
Public transport:	British Rail Lancaster

From the bus station in central Lancaster, turn left into Cable Street and then left again into King Street. Continue along this road which eventually becomes Penny Street, goes over a major road junction and then over the canal at Penny Street Bridge.

The last commercial traffic on the Lancaster Canal (a load of coal) was landed at Storey's White Cross Mill, here by Penny Street Bridge in 1947. Curiously, as recently as December 1988, an unexploded bomb was found under the bridge. It was only uncovered when the canal had been drained for routine maintenance work.

The walk to Glasson starts by turning right to walk along the left bank. This area of the canal in Lancaster was once a busy spot and the old Aldcliffe basins and wharves are still vaguely discernible, despite the DIY superstore, on the opposite bank. When the canal was operational, the company had its offices near the basin area. The Waterwitch pub, which we

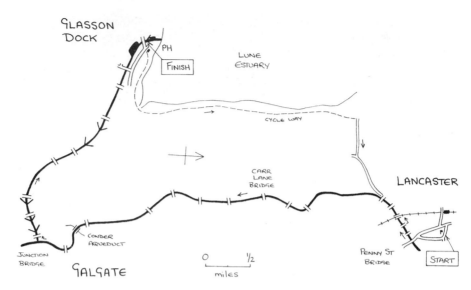

The Lancaster Canal

pass shortly on this side of the canal, has been housed in the old stables. The footbridge, however, is of much more recent vintage: January 1987. We walk on to pass a British Waterways yard on the left to reach bridge 98; a fine turnover bridge which allowed barge horses to change banks without having to uncouple from the boat. Of note here are the rope-wear marks on the protecting metals. Our route now continues on the right-hand bank. On the opposite side is another old wharf with a crane still in position and a rather derelict looking building known as the Boat House. At one time packet-boats were refitted here. The boats were pulled out of the water by means of pulleys positioned on the upper floor of the building.

Continue under the railway bridge (No. 97) and on to pass Aldcliffe Road Footbridge. Robert Swain assures us that this was once a ship's gangway. It was put here in May 1954 to replace the old pitch-pine bridge which once occupied this spot. After Haverbreaks Bridge we enter more open country although, for a short distance, a road accompanies the canal to the right. Before too long, however, it veers right and we start along the approaches to the 1 1/2 mile long Deep Cutting. It's a deep, dank, mouldy, cavernous place that in autumn, when I was here, had a peculiarly eerie feel to it. Even the ducks looked a little spooked. In the midst of the chasm, the canal passes under bridge 93: Carr Lane or Broken Back Bridge. It's notable for instead of a conventional stone parapet, the central section of both sides is railed.

After bridge 92 (Burrow Beck Bridge), the sides of the cutting subside a little although the sides of the canal remain heavily wooded and enchanting.

The turnover bridge at the junction between the Glasson arm and the Lancaster main line

The views open out and, for the next mile, we pass through open country-side, rich in herons and passing sparrowhawks. Only gradually do you notice the arrival of civilization in the form of Galgate to the left. After bridge 88 (Ellel Hall Bridge), the canal crosses an aqueduct over the River Conder before swinging left and then right under Galgate Bridge. Here is the Galgate Marina and shortly thereafter a lock-keeper's cottage and Glasson Junction turnover bridge. The line straight on continues towards Preston, we meanwhile go right to reach, almost immediately, the first lock on the Glasson branch.

The 2¹/₂ mile long Glasson branch was authorized by an Act passed on 10 May 1793. The aim was to provide the canal with a sea connection and to improve the fortunes of the Lancaster section of the waterway. Despite this relatively early enabling Act, work wasn't begun until the passage of a further Act on 14 June 1819 which allowed the company to raise the neces-sary funds. The new branch was surveyed by William Crosley who estimat-ed the cost to be £34,608. Again the work was delayed and the decision to proceed wasn't made until 1823. Part of the reason for this latter delay was the fact that Crosley was considering an alternative line from Hest Bank (north of Lancaster). Hest Bank was already being used as a transhipment point from coastal vessels, even though there was no physical connection between the sea and the canal. Crosley's estimate for the Hest Bank connec-tion was, however, roughly twice that of the Glasson branch. The branch was finally opened in December 1825, and by 1831 the Hest Bank port was closed and the Hest Bank Shipping Company transferred its business to Glasson. The line carried coal destined for North Wales, Ireland and Ulverston and imported slate, timber and grain for the growing town of Preston. The arm is fed from the River Conder. There are six locks altogether and eight bridges (including the turnover bridge at the junction). Although Miller's Canalside Tavern (by lock 6) may arrest your progress for a while, it's an easy 3/4 hour stroll with the massive nuclear power stations on the coastline ahead pulling you ever westward. Eventually we pass the diminuitive Christ Church and reach the Glasson Basin; 36,000 square yd of water, 14 ft deep.

To reach Glasson Dock, bear right along the edge of the basin to the Victoria Inn with the Lune estuary stretching out beyond. If you bear left here you can walk on to the swing bridge and the sea lock with the Glasson Dock beyond. The lock can handle vessels up to 98 ft long and 26 ft wide. The upper gates are doubled so that the lock can also protect the basin from high tides. In 1830, 16,036 tons of cargo passed through the dock, most of it remaining waterborne to continue along the canal. This trade continued to be healthy until 1846 when competition from other ports began to affect trade. There were also problems with the sandbanks in the Lune estuary which limited movement in and out of the dock. The Lune was deepened but the effect was short lived and the problem recurred regularly thereafter. Although cargo traffic has ground to a stop, the records suggest that the lock is five times busier now with pleasure boats than it was at the peak of the canal's commercial activity.

Apart from the Victoria Inn, Glasson Dock has a small post office-cum-corner shop just beyond the swing bridge. There is also a café. The stop for buses back to Lancaster is situated just outside the public conveniences next to the Victoria Inn. However, it is possible to walk back along the converted Glasson branch of the old London & North Western Railway (a distance of 5 miles). The line was opened in 1883 but by 1890 traffic was already declining in the face of competition from Preston Docks. By the 1920s it was losing money badly but it struggled on only to be closed on 7 September 1964. The start of the route can be found just after the bus-stop. Turn left down the side of the public conveniences near the bowling green. At the end, turn right to walk along a clear track which is signposted the Lancashire Coastal Way. This skirts the side of the road with the Lune to the left. The line soon bends north to go over the River Conder on a girder bridge. After passing the old station at Conder Green, there are some fine views over the salt-marshes and mud-flats of the Lune estuary. Continue past Ashton Hall and on to reach a barrier. Turn right and follow the road through Aldcliffe. If you bear left at a junction, you will reach the road that runs parallel to the canal. Rejoin the towpath (on the left-hand bank) and walk back to Penny Street Bridge and central Lancaster.

A Cawood-Hargreaves coal boat train leaves Ferrybridge Flood Lock on the Aire & Calder Navigation at Kellingley

Allbright & Wilson's chemical plant on the Sankey Brook Navigation in Widnes

The Double Arched Bridge at East Marton on the Leeds & Liverpool Canal

Lock 1 on the Huddersfield Narrow Canal in Huddersfield

Further Explorations

All of the extant sections of the Lancaster Canal are open for walking and even the dry northern line can be followed into Kendal. However, one of the major sights of the Lancaster Canal, if not the entire UK canal system, can be seen with a walk from Penny Bridge, Lancaster to Carnforth; a distance of approximately 8½ miles (13½ km) with a return by train.

You can get to Penny Bridge following the instructions from the main walk but on reaching the canal, turn left to walk along the right-hand bank. At the next bridge (Friarage Bridge), the towpath changes to the left-hand bank from where it continues its steady course through the centre of Lancaster. There are five more bridges before the canal bends left to cross two aqueducts in quick succession. The first, the Bulk Aqueduct, which crosses the A683, is a concrete affair that dates from the 1960s, the second is one of the wonders of the canal age: the Lune Aqueduct.

The aqueduct must rate as one of Rennie's finest achievements. Over 600 ft long and 60 ft high, it crosses the river by five huge arches. As with the Dundas Aqueduct on the Kennet & Avon, Rennie would have preferred to have built in brick but was forced to use stone by the canal company.

John Rennie's Lune Aqueduct

Work started in January 1794 with the canal company's own navvies building the foundations for the massive piers. By the summer the work was going on 24 hours a day with as many as 150 men engaged. A year on, in July 1795, the piers were ready and Alexander Stevens & Son of Edinburgh, were appointed as contractors to complete the structure. Although Stevens himself died in January 1796, his son finished the work in the autumn of 1797. The aqueduct had cost £48,320 against Rennie's estimate of £27,500 but all who looked on since have agreed that it was worth every penny.

The canal bends sharply left as it makes a quick tour of the suburbs of Lancaster. When the canal bends back right to head in a more northerly direction, views gradually open up over Morecambe Bay. By the time we reach bridge 118, we are firmly in the centre of Hest Bank. Here, in the early days of the canal and before the opening of the Glasson branch in 1826, was an important transhipment wharf. This was the closest point to the sea, and cargo was moved on to coasters for subsequent carriage to Ireland, North Wales and northern Britain.

The canal meanders around the countryside like an old-fashioned Brindley Canal before arriving at Bolton-le-Sands where the Packet Boat Hotel near the old wharf reminds us that this was once a stopping place for the high-speed boat. The final 1½ miles into Carnforth offers some pleasant countryside and tantalizing glimpses of Morecambe Bay. Carnforth is reached at bridge 128 where a children's playground aids towpath navigation. Walk up to the road and turn left. If you continue over a crossroad (with the A6), you will find the railway station which has regular trains back to Lancaster.

Further Information

The Lancaster Canal Trust helps both the preservation and restoration of the canal that bears its name and can be contacted at:

Lancaster Canal Trust,
16 Galloway Road,
Fleetwood,
Lancashire FY7 7BD.

The society publishes a book *The Complete Guide to the Lancaster Canal* and runs a trip boat, the *Ebb & Flow,* from Holme to Cinderbarrow on Sunday afternoons.

The history of the canal, together with walking details for the entire route, is contained within:

Swain, R., *A Walker's Guide to the Lancaster Canal.* Cicerone Press, 1990.

7
THE LEEDS & LIVERPOOL CANAL
Skipton to Barnoldswick

Introduction

The Leeds & Liverpool is the longest single canal (as worked by one company) in Britain. It was also one of the most successful. Yet it took forty-six years to complete and the two committees, one in Lancashire and the other in Yorkshire, had different views on where the canal went and even what it was for.

The L&L starts just north-east of the Royal Liver Building at a terminus basin and exit to the Mersey at Stanley Dock. From there the canal meanders north to Burscough where the Rufford branch heads towards the River Ribble. The main line now heads east to cross the River Douglas at Parbold. At Wigan the line passes the Leigh branch to Manchester before climbing through twenty-three locks and then turning north along the former Lancaster Canal to Whittle-le-Woods. The L&L now turns north-east through Blackburn, Burnley and Nelson to Foulridge where it passes through a 1,640 yd long tunnel. After Barnoldswick the line meanders around some fine Pennine country to Gargrave before turning south-east to Skipton. The canal then skirts Keighley and goes through Bingley and Shipley where there was formerly a junction with the Bradford Canal. The last stretch passes via Apperley Bridge and Rodley before forming a junction with the Aire & Calder Navigation in the centre of Leeds.

Even without the canal interest, this is a fine country walk through the southernmost reaches of the Yorkshire Dales. Towpathing at its best.

History

The Leeds & Liverpool Canal started life in 1764 as a suggestion to link the Aire & Calder Navigation in Leeds with the River Ribble at Preston. The businessmen of Yorkshire were looking to increase supplies of lime, for a cheaper means of transporting coal and for a route for their goods (especially woollen cloth) to Liverpool and the colonial market. Their counterparts in Lancashire were primarily seeking cheaper coal but were also looking to the markets in the east. Liverpool was expanding rapidly both as a port and as an industrial centre based on its import–export trade. This expansion demanded a good communication system with the hinterland and it demanded coal. The Douglas Navigation, which took the course of the River Douglas from the Wigan coalfield to the sea had been opened in 1742 but still the demand could not be met.

Precisely how the two otherwise disparate groups got together isn't clear but John Stanhope, a prominent Bradford landowner, asked John Longbotham to survey a line between the Aire and the Ribble. It was this that was discussed at a meeting in Bradford on 2 July 1766. Longbotham's proposal, which he completed in August 1767, was for a line from Leeds to Preston where one branch would go north to Poulton and the River Wyre (near Blackpool) and a second south to Liverpool. Following the circulation of a rival scheme in August 1767, Longbotham revised his plan to a more direct line from the Aire & Calder in Leeds to the Mersey in Liverpool. This proposal was presented to a group of Yorkshire businessmen on 7 January 1768 and accepted, subject to review by a more prominent canal engineer such as James Brindley. The meeting, held at the Sun Inn in Bradford, also appointed a Yorkshire committee and decided that a Lancashire committee should be formed.

By August Robert Whitworth (as Brindley's assistant) had checked Longbotham's route and a Lancashire committee had been formed. The two committees met on 19 December 1768 in Burnley. Brindley estimated the cost of the new canal (then called the Yorkshire & Lancashire) at £259,777. The route was to run from Leeds to Keighley, Skipton, Barnoldswick and Barrowford. From there the line ran through Whalley, Walton and Eccleston to cross the River Douglas (without any connection) near Tarleton and thence to Liverpool. The meeting supported the plan and agreed to put a bill to Parliament. However, there were already ructions between the two committees. The Yorkshire group was keen to build the cheapest and quickest line to Liverpool. The Lancashire committee, however, basing its main interest on the canal's ability to deliver cheap coal, preferred a route through the coalfields of south Lancashire, most notably those around Wigan. During the course of 1769 the Lancashire committee had its own survey undertaken by John Eyes and Richard Melling. As some technical faults were found in this, a second was undertaken by P.P. Burdett. His proposal took the Lancashire route from Barrowford via Burnley, Blackburn and Chorley to Wigan. From there it passed along the northern side of the Douglas Valley to Parbold where it rejoined Longbotham's line to Liverpool.

The two committees met and decided that the final route should be decided by Brindley. He reported on 11 October 1769 that the line from Colne to Liverpool was 66 miles and would cost £174,324 whereas Burdett's was 87 miles and would cost £240,881. The finances, plus the fact that the Yorkshire committee was able to out-vote their Lancashire counterparts, meant that the Longbotham line was adopted. Many of the Lancashire subscribers promptly withdrew; observing that the proposed route would not provide the benefits they had sought. Indeed, in a bout of spite, they even declared that they would not allow the new line to connect with the Liverpool Docks (which some of them owned).

Whatever the position of some members of the western committee, by 10 January 1770 a joint meeting resolved to promote Longbotham's plan for a barge canal (able to take boats measuring up to 14 ft 4 in wide) to Parliament. Why the Lancashire committee capitulated isn't evident although there may have been some discussion on a junction with the Douglas Navigation, which would thus have given access to Wigan. The Act received its Royal Assent on 19 May 1770; the main opposition coming from the Douglas who wished to protect its water supply. The Act enabled the company to raise £200,000 in £100 shares and a further £60,000 should it be needed. Longbotham was appointed engineer (Brindley having declined the post) and in November building began at both ends of the line.

Plans to forge the link with the Douglas were forwarded with all haste. Discussions with Alexander Leigh, proprietor of the navigation, were held during 1768–9 and resulted in the L&L buying the line; the majority of the shares changing hands in 1771. The strategy was to complete a line that was already being built by Leigh. This cut, started in 1753, had been designed to bypass some of the more convoluted lengths of the Douglas from Parbold (where the L&L crossed on the Newburgh Aqueduct) towards Wigan. Apart from giving access to the Wigan coalfields, the link with the Douglas would also supply the L&L with much needed water supplies. Leigh Cut was opened in 1774 and was celebrated with the ringing of bells, the playing of bands, the firing of salutes and the consumption of refreshments.

By that time most of the Liverpool section had been built, the section from Skipton to Bingley was opened and there were 31 miles of canal in Lancashire and 23 in Yorkshire. Progress was such that by March 1774 the line from Skipton to Thackley near Shipley was finished. Coal and lime traffic along the L&L and the Bradford Canal now started in earnest. By October the Yorkshire end had been further extended to Gargrave. However, funds were already beginning to run low. Work on the aqueduct at Whalley Nab was stopped and expenditure on the rest of the line curtailed. The stretch into Leeds was completed on 4 June 1777; an event that was celebrated by thirty thousand people. The only other significant work during the 1770s was that along the old Douglas Navigation: a line from the

L&L at Burscough to Rufford (the Lower Douglas Navigation) and an extension of the Leigh Cut from Dean into Wigan (the Upper Douglas Navigation) were completed in 1780 and 1781 respectively.

By the early 1780s, 75 miles of canal had been built (Leeds to Gargrave and Liverpool to Parbold) at a cost of £232,016. In addition, £53,434 had been spent to buy the Douglas Navigation. Although some toll income was forthcoming (£2,025 in 1774 rising to £7,352 in 1778 and £13,062 in 1785), finance was critical and the company was forced to borrow. With the onset of the recession caused by the American War of Independence, all building work was stopped. This period of enforced calm allowed the route through east Lancashire to be reconsidered. There had been a significant expansion of the towns of Blackburn and Burnley, and by the 1790s, when raising funds for canal building was relatively simple, re-routing the line in Lancashire became expedient. Robert Whitworth was appointed engineer in 1790 and an Act passed in June enabled the company to raise a further £200,000. The Act defined a revised line which went slightly south of the original. By 1793 the line was again altered to one similar to that suggested by Burdett nearly a quarter of a century before. Thus the Upper Douglas Navigation to Wigan was to become the main line which would then pass through Blackburn and Burnley. The new line was longer (an important consideration in view of the newly proposed – and 15 mile shorter – trans-Pennine rival Rochdale Canal) but would add some significant new trading points at not much extra cost.

Although some of the proprietors questioned the wisdom of completing the line at all, the construction of the trans-Pennine section was again under way with some of the biggest engineering projects on the line now being built: Foulridge Tunnel, which opened on 1 May 1796, was 1,640 yd long; the Burnley embankment was a mile long and 40 ft high. Although this work was slow and required a further £100,000 through a new share issue, the line, including the Burnley embankment, was open to Henfield, and thence by turnpike to Accrington and Blackburn, on 23 April 1801.

By 1804 the company had spent £554,569 (not including the purchase of the Douglas) and it still didn't have a complete waterway. Maybe it didn't mind. It was already obvious that the L&L was a success with an annual income of £51,838 enabling a dividend of £8 per share. Some 176,000 tons of coal p.a. were now being shipped into Liverpool and the canal was carrying significant quantities of wool; raw bales being carried inland and spun yarn returned for weaving. The textile industry was developing rapidly all along the line with the canal supplying the ever increasing demand for coal.

The L&L realized that its main business would be at the two ends of its line rather than cross-Pennine but it remained committed to completing the stretch from Henfield to Wigan. Work restarted in May 1805 and the section from Leeds to Blackburn was opened in June 1810. The line from there to Wigan was fraught with complex negotiations but, eventually, it was agreed that the L&L would share the southern line of the Lancaster Canal from Johnson's Hillock near Wheelton to Kirkless (a point near Wigan Top Lock). In return, the Lancaster would receive a water supply for its line from the L&L as well as tolls from the L&L's through traffic.

This final section, and hence the L&L Canal, was fully opened on 19 October 1816. Amid much jubilation the first barge left Leeds on the Saturday, spent its nights at Skipton, Burnley, Blackburn and Wigan, and arrived in Liverpool late Wednesday, again amid great celebration. It had taken forty-six years to build and was (if the shared section of the Lancaster is included) 127 miles long. There were forty-four locks on the Yorkshire side, forty-seven on the Lancashire and more than three hundred bridges

A barge is loaded from a railway wagon on a tippler at old Wigan Pier, *c.* 1891
K.C. Ward/The Boat Museum Archive

spanned the canal. It had cost £877,616 (including buying the Douglas) compared with Brindley's estimate of £259,777. The company was in debt to the tune of £400,000.

Shortly after the completion of the Wigan to Blackburn section, a link to Manchester and the rest of the canal system was made. The Leigh branch, opened in December 1820, runs from Wigan to Leigh where it meets the Bridgewater Canal. One consequence of this was that by March 1822 all the locks between Leigh and Liverpool were lengthened to permit the passage of narrow boats which could now enter the canal from Manchester.

Although overspent, the L&L was, virtually as soon as it was fully open, a success beyond the wildest dreams of the promoters. In 1824 the line earnt £94,423 allowing a dividend of £15 per share. This should be compared with Brindley's estimated annual income from tolls at just £20,000 p.a. Trade was primarily in coal which was supplied from the fields of Yorkshire and east Lancashire (especially those around Wigan) to the industrial centres that were developing all along the line. During the early nineteenth century, about 200,000 tons of coal p.a. were delivered to Liverpool; a figure that rose to over 270,000 tons by the 1830s, more than 500,000 tons by 1840 and more than 1,200,000 tons in 1858. Although lime and limestone were important cargo for the L&L, the traffic in them never reached the significance predicted by the Yorkshire committee. This underestimate was more than made up by the massive, and mostly unexpected, volumes of cargo categorized as 'general merchandise'. This traffic commanded a much higher toll rate (at 1½d. per ton) than either coal (up to 1d.) or limestone (½d.) and was thus a highly profitable business. The term 'general merchandise' covered everything from foodstuffs to night soil and manure, from flax to yarns, from snuff to cast iron, from beer to gunpowder.

They started talking about the possibility of a railway between Liverpool and Manchester in 1822 and, although several bills were defeated in Parliament, the line was opened on 17 September 1830. Meanwhile a less significant line that ran from Bolton to Leigh had already opened for goods traffic in May 1828. There were some thoughts of lowering tolls to compete but in fact it was the railway companies that struggled initially. A minimum toll agreement was made, for example, with the Bolton & Leigh in 1833; not to protect the L&L's income but to bale out the railway. In Yorkshire the rail threat arrived in 1824 with the suggestion for a line from Leeds to Hull. At first this was countered by some major improvements along the Aire & Calder Navigation. But in 1830 the Leeds & Selby Railway Act was passed. The L&S's natural extension, the Liverpool & Leeds, which would have followed the canal through the Wigan coalfield was initially lost in Parliament in 1831 but was passed in 1836 as the Manchester & Leeds Railway.

At first the railways were unable to compete with the canal for its coal and limestone traffic. Part of the reason for this was the fact that much of the industry that used these materials had been built close to the canal and thus the waterway remained the most convenient mode of transport. Gradually, of course, new works would be built near the railways but this transfer of traffic was slow. This meant that by the 1840s, when the number of railways was increasing (the Leeds & Bradford, for example, was opened in 1846), the L&L was still prospering. In 1838 the L&L carried 2,220,468 tons of cargo. In 1840 income totalled £164,908 and dividends from 1841 to 1847 topped £34. In 1847 the company finally paid off its debt.

From the opening of the canal until 1850 the company continued to improve its waterway. One enforced improvement was at Foulridge Tunnel where a partial collapse in 1824 closed the line for eighteen months. Apart from this, some stretches of the line were deepened and several new warehouses built. There were also continuous improvements at the Liverpool terminus. The most important of these was a new, direct link with the Mersey Docks that was engineered by Jesse Hartley at a cost of £42,622. The arm went by four locks to a new dock, Stanley Dock, and was opened in 1848. There was also during this period a programme of reservoir construction. Originally the line was supplied from watercourses but this was soon seen to be inadequate, and by the end of the nineteenth century there were seven reservoirs with a total capacity of 1,174 million gallons.

In 1847, when the L&L started its own carrying fleet, railway competition was beginning to bite into the profitable merchandise traffic in Lancashire (although not yet in Yorkshire). This didn't mean that the railways were thriving. They, too, found that the necessary toll cuts were eating their profit margins. But by 1850 the situation was beginning to change. In 1847 the L&L had been forced to reduce its merchandise tolls to 1d. a mile and this, plus the loss of trade to the railway, meant that the company's income in this sector fell from £58,128 to £15,333 in just two years. Although in 1850 nearly 900,000 tons of coal were still being carried annually, total income was down to just £71,523, less than half that of ten years before. The company was forced to reduce its dividend to £15. This financial pressure was also being felt by the railways, and discussions with a group of companies were held from 1848 to 1850. The result was an agreement (in August 1850) in which the railways leased the merchandise tolls on the canal for £41,000 p.a. until 1871 – later extended to 1874. The carrying fleet was also purchased outright for £13,880. This deal provided the L&L with a guaranteed income for a business it looked like losing. Despite having to pay compensation of £4,335 p.a. to the Lancaster Canal, the company must have been overjoyed with the agreement. Indeed, the L&L prospered thereafter, with dividends returning to £25 by 1856. The merchandise lease certainly had the effect of moving customers onto rail. Whereas canal merchandise traffic totalled 360,000 tons for an average of 30 miles in 1840, by 1871 it was only 282,485 for 12 miles. However, the canal's coal traffic was booming. In 1866, 1,897,000 tons were carried along the line – most on the east Lancashire section. When the merchandise lease came to an end on 4 August 1874, the canal found itself in a strong financial position and ready to compete with a railway system that was becoming increasingly inefficient.

During the course of the railway lease, the L&L made a range of improvements. Steam tugs were working the Liverpool coal trade and various capital works were under way. In 1880 steam powered cargo-carrying fly-boats were introduced, able to tow three or four unpowered boats. The efficiency of this timetabled service meant that the canal was able to attract back much of the merchandise traffic that had been lost. It was said that the canal company provided better warehousing and was actually quicker than the railway between, for example, Burnley and Liverpool. The end of the nineteenth century saw part of the old basin at Liverpool being sold to the Lancashire & Yorkshire Railway. This earnt the L&L £185,341, which it spent renovating the facilities at its western terminus.

Although trade at the end of the century was still good, the passage of the Railway and Canal Traffic Act of 1888 forced the L&L to charge toll rates similar to those of a railway competitor. In general this was about half what it had been charging. With other constraints upon the L&L's activity that were imposed with the passing of the canal's own Act of 1891, profit levels fell dramatically. A dividend of £15 in 1890 was down to just over £4 in 1900 and to nothing in 1901. This was despite the fact that the line was still carrying 1.1 million tons of coal and 600,000 tons of general merchandise annually. The 1891 Act had been introduced to restructure the company and to implement a series of improvements by raising £275,000 in share capital. The plan was to upgrade the canal into one that could handle boats able to carry 67 tons. However, the financial stringency that followed meant that much of this work had to be scaled down considerably. A bad winter in 1895, which closed the canal for two months, and a series of dry summers, didn't help. But trade remained good. In 1906 the L&L carried 2,337,401 tons of cargo to produce an income of £180,000.

The beginning of the end really came in 1917 when the line was brought under government control through the rest of the war and on to 1920. Although the government paid compensation, the company's finances were now in a sorry state and the carrying company was all but broke. Poor wages meant that boat crews were hard to find and continuing losses meant that the activity was closed on 30 April 1921 and the vessels sold. From then to the time of nationalization in 1948, the L&L steadily declined. Various

Barnoldswick Bridge and Greenberfield Lock. The old canal line is visible to the right of the cottage by the bridge, *c.* 1905

Ware/The Boat Museum Archive

parts were sold or closed. The Bradford Canal, for example, which had been jointly owned by the L&L and the Aire & Calder since 1877, was closed in 1922. The situation was partially rescued by the construction of three canalside power stations at Wigan, Whitebirk and Armley, but by the 1950s the canal was declining rapidly and many locks and bridges were in a poor condition. Income in 1956 was £127,500 and the canal lost £132,500. Commercial traffic effectively ended with the great freeze of 1962–3 when traders were unable to move boats for many weeks. The last commercial traffic on the waterway was the delivery of coal from the Plank Lane colliery to Wigan power station in 1973.

The L&L was not viewed enthusiastically by the Board of Survey in 1955 but at least they considered it worthy of retention. The canal never faced the same kind of threat that closed its fellow cross-Pennine routes via Rochdale and Huddersfield. Later, in the 1968 Transport Act, the line was classified as a cruising waterway and, apart from a short section at the Liverpool terminus, is maintained as such today.

The Walk

Start:	Skipton (OS ref: SD 988515)
Finish:	Barnoldswick (OS ref: SD 882474)
Distance:	12¹/₂ miles/20 km
Map:	OS Landranger 103 (Blackburn & Burnley)
Return:	A number of buses run between Barnoldswick & Skipton even on Sunday (enquiries: (0282) 831263)
Car park:	Several well signposted in Skipton or on-street near Rolls Royce Bankfield site, Barnoldswick
Public transport:	British Rail Skipton

The walk starts at the bus station in Skipton. This is a short walk from the centre of town and is well signposted. Cross the small footbridge which leaves the far side of the bus area to go over the canal and then turn right to walk along the left-hand bank.

The canal reached Skipton from Bingley in April 1773 (it didn't open to Leeds until June 1777) when two boatloads of coal were sold to some lucky punters at half the previous price. The occasion was greeted with the ringing of bells, the lighting of bonfires and what the *Leeds Intelligencer* described as 'other demonstrations of joy'. From then on the town became an important canal centre. It was, for example, the headquarters of one of the largest private carrying fleets on the L&L. The canal company itself also had a depot

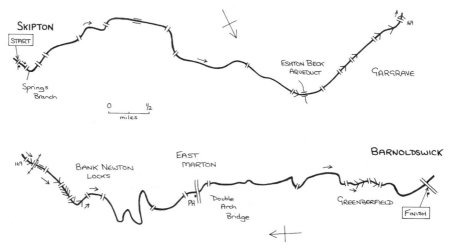

The Leeds and Liverpool Canal

here. Today there are still a number of old canal buildings including a warehouse that has been converted into a pub.

A short way along the towpath we go under Belmont Bridge to where the Springs branch enters the L&L on the opposite bank. This short (770 yd) arm heads along a narrow channel towards Skipton Castle. It was built for the Earl of Thanet, the owner of the castle, to carry limestone from his quarry at Skipton and, later, that at Haw Bank near Embsay. It was probably opened in 1773–4 and was extended in 1797 from the Watermill Bridge through the deep rock cutting. Chutes were constructed at the terminus to drop rock into waiting boats. The Haw Bank quarry, which was some 3/4 miles from the branch, was connected to the line by a horse tramway. When opened, the quarries were being worked by the Mercer Flatts Lime Company but from 1785 the canal company took over the lease which it held through to nationalization. In the 1830s annual production was about 80,000 tons p.a. Most of it was shipped down the canal for use in ironworks or for burning in the many limekilns along the line. The branch gradually declined in importance after Haw Bank was connected to the Midland Railway in 1889. However, an attempt to resurrect the canal trade was made in 1964 when small amounts were shipped out from Haw Bank, which is still being worked. If you have the time, it is well worth a quick stroll along the arm to the terminal basin.

The canal leaves Skipton heading due west, passing the backs of several terraced houses. The line goes over a small aqueduct and then under two swing bridges: Brewery and Gawflat. After passing under the new(ish) A629(T) fly-over bridge, we reach Niffany Swing Bridge where the towpath moves on to the pavement of the road before re-establishing itself through a wooden gate. Already the countryside is becoming more remote, and after another small aqueduct the line goes under the second of the new(ish) flyovers. After passing two more swing bridges (Thorlby and Highgate), we go under Holme Bridge and on to Holme Bridge Lock. Here, for a short distance, the canal acts as the southern boundary of the Yorkshire Dales National Park. In 1882 it was planned to build a reservoir on the northern side here but the plan was dropped following objections from the local landowner Lord Hothfield. The water would have come from Eshton Beck, which the L&L crosses via aqueduct shortly after the lock.

Like many of the bridges along the L&L, Ray Bridge (No. 172) has a white band painted around its arch together with a vertical stripe on the side of the arch to mark the centre of the navigable channel. These markings were painted to help with night navigation. We reach Gargrave at Eshton Road Bridge where we are forced to rise up to and then cross the road. On its westward march, the Yorkshire end of the canal reached Gargrave in March 1774 and there it stayed until October 1790 before the building work recommenced. The construction of the line to here was important as the main water supply for the Yorkshire end fed into the canal from Eshton

Beck. Once the through line was fully operational, the town became an important staging post and it boasted both a warehouse and stabling.

The canal goes through Eshton Road Lock and past the old coal wharf. At Higherland Bridge, the Pennine Way crosses the canal. We, meanwhile, move on to Higherland Lock. One of the more curious things about the locks on the L&L is the variety of ground paddle gear, and Higherland Lock exhibits one of the simplest. Here, just beyond the upper gates, are two wooden structures which look like an extra pair of ill-fitting gates. They're not! These simple wooden barriers are moved sideways to cover or uncover the ground holes.

Continue under Anchor Bridge (cross here for The Anchor Inn) and on to Anchor and Scarland Locks. Stegneck Bridge is followed by Stegneck Lock and the bridge that carries trains from Leeds to Settle. After Priest Holme Aqueduct the towpath changes sides at Priest Holme Changeline Bridge. It is necessary to walk along the road for a short distance before being able to regain access to the towpath via a gate. The canal now makes a straight course, past a lock-keeper's cottage, to the first of the Bank Newton Locks. There are six altogether and the canal twists and turns in order to gain height in as short a distance as possible. After the first lock is Carpenters Yard Bridge and the old canal company yard where maintenance boats were built. The yard acted as the store for the company's legal papers during the Second World War. After the fourth lock is Plantation Lock Bridge. On your way up the flight it's worth noting another piece of canal architecture which, at first, tests the intellect a little. I'm only able to explain it following the advice of Mike Clarke (see Further Information). At the far corner of each lock is an upright metal hook. When the barge was ready to leave the lock, the towrope coming from the horse was passed through a pulley on the boat's towing mast and then forwards on to the hook. The pulley system thus effectively provided the horse with a first gear with which to get the boat moving. After a certain distance a short stick attached to the rope would get trapped in the pulley, the rope would become detached from the hook and the system would then revert to normal towing without any attention from the boatman.

Above the sixth lock the canal reaches Newton Changeline Bridge where the towpath changes sides. Again, we have to walk along the road for a short section before being able to regain access to the towpath. After Newton Bridge the canal enters into a wonderfully convoluted section. For the next half an hour you will approach the TV mast on the hill ahead from virtually every conceivable angle. Although built during the early part of the nineteenth century, the highly circuitous route between here and Barnoldswick is almost a throwback to the age of Brindley. Here we see an unashamed contour canal that follows the line of the country rather than beating a bold straight course over the top of it. Amateur canal engineers can work out whether the line should have been built on the other side of the valley. Some of the bends of the actual line are so acute that the company installed large vertical rollers on the inside corners for the tow lines to run on. None of these are still in position although one of the roller stands remains (examples still exist in Wigan – see Further Explorations).

Eventually the route straightens and heads south-west via Langber Bridge and then Williamson Bridge to reach East Marton. The next bridge on the line is the famous Double Arched Bridge where a second arch has been built half-way down to strengthen the structure. Those more interested in quenching their thirst than the structure of bridges should go underneath and up the side. At the road turn left for the Cross Keys pub.

There is now a quiet 2 mile walk that goes under Old Hall and South Field Bridge before reaching the newly (and splendidly) built Greenberfield Changeline Bridge where the towpath changes sides. Greenberfield is one of those little hot spots of canal interest. The main focus is in the fact that at one time there was a three-lock staircase here. This was replaced by three separate locks in 1820 in order to save water. As you approach Greenberfield Bottom Lock, a slight depression on the right marks the course of the original line. Although the channel disappears after Greenberfield Middle Lock, look right to see the old route going through a garden and on to a redundant canal bridge. Continue past Greenberfield Lock Bridge and up to Greenberfield Top Lock. The old canal bridge is now just to the right and the site of the former staircase is buried under a well-manicured lawn. Both the old route and the new bring the canal up to the summit pound, 487 ft above sea-level. The next lock west is at Barrowford, approximately 6 miles on. Further along there is a lock-keeper's cottage to the right. An information notice informs its readers that lock-keepers have lived here since 1816 and that the first occupant was a retired army captain by the name of Isaac Jones. Just beyond the cottage is another small building dated 17 August 1893. This marks the arrival of the feeder pipe from the Winterburn reservoir which is 9 miles to the north (just south-east of Malham).

It is now just a short stroll into Barnoldswick. After Greenberfield Bridge the canal bends right and then left to pass a redundant bridge (No. 155) and the Rolls Royce Bankfield site to right. The next bridge is Coates Bridge. Go underneath and turn right to reach the road. Turn left and cross the road to reach the bus-stop for transport back to Skipton.

Further Explorations

Virtually all 127 miles of the L&L are open for walking and much of it has been developed with walkers and cyclists in mind. It would make a fine long-distance walk over the course of, say, seven or eight days.

For a single walk of about 12 miles in the Wigan area, park the car at

Burscough, near Burscough Bridge railway station (OS Landranger 108, ref: SD 444124). From the station turn right and walk for about 250 yd to the canal. Turn left to go along the left-hand bank and under Burscough Bridge. Before setting out, it's worth bearing in mind that although cyclists need a permit to ride on BW towpaths, the stretch from Burscough to Wigan was upgraded in 1984 in an attempt to encourage wider use by those with bicycles. You should therefore keep your eyes and ears open as, on summer weekends, cyclists hurtle along the canal and expect towpathers to get out of the way with minimal warning.

The town of Burscough was once an important staging post on the Wigan–Liverpool packet-boat run. The L&L had a yard here where horses could rest and where the company's store of horse provend (i.e. feed) was kept. By the second half of the nineteenth century, Burscough was the centre of the canal's merchandise trade; gaining further importance after the provend depot and vet centre was established here in 1888. Houses, often including stabling, were built for the boatmen alongside the canal from New Lane to Parbold. There was even a mission in the hay loft of the shop at Crabtree Lane. Perhaps most notoriously, Burscough was the point to which manure and night soil were delivered from the cities prior to distribution to local farms. The smell from the accumulated heaps was, perhaps not surprisingly, objected to rather strongly by the local residents. Just before a railway bridge (the Preston to Liverpool line) is Ainscough's Mill with its fine, if dilapidated, canalside awning. The mill received its grain supplies from Liverpool Docks. It also received coal via the canal and the mill owners worked the last horse-drawn boat along the L&L, the *Parbold*, which ended service as recently as 1960. After going under the railway the towpath goes over the roving bridge that marks the entrance of the 7¼ mile long Rufford branch. Opened in October 1781, the line falls by seven (72 ft by 14 ft) locks to a tide lock into the River Douglas at Sollom and then on along the old Douglas Navigation to the Ribble. At the junction are a series of canal cottages, including boatmen's houses on the right, together with a dry-dock. The first two locks along the branch can be seen from the roving bridge.

Along the main line the canal passes through the low-lying, market-gardening farmland and past Glovers Swing Bridge, Ring-o-Bells Bridge, Moss Bridge and Spencer's Swing Bridge. On several occasions the line cuts across minor roads by means of one-arch aqueducts. At Parbold the canal crosses the River Douglas on the Newburgh Aqueduct. This aqueduct was nearly destroyed in June 1838 when a sudden storm caused the river to flood to such an extent that the structure was nearly washed away. As this was the busiest, and hence financially most important, stretch of the L&L, emergency repairs continued 24 hours a day for almost a fortnight. During this time a wooden aqueduct was put into position to allow boat movement.

A little further on is Parbold Bridge. This was formerly called Windmill Bridge because of the old mill which stands on the southern bank. The mill received its grain supplies and exported its flour via the canal.

Beyond Windmill Bridge, look for a blocked arm that leaves the line on the opposite bank. The route between Parbold and Barrowford, as defined in the 1770 Act, was to go through Whalley, Leyland and Eccleston. The towns of Blackburn and Burnley were to be served by branch canals and Wigan by the Douglas Navigation. This blocked arm is all that remains of a length of the original route that was built before the change of course was decided. At one time it was used as a graving dock where the bottoms of boats were burnt clean and then tarred.

We now pass Alder Lane Bridge and Chapel House Bridge to reach Gillibrand Bridge. It is interesting to note the wartime defences that were built along this section of the L&L. Keen-eyed towpathers will already have spotted the pillboxes but at Gillibrand Bridge, among others, there are large concrete anti-tank obstacles that were designed (if such a word can be used) to block the bridges over the canal. After Hand Lane Bridge the canal splits into two channels and the three Appley Locks. The original line (that nearest the towpath) bears just one 12 ft deep lock which almost immediately caused problems. As the locks above Appley (into Wigan) were built with less fall (typically 6–7 ft), there were water shortages. As a consequence, a new channel was built with two locks of more conventional fall and this appears to have improved the situation. At one time both channels were operated with one acting as a sidepond to the other in order to save water. This also had the advantage of speeding traffic flow. Nowadays the original, deeper lock is used. All the locks along this stretch were originally built to take the Yorkshire barges and were 61 ft long by 14 ft 4 in wide. However, when the Leigh branch from Wigan to Manchester was opened in 1821, it was decided to extend all the locks between the branch and Liverpool so that conventional Midland narrow boats could navigate the line. Thus the locks along here are all 72 ft long.

Appley Bridge is marked by an old stone quarry wharf and a series of canal cottages. One of the latter has been converted into the Kettle's On tea shop while, on the other side of the canal is The Railway pub. From here the canal winds through some pleasant open country with the River Douglas winding around the fields to the right. We pass three swing bridges (Finch Mill, Ranicar's and Fisher's) before reaching Dean where there are two extant locks and the remnants of a third in the shadow of the M6. The first lock on this site was the old river lock which ran from the Upper Douglas Navigation down to the River Douglas. This can be seen blocked off to the right (it was closed in the 1880s). This was the original terminus of the Leigh Cut to Wigan which was completed in October 1774. The section from Dean into Wigan was opened in 1781. However, the lock between the river and the new navigation was kept in use so that boats could reach the coal staithes on the opposite side of the river which were served by a number of tramways. The next Dean Lock (the central one) was opened when the Upper Douglas Navigation to Wigan opened in 1781. Later the locks were doubled up to enable two-way passage.

After going under the M6 and the railway bridge, the canal reaches Gathurst where, near Gathurst Bridge, is The Navigation pub. The outskirts of Wigan start at Crooke and soon we bid farewell to the pleasant rural scenery. Here, into the mid-1950s, the John Pit colliery delivered its product via a tramway to the canal edge. It was then tippled directly into waiting barges for delivery to Wigan power station, Ainscough's Mill, Liverpool gasworks, Tate & Lyle's sugar refinery and elsewhere. The staithe was positioned on the Gathurst side of the Crooke Hall Inn. Just on the other side of Crooke Bridge, on the northern bank, was another conduit of coal, this time a branch canal known as the Tunnel Canal. This arm, which was active from 1798 to around 1850, went north into an underground colliery near the Standish and Follient Woods. It was worked by narrow boats similar to those used on the Bridgewater Canal at Worsley.

After Grimshaw's Bridge the canal passes the remnants of the former Crooke Lock (the lock chamber and the gate recesses can still be seen) and under Martland Mill Bridge to reach Hell Meadow Lock. With increasingly dreary surroundings, the line goes under the railway and by Pagefield Lock to continue towards central Wigan. Although the Yorkshire committee was never keen to have the L&L running to Wigan, the coal-hungry, Lancashire committee eventually had its way. The first line from Liverpool, which utilized the Leigh Cut and then the Douglas Navigation for the final stretch, opened in October 1774. The L&L itself arrived in 1781.

The line enters central Wigan by passing under a footbridge, a railway bridge and Seven Stars Bridge. The line then bends sharply left to reach Wigan Pier. The canal basin at Wigan has been developed into a major tourist attraction with many of the restored canalside warehouses and mill buildings open to the public. One part, for example, includes an exhibition entitled 'The Way We Were'. The pier itself, made famous by various sources including George Orwell, was the end of a tramway which brought coal to the canal from local collieries. The pier was built in 1822 and, at its peak, tippled 50,000 tons of coal a year into waiting canal barges. The tippler was dismantled in 1929. The precise site of the pier is marked by the upturned railway lines that hang over the canal. On the opposite bank, where there are now gardens, was also a coal loading wharf. Further along on that side is the first of the canal warehouses. The nearest was built in the 1890s and now houses the exhibition centre. The brick warehouse beyond it, with the covered hoist, is ten years older and now houses a pub/restaurant (The Orwell). This once sported a similar canopy to the exhibition centre

Wigan Pier: the tourist resort

but it was removed during the Second World War. This latter warehouse and the one opposite it on the towpath side were both used to store cotton prior to delivery to the many local mills. The stone warehouse which adjoins the pub and the terminal warehouse (the one that incorporates two entrances for canal barges) are earlier than the others, being built in the 1790s and 1770s respectively. All these buildings were scheduled for demolition in 1973 but, thankfully, British Waterways refused. By 1979 even BW thought that they should be knocked down but by then the terminal warehouse had been listed and thus it was the local authority's turn to do the refusing. In 1982 the County Planning Officer recognized the potential of the site, and by 1984 the restoration progress was complete.

Continue over Pottery Changeline Bridge and turn left to descend to the left-hand bank. Underneath the bridge are restored examples of the rollers that protect the bridge arches from towline wear. On the other side of the bridge there are a group of canal cottages followed by Trencherfield Mill on the left. The building presently on this site was peculiarly old-fashioned in being built next to the canal rather than the railway. A little further on, the towpath crosses the mill's own canal arm and dock to which coal and cotton were delivered. On the opposite bank is a hire boatyard that was formerly the site of James Mayor & Co., boat-builders. Within a short distance we reach Wigan Bottom Lock. The twenty-three Wigan Locks raise the canal just over 214 ft in approximately 2^1/$_2$ miles. The second lock of the flight is reached after crossing the busy (and tricky to cross) Chapel Lane at Henhurst Bridge. This bridge was important because the towpath for boats heading for Manchester changed to the right-hand bank whereas those going to Leeds stayed on the left. The cottages on the towpath side of the lock formerly housed the lock-keeper and toll collector. The toll office was later transferred into the smaller red-brick building to the right.

The line crosses a small aqueduct over the River Douglas and goes on to a small bridge that was once the canal entrance to the Wigan Electric Light and Power Company. The opening is now blocked but a small culvert a little further on is open and delivers water into the pound from the River Douglas. From here the main line of the canal goes straight ahead on its way to Leeds, some 92 miles to the east. The right turn, meanwhile, heads towards Manchester, 23 miles to the south-east. This is the Leigh branch which runs for 7^1/$_4$ miles to the town of Leigh where it forms a junction with the Bridgewater Canal and hence provides a through route to central Manchester and the rest of the canal system. The line was originally proposed in 1803 but it took some time to reach agreement with the proprietors of the Bridgewater and to fight off the opposition of the Rochdale and the Manchester, Bolton & Bury Canals. It wasn't until June 1819 that the line received its Royal Assent. It was opened in December 1820 at a cost of £61,419.

To complete the walk, return to Pottery Changeline Bridge. Go underneath and bear right and then left to cross a new bridge in front of the terminal warehouse. Turn left here to visit the Wigan Pier exhibitions. To reach Wigan Wallgate station, go straight on and cross the road. Turn right and walk towards the town centre. Go under a railway bridge and continue on to pass the North Western train station on your right. A hundred yards further on, the Wallgate station is on the left. For enquiries on trains back to Burscough Bridge ring: (0942) 42231.

One of the great short strolls on the canal system, let alone the L&L, is at Bingley, a few miles north-west of Bradford on the A650 Keighley road (OS Landranger 104, ref: SE 110392). You can park near Bingley station (train enquiries: (0532) 448133) in a huge car park by the canal. Go through the gate and turn left to walk along the left bank. Almost immediately, you pass the waterbus stop (for enquiries about trips phone: (0274) 595974) and go under Park Road Bridge. The canal hereabouts is being shifted right at a cost of £4.5 million to accommodate some road-works. Luckily it shouldn't alter the character of the spot too much and the canal will still sit in the shadow of the massive Bowling Green Mill that is now home to Damart thermals.

Continue around the canal bend to the first lock rise: the Bingley Three Rise. This is a staircase of locks in which the top gates of one lock act as the bottom gates of the next. Such structures were used to overcome sharp height changes. The problem is that a boat has to pass up all three locks

Bingley Five Rise

before another can come down. Above the flight, the canal bends gradually left through a pleasantly wooded stretch and on to the main course for this trip: the Bingley Five Rise.

This magnificent set of locks is based on the same principle as the three rise. However, the extra two locks plus an altogether more grandiose architecture make this a real tourist spot. The locks were designed by John Longbotham and were completed in 1770. Even then the five rise was seen as something extraordinary and it had its own opening ceremony which included bands playing, guns firing and local church bells ringing. Each lock measures 62 ft by 14 ft 4 in and holds 90,000 gallons of water. Altogether they take the canal up 60 ft. I'm told that this can be done in thirty minutes although when I was here, a stream of boats seemed to arrive at the bottom, took one look at the climb and returned back the way they came.

You can walk up the flight either via the path to the left or on the opposite bank. At the top there is a swing bridge and some canal buildings. If you're the lucky sort, you may also find the ice-cream stand open for refreshment prior to your stroll back to the car.

Further Information

There is no Leeds & Liverpool Canal society as such but the Inland Waterways Association has local branches that cover the waterway. For further information contact the London office (see Appendix B).

For information on the Leeds & Liverpool Canal look no further than:

Clarke, Mike, *The Leeds & Liverpool Canal.* Carnegie Press, 1990.

8
THE MACCLESFIELD CANAL
Macclesfield and Bollington

Introduction

Wedged between the busy Trent & Mersey and the scenic Peak Forest, the Macclesfield Canal is almost overlooked as it skirts the Peak District National Park on its route from the Potteries to Marple. It doesn't deserve to be.

At its southern end, the Macclesfield leaves the Trent & Mersey Canal just north of the Harecastle Tunnels at Hardings Wood Junction, Kidsgrove. From there it crosses the T&M at Pool Lock Aqueduct and heads north through open country to Congleton. The canal fits its twelve functional locks into a little under a mile at Bosley before continuing along a meandering course to reach the eastern edge of Macclesfield. From there, the line passes through Bollington and High Lane before reaching the Peak Forest Canal at Marple Junction.

Although there is a short section along the busy streets of Macclesfield, this circular walk has bags of interest including a return stroll along the Middlewood Way.

History

During the latter half of the eighteenth century there had been several attempts to raise interest in a waterway between the Peak Forest Canal at Marple and the Trent & Mersey at Hall Green near Kidsgrove. A canal from Macclesfield to Congleton and Northwich had been proposed in 1765 and a correspondent to the *Derby Mercury* suggested a Caldon Canal–Leek–Macclesfield line in December 1793. In 1795 a group of promoters planned a canal from the Poynton collieries to Stockport. This plan

A LNER trip boat at Bollington in the 1930s

R.W. Lansdell/The Boat Museum Archive

was later extended south through Macclesfield to the Trent & Mersey near Kidsgrove and to the Caldon at Leek. Yet another group of enthusiasts sponsored Benjamin Outram to carry out a survey for a line from the Peak Forest to the Caldon at Endon via Rudyard with a branch canal or railway from Poynton and Norbury to Stockport. This scheme was reported to a meeting on 11 March 1796 at Macclesfield. The cost was estimated at £90,000 with a revenue of £10,175 p.a. At a follow-up meeting in April it was agreed to proceed. However, following some concern over the likely profitability of the new route, and with opposition from the Trent & Mersey, which was promoting its own line from the Caldon to Leek, the plan was dropped. Other schemes were proposed by the Peak Forest Canal Company. In 1799 it considered building a line from Marple to Poynton and Norbury collieries, and in 1805–6 it evaluated the potential for a line via Macclesfield to the Trent & Mersey near Kidsgrove.

Despite all these valiant attempts, it wasn't until 6 October 1824 that the form of what is now the Macclesfield Canal was first proposed. A meeting in the town was held to consider the issue and, almost immediately, some £60,000 was subscribed to the project. It must be said that not all those present were keen on the idea and one far-sighted individual, a Mr Wakefield, was audacious enough to have suggested that a railway might be better. The committee promised to assess both ideas but the mood of the meeting held sway. In November a group of Macclesfield promoters visited the Peak Forest committee. The two parties got on well and agreed that the prospects for the new line would be highly beneficial to both concerns. With the Peak Forest's support, the committee asked Thomas Telford to carry out a survey. He produced two reports and 'most unequivocally' declared in favour of a canal which was to be highly beneficial for the silk and cotton mills and the many other manufacturers in and around the towns of Macclesfield and Congleton. He estimated the cost of construction as £295,000.

The Macclesfield Canal Act was passed in April 1826. Among the shareholders were many of the Peak Forest's promoters, including Samuel Oldknow (see Chapter 9). Even Thomas Telford subscribed £1,000. Despite carrying out the original survey and evaluating the various contractors who had tendered for the work, Telford did not become involved with the construction work. Instead William Crosley, formerly one of the resident engineers on the Lancaster Canal, was appointed as chief engineer. By all accounts the construction work went smoothly and quickly. At the southern end the line ended at Hall Green about a mile from the junction at Hardings Wood, Kidsgrove. That final mile was built by the Trent & Mersey who felt that they needed to exert some kind of control over the traffic travelling north along the new route. As the new line was marginally higher than the Trent & Mersey level, the Macclesfield company built a stop lock at the junction between the two to prevent water loss into the Trent & Mersey.

The new canal opened on 9 November 1831 at a cost of about £320,000 raised in 3,000 £100 shares and a small amount of borrowing. The line is 26^1/8 miles long from Marple, where it joins the Peak Forest at the top of the Marple flight, to the junction with the Hall Green branch on the summit level of the Trent & Mersey. The only functional locks are the twelve at Bosley which carry the canal up 120 ft to the 518 ft summit level. The line is fed by reservoirs at Bosley and Sutton. The new canal provided a waterway route to Manchester that was 10 miles shorter than that by way of the Trent & Mersey to Preston Brook and then the Bridgewater into Manchester, but with twenty more locks. The opening was celebrated by two processions of boats, twenty-five from the north and fifty-two from the south. The leading boats contained various dignitaries, while following boats were loaded with bands and artillerymen or simply with onlookers. Merchant vessels came shortly thereafter loaded with a wide variety of cargo such as coal, grain, salt, iron, timber, lime, coke, cotton bales and groceries.

Almost from the off, the Macclesfield had to compete with rival routes: the Trent & Mersey for goods going between Manchester and Staffordshire, and the new Cromford & High Peak Railway, which had also opened in 1831. This latter line joined the Peak Forest Canal with the Cromford Canal

A snake, or roving, bridge on the Macclesfield Canal. The design allows the horse to change towpaths without unhitching. Probably 1950s

R.W. Lansdell/The Boat Museum Archive

and was thus a fast route for goods being shipped between Manchester from the Nottingham–Leicester–Buxton area and on towards Chesterfield and Sheffield. The Cromford & High Peak had been authorized in May 1825 and was an important source of trade for the Peak Forest and a potential threat to both the Trent & Mersey and the Macclesfield. In 1837 the new canal faced additional railway competition with the opening of the Grand Junction Railway. The company responded in the only way open to it. It reduced the rates. The company also successfully convinced the Peak Forest to reduce its rates in an attempt to kick-start the business. By July 1833 further reductions were needed. The Trent & Mersey had flexed its muscles and reduced its own tolls in an attempt to keep its Manchester via Preston Brook trade. The vigour of the Trent & Mersey's attack can be seen by the fact that toll reductions were not applied to the Hall Green branch. With these reductions, limestone was being carried for 1/4d. per ton per mile, lime for 1/2d., coal for 3/4d., and merchandise for 1d.

Some trade must have been forthcoming for in 1833–4 the company paid a dividend of 1 per cent. This figure was increased to 2¹/₂ per cent in 1836 and was maintained until 1839. From 1837, however, competition from the Grand Junction Railway was being felt and the Macclesfield's shareholders received the last of their pay-days in 1840–1 with just 1¹/₂ per cent. Any income after that was used to reduce the company's debt. The opening of the GJR stimulated a meeting of the canal companies and carriers on the Manchester–London route to consider the use of steam power and to appraise the situation regarding the Manchester & Birmingham Junction Canal. This 16 mile long canal would have run from the Bridgewater at Altrincham to the Ellesmere & Chester at Middlewich and promised a shorter line from Manchester to the Potteries and Birmingham with less lockage and tunnels. The Manchester & Birmingham Junction was surveyed by W.A. Provis with a proposed capital of £500,000. The Macclesfield, Peak Forest, Trent & Mersey and other companies fought the new line vigorously and successfully with the plan failing to reach Parliament.

There was further concerted action at the beginning of 1839 when the Peak Forest suggested that unless toll reductions were made the railway would take virtually all of the London–Manchester traffic. In particular, accusing fingers were pointed at the Coventry and Oxford companies who were proving obstinate to the idea of lowering tolls. The rates at the time were: 1d. per ton per mile on the Trent & Mersey and Macclesfield, 1¹/₄d. on the Peak Forest, 1¹/₂d. on the Grand Junction, Coventry and Birmingham, and 2d. on the Ashton and Oxford. This meant a total of 28s. 5d. on the London-Manchester line via the Macclesfield Canal. The report convinced the southern companies and later in the year the tolls on the route were down to 21s. 6³/₄d. A further meeting in March 1839 heard that Pickford's had experienced a sudden drop in trade along the three associated canals (the Macclesfield, Ashton and Peak Forest) with the railway companies taking a significant proportion of the light goods traffic. There was also an increase in competition for the Staffordshire iron trade from a newly introduced coaster service importing iron from Scotland. It must be said that the Macclesfield was less energetic in its attempts to fight off railway competition than its neighbours at the Peak Forest. The company had started operating steam boats and, from 1842 (until at least 1846), was operating a fast passenger craft. But it did little else. It was almost as if it was resigned to the inevitable. Its strategy was to cut staff and reduce wages. This allowed it to decrease tolls to the point where much of the merchandise was passing at ³/₄d. or ¹/₂d. per ton per mile. The consequent reduction in income meant that no dividends were paid after 1840.

In July 1845 the North Staffordshire Railway was promoted and this stirred the Macclesfield into some kind of response. The company threatened to build a railway between the Manchester & Birmingham Railway at

Macclesfield, and the lines of the proposed railways at Kidsgrove. Whether or not the plan was of serious intent, the result was that the NSR, who by that time had already agreed to buy the Trent & Mersey Canal Company, offered the Macclesfield shareholders 1,000 shares in return for abandoning their scheme and helping the railway get its Act. The Macclesfield committee supported the scheme but not necessarily the terms. It recommended that its shareholders should agree to lease the Macclesfield to the NSR only if the railway company would offer £2 per share p.a. In October the NSR made an alternative proposal in which it was suggested that the Macclesfield company should subscribe £40,000 to its company and appoint a director, and should give the railway an option to buy the waterway within five years at 50s. a share plus payment for debt. The Macclesfield's shareholders rejected these terms but not the idea. The committee was asked to seek an alternative agreement with any other railway company, or even to seek powers to make a line of its own.

In December 1845 the committee was able to report that the Sheffield, Ashton-under-Lyne & Manchester Railway was prepared to buy the Macclesfield at a perpetual yearly rent of £6,605, equivalent to 50s. a share on 2,642 shares. The deal was to include a payment of £60,000 to cover any outstanding debt. The shareholders agreed at a time when the Trent & Mersey was fighting hard for the Manchester traffic by reducing tolls; a fact which may well have prejudiced the minds of the shareholders. With the enabling Act passed, the Macclesfield, together with its neighbour the Peak Forest, became the property of the Sheffield, Ashton-under-Lyne & Manchester Railway, though the canal company continued as a separate entity, collecting rents, until 1883. The last meeting of the canal company was held on 15 July 1847.

About the first thing the Sheffield, Ashton-under-Lyne & Manchester did with its new possessions was to obtain an Act which allowed it to sell water from either line. Shortly thereafter, the company was approached by the manager of the Navigation Department of the North Staffordshire with a view to the NSR taking over the waterways. However, it later transpired that the approach was made without authority and negotiations were not started. At the time, the Macclesfield was returning annual toll receipts of just over £9,000. In 1848 the Sheffield, Ashton-under-Lyne & Manchester started its own carrying business along the Ashton, the Peak Forest and the Macclesfield. The trade was linked with its own goods yard at Guide Bridge near Ashton-under-Lyne. In 1854, 6,894 tons of the trade along the Macclesfield were carried by the new venture, out of a grand total of 214,445. The traffic was primarily in raw cotton. The company operated the carrying service until 1894 by which time back-carriage was becoming hard to find and the business was closed.

By 1905 traffic had diminished considerably. The prime business along the Macclesfield by this time was still in coal but many of the local collieries wcrc closing. Othcr cargo includcd raw cotton, grain (carried from Manchester and other docks) and stone. The decline continued through the two wars and the Stoke-Marple coal trade finally ended in 1954. By then the canal had been nationalized for six years although without great prospect. But the line was not to suffer the almost terminal decline of its Peak Forest and Ashton neighbours. Although the Macclesfield was listed in the Report of the Board of Survey in 1955 as having insufficient commercial prospects to justify its retention, the Transport Act of 1968 listed it as a cruising waterway. On this basis it exists under the guardianship of British Waterways and is a justifiably popular route for holiday cruisers.

The Walk

Start and finish:	Macclesfield railway station (OS ref: SJ 919736)
Distance:	7^1/$_2$ miles/12 km
Map:	OS Landranger 118 (Stoke-on-Trent & Macclesfield)
Car park:	Several signposted in Macclesfield
Public transport:	British Rail Macclesfield

The walk starts at Macclesfield railway station which is east of the town centre not far from the Buxton Road (A537) and across the street from the bus station. From the railway station turn right and then right again to go under the railway. Our way now bears right to take the road to Buxton. Pass The Bull, then Victoria Park and carry straight on at the next roundabout. Just before the Puss In Boots, the road crosses the canal. On the right is the canalside Hovis Mill. We, however, cross the canal and turn left to take the steps which go down to the towpath.

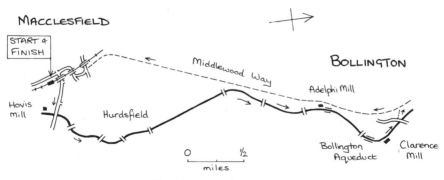

The Macclesfield Canal

The walk starts along the right-hand bank and follows the canal as it bends right past Peak Forest boats and goes under a bridge into open country. In just half a mile, the bustle of Macclesfield gets left behind and all we are left with are the sounds of the birds. The canal now bends back left to go under a skew bridge (No. 35) which, like many along the way, was built of stone which has a pink tinge to it. The Cheshire Ring Canal Walk book (see below) tells us that this highly individual stone comes from the millstone grit quarries at Teg's Nose, a country park about 2 miles east of Macclesfield.

From here the canal goes through Hurdsfield and although we pass close to the houses, the line still has a quiet relaxing air. We pass under an expanded road bridge (No. 34) and on to the first milestone which informs us that Marple is 10 miles off. These stones were removed during the Second World War in case the enemy invaded and used them to find their way about; a move, which with the great aid of hindsight, seems a little ludicrous. Luckily these impressive stones were restored to their former positions by the Macclesfield Canal Society; no mean feat as apparently each weighs about half a ton.

The canal goes under the next bridge (No. 33) and then expands to form a winding hole. There is now a long straight section of canal which has three distinct 'narrows' marking the sites of former swing bridges. Interestingly only two of these are allowed for in the bridge numbers. The site over the fence is a chemical works which, perhaps fortunately, is obscured from our view. The next bridge (No. 30), an accommodation bridge, has a plaque on it which commemorates its reopening by the local MP on 3 August 1991 following reconstruction by the Macclesfield Canal Society. The rope grooves on the sides of the bridge suggest a busy past. There are stop plank grooves underneath the bridge and a rack of planks on the far side.

Shortly we pass another milestone (Marple 9 miles; Hall Green 17^1/$_4$) and walk on to a fine turnover bridge: Clarks Change Bridge. The design of these wonderfully elegant bridges was immensely practical as they enable the towpath to change sides without the boatman needing to unhitch the horse. Again there is a stop plank rack on the far side; seemingly being embraced by the arms of the turnover bridge. We cross the canal here to continue the walk along the left-hand bank. The path now goes on past Kerridge dry-dock (to the right). This was formerly the dock for a gritstone quarry up on Kerridge Hill beyond. The two were connected by means of a tramway. After curving left to go under bridge 28, the outskirts of Bollington can be seen away to the right and we walk on to Adelphi Mill with the Anglo Welsh boatyard opposite. Adelphi Mill was built in the 1850s for George Swindells, a local cotton manufacturer. He had it built here to take advantage of the canal transport. However, within twenty years, the Macclesfield, Bollington & Marple Railway was opened and the canal was superceded. The mill produced cotton and silk and, after the Second

World War, rayon and nylon for Courtaulds. The mill was closed in the 1970s and has since been converted into smaller industrial units and offices. Near the end of Adelphi Mill is the Macclesfield & Vale Royal Groundwork Trust Discovery Centre and shop. This is open from Tuesday to Sunday in the afternoons. Admission is free. There is also a cycle hire shop here.

At the end of the bridge is the Barge Inn. Central Bollington, which has numerous hostelries as well as a reasonable range of shops, can be reached by descending the steps just after the mill and turning right to go under the aqueduct. For those uninterested in such temptations, continue over this small aqueduct, on through the pleasant Bollington suburbs and under the next bridge (No. 27). The canal crosses the Dean Valley by means of a 60 ft high embankment and a short stone aqueduct: the Bollington or Dean Aqueduct. From the embankment there are fine views of Bollington to the right and of the playing fields and the Bollington railway viaduct to the left.

On the other side of the valley is Clarence Mill. This former cotton mill was built in the 1820s (i.e. at a similar time to the canal) again for the Swindells family. It is firmly canal oriented and has a fine unloading bay where the raw cotton and coal were delivered. In the 1820s the first section

Clarence Mill at Bollington

with the tower was built. It was later extended to the left in the 1850s. The section with the chimney was built in the 1890s and 1900s. The mill produced fine cotton for high quality clothing and table linen. It was closed in the 1960s and is now divided into offices and small industrial units. A little further on a small railway track crosses the towpath. This was used to carry skips loaded with boiler ash from the mill to a dump over the bank.

After going under a fine skew bridge, turn left to go up to the road. Turn right. After 100 yd, turn right along the drive of Lodge Farm. On reaching a fence on the left, cross a stile and then descend the bank on to the tarmacked Middlewood Way. This was formerly the line of the Macclesfield, Bollington & Marple Railway that has now been converted into a path and cycleway. The line was opened on 2 August 1869 and closed on 5 January 1970. It was reopened for walkers and cyclists in 1985. A full description of the Middlewood Way appears in *Railway Walks: LMS* by Jeff Vinter (Alan Sutton, 1990).

Turn left to walk over the Bollington viaduct. It should now be possible to walk on along the Middlewood Way all the way back to central Macclesfield from where local signposts will point you to the station. However, when I was here the route was blocked at Clarke Lane Bridge. If this is still the case (notices will indicate if it is), continue along the course of the line until it reaches a road. Here turn left to walk up to the aqueduct near the Adelphi Mill. If you ascend the steps, you can return to the towpath. Turn right to return to Macclesfield via the canal.

Further Explorations

The whole of the Macclesfield Canal is open for walking and all of it is worth a look. At 27^1/$_2$ miles, very fit towpathers could walk the entire length from Marple to Kidsgrove on one long summer's day. Mere mortals could consider making a weekend of it; spending the Saturday night in Macclesfield itself (for accommodation enquiries ring the tourist information office: (0625) 618228). If you started at midday in Marple, afternoon tea could be had in Bollington before reaching Macclesfield by early evening. A late start on Sunday could permit afternoon tea at the National Trust's Little Moreton Hall, just south of Congleton (about a mile from the canal and well signposted) before walking on to Kidsgrove. Both ends are served by British Rail (enquiries for Marple ring: (061) 228 7811; for Kidsgrove ring: (0782) 411411).

For a simpler afternoon's stroll of approximately 7 miles, start at Kidsgrove railway station (OS Landranger 118, ref: SJ 837544) where there is a car park (fee payable). Almost directly opposite the station itself, a series of steps descends to the Trent & Mersey Canal where the iron-enriched waters shine a brilliant yellow. For those who are not familiar with the T&M at Kidsgrove, you might wish to start the walk by turning right to go along the right-hand bank. Within a short distance the canal reaches the Harecastle Tunnels that take the line south to Stoke-on-Trent and beyond. There are two tunnels here. James Brindley's was the first and is to the right. It is 2,880 yd long and was opened in 1777 having taken eleven years to complete. There were constant problems with a poorly built tunnel through a series of mine workings which led to numerous collapses. As a consequence a 'new' tunnel was built by Thomas Telford. This is the one that is open today. It is 2,926 yd long and was first opened in 1827 having taken just three years to build. It is not without its problems and was extensively renovated during the 1970s. For more information on the T&M, the tunnels and walks along the canal, readers should seek out *Canal Walks: Midlands*.

Return to the steps from Kidsgrove station and walk on along the left-hand bank of the canal. The towpath goes under a road bridge (the entrance to the station) and then the railway itself before reaching a handsome roving bridge at Hardings Wood Junction: the start of the Macclesfield Canal. Cross the bridge and turn left to walk along the right-hand bank. The route creeps around the back of some houses, under another bridge and on to Pool Lock Aqueduct that takes the Macclesfield over the Trent & Mersey in a kind of canal flyover. This is made possible by the fact that the T&M has descended by two locks since Hardings Wood Junction whereas the Macclesfield has stayed at the same level.

Cross the canal at the bridge just before the aqueduct to continue along the left-hand bank. Within a short distance the line goes over a second aqueduct, Red Bull Aqueduct, which spans the busy A50. Already the canal is into the open country that makes this stretch of the Macclesfield so attractive. The canal goes under two bridges to reach the first lock: Hall Green Lock. It is interesting to note that, contrary to expectations, this was the official start of the Macclesfield Canal. The stretch we have walked along so far actually belonged to the T&M who built it in order to have some control over this new short cut to Manchester. The lock, which has a rise/fall of just 12 inches, was built by the Macclesfield company as a stop lock to limit the flow of water into the ever water-hungry T&M. The T&M, in some kind of petty retaliation, then added an extra lock of their own in order to be able to charge a toll for those entering their line. The T&M lock-gates have been removed but the chamber, together with the recesses for the gates, can still be seen on the Kidsgrove side of the present lock.

The walk now passes under a footbridge and then on under six bridges. After passing a milestone (Marple 25 miles), the canal appears to pass

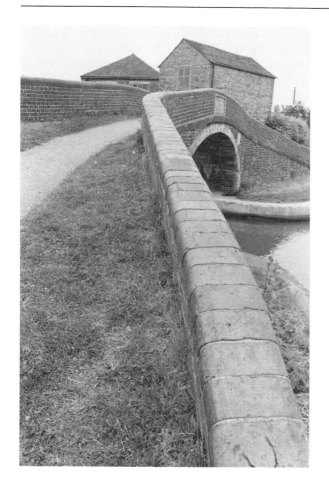

Hardings Wood Junction
where the Macclesfield
Canal joins the Trent &
Mersey

through the grounds of Ramsdell Hall. Curiously the owners of the hall welcomed the building of the commercial waterway through the grounds of
their estate and they even had their own wharf built. For those who wish to
visit the National Trust's Little Moreton Hall, go up to the next bridge (No.
86) and turn left to walk along the lane for about a mile.

From here the canal continues under five bridges and then over Watery
Lane Aqueduct. Three bridges further on and the canal reaches the outskirts of Congleton. Bridge 77 is a splendidly graceful turnover bridge which
takes the towpath to the right-hand bank. Our way passes the warehouse on
Congleton Wharf, the canalside Moss Inn, and then goes over Congleton
Aqueduct. This is not the first aqueduct on this site. The original had to be
pulled down and the present one was built by Thomas Telford. It is a typical Telford design being constructed of an iron trough. Those who are

familiar with the Wootten Wawen Aqueduct on the Stratford and the Nantwich Aqueduct on the Shroppie will recognize the style.

The towpath changes back to the left-hand bank at the next turnover bridge and then enters a deep chasm made by the railway and road bridges and Oakes Millers factory on the right of the canal. Go under the road bridges and turn left to go up some steps to Congleton station where trains will take you back to Kidsgrove.

For a glimpse of the northern terminus of the Macclesfield Canal and the junction with the Peak Forest Canal at Marple see Chapter 9.

Further Information

The address for those interested in joining the Macclesfield Canal Society is:

Macclesfield Canal Society,
'Hawkesbury',
72 Blakelow Road,
Macclesfield,
Cheshire SK11 7ED.

For more details of the history of the waterway:

Hadfield, C., *The Canals of the West Midlands*. David & Charles, 1985 (3rd Edition).

9
THE PEAK FOREST CANAL
Newtown to Whaley Bridge

Introduction

It's a wonderful little canal the Peak Forest and a strangely well-kept secret. It doesn't have the historical significance of the Bridgewater or the sheer effrontery of the cross-Pennine lines like the Rochdale or the Huddersfield Narrow. It was never as busy as the Aire & Calder nor is it ultimately as scenic as the Leeds & Liverpool. Yet it can compete on most of these things and many will argue that it wins hands down. And to think that it very nearly disappeared forever!

The 14³/₄ mile long Peak Forest Canal starts at a junction with the Ashton Canal and the Huddersfield Narrow Canal at the Portland Basin in Ashton-under-Lyne. From here the line goes south through Dukinfield to Hyde. After passing through the Woodley Tunnel the canal continues via Romiley and the Hyde Bank Tunnel to go over the famous Marple Aqueduct. Here the canal passes through the town of Marple itself, ascending the hill by sixteen locks to a junction with the Macclesfield Canal. The 'upper' level of the PFC now enters rural Cheshire where it runs south-east via New Mills to another junction near Buxworth. Here one arm continues to a terminus at Whaley Bridge, where the Cromford & High Peak Railway formerly offered a route to the Cromford Canal and all points south and east. A second arm goes east to Bugsworth Basin, once the busy terminus for a series of tramroads which delivered limestone from the quarries of the northern Peak District.

You can walk the whole thing or take just a couple of short strolls along the highly accessible towpath. A fine walk on any occasion and the stretch through Marple is as good as you'll find anywhere.

History

The Ashton Canal, a line from Piccadilly in Manchester to Ashton-under-Lyne, was enabled by an Act passed in June 1792. The main purpose of the new canal was as a conduit for the many collieries around Ashton and Oldham but the line was also seen as being a possible first step in a new cross-Pennine route (later to be completed via the Huddersfield Narrow Canal – see Chapter 5). It was only later that the Ashton's shareholders began planning a branch to continue their line on through Marple to the limestone quarries of the northern Peak District.

At a meeting on 8 May 1793 the Ashton promoters agreed to introduce a bill for what would be the Peak Forest Canal. However, when it came to it, it wasn't the Ashton company that made the necessary moves. Instead a wholly independent group held a meeting in July 1793 to review the plans for a scheme to build a line from what is now Portland Basin, Ashton-under-Lyne, through Hyde and Marple to Chapel Milton. The prime purposes for the new canal were firstly to access limestone for burning to lime and secondly to serve the mines and industry between Hyde and Dukinfield. As part of this plan a tramroad was to run from the canal to the limestone quarries at Doveholes, about 4 miles north of Buxton. One of the key movers of the scheme was the cotton baron and chum of Richard Arkwright, Samuel Oldknow. Oldknow had extensive interests along the route and had plans to build limekilns at Marple which would take advantage of the new waterway for supplies of both limestone and coal.

The Peak Forest's Act was passed, with the support of the Ashton company, on 28 March 1794. The Act enabled the company to raise capital of £90,000 and a further £60,000 if necessary. The narrow boat canal was to run for 14³/₄ miles from Ashton to Bugsworth (near Whaley Bridge) through three short tunnels and over a substantial aqueduct (at Marple). The line included sixteen locks at Marple rising 209 ft to a summit pound that was 518 ft above sea level. From Bugsworth, the Act enabled the construction of a tramroad to Chapel Milton and a ¹/₂ mile long branch to Whaley Bridge. Benjamin Outram was appointed as engineer with Thomas Brown his resident assistant.

Work began at the Ashton Junction at Dukinfield and on the Marple to Bugsworth length. Curiously, although Outram had been appointed engineer, the canal committee spent some years flirting with an American engineer called Robert Fulton. Fulton, together with Charles McNiven, had been contracted to cut part of the line. Although Outram was unhappy with the contractor's work rate, the committee was impressed by Fulton and began preferring his advice to that of its appointed engineer. For a time, for example, it considered Fulton's proposal for an iron-arched aqueduct at

Marple rather than Outram's masonry design. The committee then had Outram following up Fulton's suggestions for converting the canal into one suitable for tub-boats and for the replacement of the planned Marple locks with an inclined plane. The company even paid for the printing of a book by Fulton which expanded on his ideas: *A Treatise on the Improvement of Canal Navigation* (published in 1796). Curiously, Fulton's influence disappeared virtually as quickly as it had arisen. The canal was built as a narrow boat line, the Marple Aqueduct was constructed using masonry and any thoughts of an inclined plane at Marple were quietly forgotten.

By July 1796 work was proceeding apace but already funds were running low and the company found itself with a £4,000 overdraft that it was unable to pay off. Members of the committee were forced to guarantee the monies owing and to advance a further £4,000; a sum that was needed to complete the summit level and the tramroad. Both were opened on 31 August 1796 and limestone from the quarries near Doveholes was transported into Marple. Samuel Oldknow's limekilns were now operational and coal and limestone were shipped to the kilns along two arms that lead from Marple Basin, while lime and lime ash were loaded on the short arm which ran from Posset Bridge in central Marple.

Despite the opening of the 'upper' canal, the company was still short of funds. In March 1797 work was stopped on the 'lower' canal while the committee arranged for a new issue. Even with this there was still insufficient money available and it was decided not to build the proposed flight of sixteen locks at Marple but to install a single-track tramroad instead. The company also tried to encourage trade by reducing toll rates on lime, limestone, building stone and coal destined for limekilns. With the tramroad replacing the locks, the canal, including the Whaley Bridge branch, was open for traffic on 1 May 1800. The work to that point had cost £117,140, £36,540 of which had been borrowed. An Act in June then allowed the company to raise further funds (to a total of £150,000) by new shares or promissory notes. This allowed the company to build an 84 acre reservoir at Chapel-en-le-Frith (a second reservoir, Toddbrook, was built at Whaley Bridge in the 1830s).

With the canal now fully operational, the problems in having to tranship lime and limestone on to the Marple tramway and back off again were beginning to cause congestion. Some 1,000 tons of limestone a day were being handled and night working was introduced in December 1800. The situation was only partially resolved when the tramway line was doubled. As a consequence, in 1801 the company sought to raise the necessary funds to build the locks they had originally planned. In August 1803 Richard Arkwright (presumably at Samuel Oldknow's request) agreed to loan the company £24,000 (he was repaid in 1813) and Thomas Brown was engaged to engineer the work. The locks were opened in October 1804 at a cost of about £27,000. The tramroad, however, wasn't closed until

Ice breakers at work on the canal near Romiley in *c.* 1903

Ware/The Boat Museum Archive

February 1807 which may suggest just how busy things were. The company funded this work through a further share issue authorized by the passage of an Act in 1805. This brought the final cost to £177,000.

The Peak Forest company now set about increasing its lime and limestone business. Drawbacks on limestone and coal were offered to investors who built kilns on the Peak Forest or neighbouring canals. They even offered to financially assist the construction of new kilns if Peak Forest limestone was used exclusively in them. Premiums and loans were offered for boats built before a certain date on the upper level. Special promotions of Peak Forest limestone were made to sites along the Bridgewater Canal, the Leeds & Liverpool Canal, the Huddersfield Canal and the summit levels of the Rochdale and the Manchester, Bury & Bolton Canals. They bought additional quarries, rented kilns (which they operated themselves) and maintained a wharf in Manchester from where limestone was sold for road-building. They drew up agreements with the new Liverpool & Manchester Railway; granting drawbacks on stone moved by canal to Manchester and then transhipped on to rail for sales elsewhere. This vigorous promotion was highly successful and the trade in limestone and lime was a busy one. In 1808, 50,000 tons of limestone were moved from Bugsworth and in 1811 a maiden dividend of £2 per share was paid. By 1824 the company loaded 291 narrow boats in just four weeks and in 1833 it was carrying an average of

1,743 tons on 279 boats every week. Furthermore, the prospects for growth were good especially as the much discussed Macclesfield Canal (which would link the Peak Forest with the Trent & Mersey and provide a shorter route from the Potteries to Manchester) now looked like becoming a reality (it was finally opened in 1831). In addition, the Cromford & High Peak Railway, which would connect the Peak Forest with the Cromford Canal, was about to be built. This latter line, which also opened in 1831, offered the potential for a short cut between Manchester and the East Midlands and possibly even to London. Thus by the 1830s the canal company was in good shape. In 1832 the dividend was up to £3 10s. By 1833 it was £4 and by 1835, £5. In 1838 the canal carried 442,253 tons of cargo and the total revenue, which includes the company's own lime amd limestone business, was £19,169.

Inevitably by the late 1830s railway competition was beginning to raise its head. In 1838, the through line from Manchester to London was open and the first rounds of toll cutting began. Competition was also developing between the canal companies for traffic from Manchester to Sheffield, Manchester to Cromford and Manchester to Nottingham. The Peak Forest company supported the Cromford & High Peak Railway route to the East Midlands and reduced both its toll and its wharfage charges in order to maintain and encourage traffic. However, by April 1843 the company was beginning to feel the affects of competition for its lime trade and was forced to increase the drawbacks offered.

These aggressive actions meant that by 1844 traffic levels had increased but there was no concomitant increase in receipts. Dividends were reduced to £3 and the writing was firmly written on the canalside wall. In 1845 the Sheffield, Ashton-under-Lyne & Manchester Railway offered to take over the canal and its debts on a perpetual lease for £9,325 p.a. (equivalent to a guaranteed dividend of 5 per cent per share after the payment of loan interest). This deal was completed on 25 March 1846 and the canal passed into railway control on 27 July. At the same time, the SA&MR (which was soon renamed the Manchester, Sheffield & Lincolnshire Railway) also leased the Ashton and Macclesfield Canals.

With the transfer of ownership the canal began to lose trade. In 1838 the canal carried 442,253 tons. In 1848 this had dropped to 343,549 tons, and by 1855 it was down to 187,189. Traffic stayed at this level into the 1860s. These figures, however, mask the significant loss of income that resulted from the ever-falling toll rates. Such reductions were needed even more when the Stockport, Disley & Whaley Bridge Railway was opened to the Cromford & High Peak in June 1857. Any tentative agreements with this new company on competition must have disappeared when it was absorbed by the London & North Western Railway in 1866.

In 1883 the Peak Forest company was officially dissolved and all its interests incorporated into the railway company, which in 1897 became the

Great Central Railway. By then traffic along the canal had declined significantly, carrying just 136,148 tons in 1905 to yield revenue of £4,138. There was thus an operating loss of £976. Cargo at this time included lime and limestone, coal, stone, cotton and grain. Activity at Bugsworth Basin came to a halt in 1922 and the tramroad line was abandoned in 1925. With the basin shut, trade along the upper level of the canal virtually ceased. On the rest of the line some through traffic continued to and from the Macclesfield Canal, and a small amount of coal and raw materials was shipped towards Manchester and the Ashton Canal. Most of the traffic, however, was short haul with cargo comprising of coal, timber, cotton, shale, soda ash and acid. Olive Bowyer reports that most of the boats that remained in the 1930s were still horse-drawn although some were actually man-powered with the ropes either tied around the puller's waist or slung over his shoulder.

Through the Second World War low levels of traffic continued on both the Ashton and the Peak Forest Canals but, once the war was over, the business collapsed. With nationalization in 1948 little or no maintenance work was undertaken and it was evident that the Ministry of Transport saw no future for its Peak Forest line. The top pound and its link with the

Bugsworth Basin after traffic from the quarries had ceased, *c.* 1930
Ware/The Boat Museum Archive

Macclesfield could be kept open for pleasure boats but all commercial traffic along the canal ceased in 1958. In the long, cold winter of 1961–2, Marple Aqueduct was badly damaged when the water that had seeped into the stonework froze and caused the masonry to be damaged. Stop planks were inserted and there was a grave danger that the whole line would be lost forever. The estimate for repairing the structure was £35,000 compared with one of just £28,000 to demolish it and carry the water across the valley by pipe.

It was at this juncture that the town clerk of Bredbury and Romiley council called a meeting of local authorities and interested parties to discuss the situation. It was agreed that the extra money must be found, that the aqueduct should be listed as an ancient monument and that it carry a fully navigable waterway and not a rubbish tip. Despite these fine words, the situation cannot have been helped when further damage during the following winter increased the estimated cost of the repairs. On top of this, the sixteen Marple Locks, which hadn't been used for some time, were beginning to disintegrate.

In 1964 the Peak Forest Canal Society was formed with the aim of restoring the canal. At first British Waterways was unwilling to let the society's members work on the line. However, volunteers appear to have replaced broken lock beams and to have mended sluices, and by 1968 the restoration of both the Peak Forest and the Ashton Canals was in full swing. The passage of the Transport Act was then a great help to the cause. BW was gradually convinced of the efficacy of the project and, thanks to the Act, it was allowed to work with interested parties to further the redevelopment of waterways under its jurisdiction. Thus, with the help of the society and the local authorities, it was agreed that the canal would be restored. With the considerable assistance of numerous volunteer navvies, the restoration work was completed in March 1974 and the line was officially reopened on 13 May. Today the canal is a justifiably popular part of the Cheshire Ring.

The Walk

Start:	Newtown (OS ref: SJ 995847)
Finish:	Whaley Bridge (OS ref: SK 013816)
Distance:	4¹/₂ miles/7 km
Maps:	OS Landranger 109 (Manchester) and 110 (Sheffield & Huddersfield)
Outward:	Train from Whaley Bridge to Newtown (enquiries: (061) 480 4482)
Car park:	At Whaley Bridge station
Public transport:	British Rail Whaley Bridge

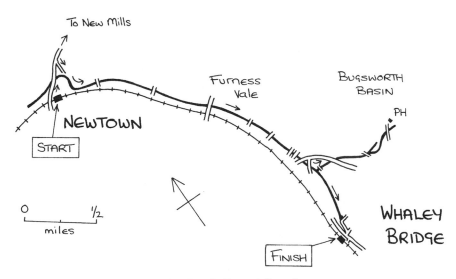

To New Mills

Furness Vale

BUGSWORTH BASIN

PH

NEWTOWN

START

WHALEY BRIDGE

0   ½
miles

FINISH

The Peak Forest Canal

From Newtown station turn left towards New Mills. After 250 yd the road goes over the canal at Thornsett Road Bridge. Although there is no immediate access to the towpath, you get fine views of the mills to either side of the bridge. To the left (towards Marple) is the Albion Mill (left) and Brunswick Mill (right). Both were cotton mills that used the canal for supplies of coal as well as for delivering raw materials and taking away finished product. To the right (towards Whaley Bridge) is the, now closed, Canal Foundry (left) and, another cotton mill, Warksmoor Mill (right). To reach the towpath continue along the road and turn right in to Victoria Street (a signpost for the canal points the way). This reaches the PFC opposite New Mills Marina.

Turn left to walk along the left-hand bank with the Goyt Valley falling away to a fine railway viaduct which carries the Manchester to Sheffield railway. At one time there were further cotton mills along the towpath side of the canal just here. One, the Victoria Mill, was destroyed by fire in 1986. The other, Woodside Mill, suffered a similar fate in 1961. The outbuildings of both are now occupied by small businesses. The canal bends right and then left to pass a small summer-house on the right bank with a well-tended garden and then on under Bankend Bridge. We now walk through a narrow wooded section which leads to Mellor's or Carr Swing Bridge. This whole section feels so remote that it's hard to believe that the line once saw forty boats of limestone a day chundering down the line towards Marple and, presumably, forty boats a day returning to Bugsworth Basin.

After passing Furness Vale Marina (and dry-dock) and a small row of terrace cottages (built for local mill workers), the canal goes under Furness

Bridge. A couple of hundred yards after the bridge, on the far bank, was once the site of a stone wharf which received its stone via a tramway from Yeardsley Quarries. The tramway passed through a tunnel under both the A6 and the railway. From here the canal passes under a lift bridge and a footbridge with the sewage works to left. This is the wonderfully named Bongs Bank Bridge. The next bridge is number 33, Greensdeep Bridge.

Walk on to pass a roller mechanism (used in times of maintenance to drain the canal by winding up a chain attached to a plug in the bottom of the canal), some houses to the right to reach a metal footbridge, Bothomes Hall Lift-Up Bridge, then a major road bridge. This is Bridgemont. If you cross the footbridge you can reach the Dog and Partridge pub.

Continue under the next major road bridge to reach the junction with the Buxworth arm. Our route to Whaley Bridge bends right here but before continuing take a short diversion along what was once the main line by turning left to walk along the left-hand bank. The Buxworth arm was opened in August 1796 and effectively abandoned in 1927. It was left to nature until the 1960s when the Inland Waterways Protection Society cleared the channel and restored it for navigation. Start by going under the footbridge, over the River Goyt Aqueduct and under the A6(T) road bridge. Stop planks mark the end of the rewatered section and beyond here the bed of the canal becomes increasingly overgrown. In 1972 the section right up to the basin was navigable but alas no longer. We pass some small cottages (known as Teapot Row) to the left and then on to a lock with Canal House and a small outhouse just a little further on. We are now at Bugsworth Basin.

Bugsworth Basin is classified as an ancient monument and is steadily being restored by the Inland Waterways Protection Society. Here is a complex of transhipment arms and basins that were once the whole rationale behind the existence of the PFC. Limestone was delivered to the basin from the various quarries around Doveholes by means of tramways and then shipped out, either as stone or as lime, along the canal. The lock at the entrance to the basin was a gauging lock, where boats were assessed for the weight of cargo carried by their height in the water. To the left is the wharfinger's office, where all the relevant gauging details were logged. The small building a little further on was the stables.

If you walk on to the newly erected footbridge, you can get a good view of the first part of the basin. There were three canal arms. The right-hand one went on to the upper basin whereas the first two in front of us were served by tramways which have been picked out by IWPS in stone chippings. Limestone was originally moved here by cart but work started on the Peak Forest Tramway in 1795. It was opened in 1796. The main quarry tramway was 4 ft 2 in gauge and ran for 6½ miles from here via Chapel-en-le-Frith to the quarries at Doveholes with branches to other quarries. There was an 85 yd tunnel at Chapel Milton (reputedly the world's first railway tunnel)

Bugsworth Basin, 1992

and a 209 ft inclined plane at Chapel-en-le-Frith. It reached a summit level of 1,139 ft. To cope with the traffic the track, apart from in the tunnel, was doubled in 1803. The stone was shipped down the tramway in iron boxes on wagons which could carry 2–2^1/$_2$ tons each. At Bugsworth the containers were either lifted into the waiting boats or taken straight to limekilns which bordered the basin area. Boxes were used in order to make the transhipment from the tramroad to the canal and from the canal to the Marple tramway easier. Following the opening of the locks at Marple in 1805, use of the boxes was abandoned and wagon tipplers were installed. The trucks made their descent under the force of gravity and were hauled back up the hill by horses. Activity at Bugsworth Basin came to a halt in 1922 and the tramroad line was abandoned in 1925. With the basin shut, trade along the upper level of the canal virtually ceased.

If you walk across the whole length of the footbridge and follow the course of the tramway (keeping to the left of the low wall), you cross a bridge with a view down right to the right-hand canal arm as it moves on to the upper basin. On the far bank are the remains of a series of limekilns. Eventually the path leads round to the Navigation Inn with the upper basin to the right. The tramway to Doveholes went straight on from here, whereas that to Barron Clough quarry bore right. Return back along the tramway towards the wharfinger's house. Just before the tramway bears left over the bridge, go right down some steps to a small basin that once served another, earlier, series of limekilns. Go right under the bridge to walk alongside the canal arm and then right again under a second bridge tunnel. This path passes to the right of the site of a building that housed a stone-crushing machine and returns to the wharfinger's house.

After this brief sojourn, return to the junction with the PFC main line. Cross the small footbridge at the beginning of the arm and turn right to walk along the left-hand bank of the canal. This, the route to Whaley Bridge, was originally built as a branch line. It is about half a mile to Whaley Bridge Basin where we cross an overflow weir and follow the path round the edge of the triangular basin to the terminus building.

This was formerly an important transhipment point on to the Cromford & High Peak Railway. The C&HPR provided a route to the Cromford Canal at Lea Wood near Cromford and thus to the River Trent and the rest of the canal network. The route was originally conceived as a canal navigation. Indeed, John Rennie had carried out a preliminary survey and presented an estimate of £650,000. The line was thought by many, including the highly supportive Grand Junction Canal Company (now known as the Grand Union), to be a way of shortening the route from London to Manchester. Naturally the Trent & Mersey, whose business was most under threat, was antagonistic and it appears to have won the argument. The concept, however, stayed alive and it wasn't long before thoughts had switched to a railway. Following an Act of 2 May 1825, Josias Jessop was appointed engineer and the work began. The line, which opened on 6 July 1831, included stationary engines to haul wagons up as many as nine inclines to the more conventional railways that ran along the summit to the PFC. The line was an important route for various cargoes going north–south, including agricultural produce and even water. The company faced severe competition by the late 1840s from other railway lines but the route remained open until 1967. It is now part of a long-distance path, the High Peak Trail. If you go around the back of the terminus warehouse, you can still see the loading bays for the C&HPR as well as the point where the feeder from the nearby Toddbrook reservoir enters the canal. The first stretch of the C&HPR left the terminus building and bore left to pass what is now the car park and over the river via a girder bridge. A little further on is the site of the first of the railway's inclines.

To complete the walk from the warehouse, turn right and follow the road left. The railway station is on the right.

Further Explorations

All 14³/₄ miles of the Peak Forest towpath are open to walkers: from Portland Basin, Ashton-under-Lyne to Whaley Bridge Basin. Public transport between the two ends is awkward and towpathers wishing to walk the entire route in one day are best advised to go with a like-minded car driver for a 'two-car trick'. For those less fortunate, trains can be used if you're prepared for the outward journey to be a little time consuming and you are able to walk an extra 1¹/₂ miles. If starting at Whaley Bridge, take the train to Newtown. Leave the station and turn right. After ¹/₂ mile turn left into central New Mills. At the top of the main street, bear left to reach New Mills station from where you can take a train to Guide Bridge. Leaving this station, turn right to reach the Ashton Canal. Turn right to walk along the right-hand bank for about ³/₄ mile to Portland Basin where the PFC goes south while the Huddersfield Narrow Canal goes straight on. Turn right to walk along the right-hand bank of the PFC. If attempting the full walk, arm yourself with Olive Bowyer's excellent towpath guides (see below).

Those who prefer a bit of a mooch rather than a walk, should head for Marple where you can park in the 'shoppers' car park in the centre of the town. Return to Stockport Road and turn left to go past the cinema and the Liberal Club to reach the canal at Posset Bridge (or New Mills and Stockport Road Bridge). The bridge get its name, so the story goes, because Samuel Oldknow supplied the navvies with a locally brewed posset of ale to encourage them to build the canal to schedule. They did.

Turn right to walk up Lockside. Here, in a short, steep 200 yd, we pass four locks each with a massive side pound that seem to invade neighbouring gardens on the far side of the waterway. They were built to ensure that there was sufficient water to allow free flow along the line without draining each pound every time a lock was used. At the top of the slope we pass the fourth lock of this sequence and reach the summit pound; 518 ft above sea-level. Here on the right is the old canal toll-house. Olive Bowyer informs us that the building doubled as the paymaster's house, so this must have been a busy spot on a Friday night. It is also the spot where Denis Howell proclaimed the restored line open in 1974. Part of the rationale for the position of the toll-house here is that the nearby roving bridge marks the junction

The horse tunnel at
Posset Bridge, Marple

with the Macclesfield Canal (see Chapter 8). Looking right we can see the former stop lock that separated the two canals and an old stone warehouse. Looking left we see the upper Peak Forest on its way to Whaley Bridge.

We now turn round to walk back to Posset Bridge where you can either go under the horse tunnel or the even more unusual pedestrian tunnel to the right of it. Looking back at Posset Bridge you can see that there is a redundant arch to the left of the currently used channel. This marked the course of the Limekiln arm which followed the Strines Road to a group of kilns about a 1/4 mile to the south. It was infilled some time after the Second World War. A second canal arm left the main line here and headed left towards a basin that was situated where the town centre car park is today. This was the Hollins Mill arm which served the cotton mill which was demolished in the 1950s. The canal arm was infilled at the same time.

The walk continues on to lock 12 with Memorial Park to the left. This popular and pleasant stretch passes two more locks before reaching a fine building to the right: Oldknow's warehouse. One interesting feature of the building is the boat entrances on the far side that enabled covered loading and unloading. The warehouse has been successfully converted into office space and is now called Lockside Mill. The small building to the right of the warehouse is Tollgate Cottage where tolls were collected from boats going to and from the many wharves between here and Posset Bridge.

The towpath changes sides at Station Road or Brabbins Change Bridge. Before crossing, however, it's worth noting the roller which still remains here to protect the stonework from towrope abrasion. Continue along the right-hand bank and past the next series of locks (Nos 1–8). There are sixteen altogether in the Marple flight and they raise the level of the canal 209 ft. Thus, each is about 14 ft deep compared with an average of just 7 ft on most canals. After passing Bottom Lock House we reach the end of the flight and a broad expanse of water that was, before the completion of the locks, the southern terminus of the lower canal. Loads of limestone were transhipped from here to the tramway that once ran up the hill on to waiting narrow boats. The tramway took a route to the west of the current line (i.e. to the left as we are presently walking) and then across the canal at lock 10 just above Oldknow's warehouse. The upper terminus was at a point just beyond the Macclesfield Junction.

The towpath again changes sides and we now walk under a railway viaduct and on to the most famous engineering feature on the PFC, the Marple (or Goyt) Aqueduct. This takes the canal over the River Goyt, some 97 ft below, by three arches. In designing the aqueduct, Outram copied the ideas of a Welsh bridge builder, William Edwards, by perforating the shoulders of the arches with hollow cylinders or roundels thereby reducing the weight of the whole structure. The aqueduct was opened in 1800 having taken seven years to build. The neighbouring railway viaduct, which overlooks the whole scene, was opened sixty-five years later. If you cross the aqueduct and turn left down a signposted path, you can reach its base. It's a densely wooded spot and wonderfully emotive. The view of the aqueduct, however, is obscured in summer by the overhanging foliage and some may view the steep climb back up the valley as somewhat unnecessary.

To complete the walk, return along the towpath to Posset Bridge and turn right to reach the town centre.

Further Information

The Peak Forest Canal Society is presently not operational. Enthusiasts for this wonderful line should therefore seek comfort in the arms of the Inland Waterways Association which has a Manchester group. The address of the head office in London can be found in the appendix.

Information about the canal can be obtained from the New Mills Heritage Centre, Rock Mill Lane, New Mills, Derbyshire. The centre sells various books on the line including those published by The New Mills Local History Society. Most pertinently, these include three by Olive Bowyer:

The Peak Forest Canal: Lower Level Towpath Guide;
The Peak Forest Canal: Upper Level Towpath Guide;
The Peak Forest Canal: Its Construction & Later Development.

Alternatively, the history of the PFC can be traced in:

Hadfield, C. and Biddle, G., *The Canals of North West England*, Vols. I and II. David & Charles, 1970.

Those interested in the Bugsworth Basin and tramway will find the following of interest:

Ripley, D., *The Peak Forest Tramway (including the Peak Forest Canal).* Oakwood Press, 1968.

10
THE RIPON CANAL

Ripon

Introduction

The fine city of Ripon is the most northerly point on the English waterways network. As you stand admiring the cathedral in a city well to the north of York and Harrogate, it is almost inconceivable that it was once possible to navigate inland all the way from here to London or Bristol. And for most of the distance, you still can.

The Ripon Canal is just 2¹/₄ miles long. From Bondgate Green Basin, almost literally in the shadow of Ripon Cathedral, the line heads south-east to Rhode's Field Lock. Here it takes a more southerly course through Bell Furrow's Lock and alongside Ripon racecourse to Oxclose Lock. Just a few hundred yards further and the canal ends as it enters the River Ure.

It's not very long, the Ripon Canal, but it makes for a pleasant stroll and Oxclose Lock is as peaceful a spot for a picnic as you'll find anywhere.

History

By the middle of the eighteenth century the River Ouse was a thriving navigable waterway and the people of Ripon began to see the desirability of forging a link with York, Hull and the coalfields of South Yorkshire. The idea of extending the navigation of the River Ouse via the River Ure appears to have arisen in 1766. The mayor of the city, Christopher Braithwaite, called a meeting at the York Minster Inn at which plans and estimates prepared by the engineer John Smeaton were presented. The sum of £15,000 was raised by public subscription and a petition to Parliament was made in early 1767. The necessary Act received Royal Assent on 15 April.

The Act enabled making part of the River Ure navigable from its junction with the River Swale upwards past Boroughbridge to Oxclose, with a cut to Ripon. There were to be two locks (Milby, below Boroughbridge, and Westwick) on the river section, an entrance lock to the canal at Oxclose, and two others at Ripon. This route was marginally different to that originally discussed. The 1766 Ripon Canal line left the Ure at Westwick Lock. Pat Jones suggests that this may well have been seen as a way of avoiding conflict with the owners of Newby Hall. In the event it seems likely that there was no objection and so the river was used to Oxclose. The canal was designed to take Yorkshire keels measuring 58 ft by 14 ft 6 in. Smeaton's estimate for the entire route from Linton-on-Ouse into Ripon was £10,843 11s. plus the cost of the land.

Commissioners were appointed and empowered to borrow whatever money was necessary. The line seems finally to have been surveyed by the young William Jessop under Smeaton's supervision. Once John Smith was appointed resident engineer, with Joshua Wilson of Halifax as masonry contractor, the work began. However, by this time the length of the Ure Navigation from Linton Lock to the River Swale had been ceded by an Act to the Swale Navigation who were now responsible for the construction work on that stretch.

By 31 October 1769 the Ure Navigation was open to Westwick Lock just south of Newby Hall. Cutting of the Ripon Canal itself seems to have begun early in 1770, and by 28 September 1771 all the works were complete. However, it was found that certain parts of the line had silted up and needed to be dredged before use. History seems uncertain about the first commercial passage but it seems likely that by the summer of 1773 a regular traffic was operating along the line. When completed, the navigation was 10¼ miles from Swale Nab, including a 1,105 yd cut from Milby to Boroughbridge, one of 616 yd at Westwick, and the 2¼ miles of canal to Ripon. The final cost was £16,400 with the Ripon Canal alone costing approximately £6,000. The commissioners raised the additional funds by calls on shares and by giving landowners rent-charges on the tolls.

In February 1773 three firms advertised that they were starting a two-boats-a-week service from Ripon to York. In 1777 a new service through from Hull to Boroughbridge was announced. By 1781 warehouses were available at both Boroughbridge and Ripon. Although the main canal business was in coal, the navigation to Boroughbridge also had a healthy trade in flax and timber, and Ripon was an important port for lead brought by road from the mines near Greenhow Hill west of Pateley Bridge. In 1824 it was claimed that Ripon lead was being conveyed by water to all parts of the country. The coal traffic mostly came from the area covered by the Aire & Calder Navigation. At Ripon there were at least three coal merchants in 1788, and in 1789, the A&CN bought its own wharf. Humber keels couldn't navigate beyond Boroughbridge which became a transhipment port. Trade to Knaresborough was also unloaded at Boroughbridge; it was then sent on by road. Sometimes, if water levels were low, such transhipment occurred at Swale Nab. Although well used, it is evident that movement along the navigation could be desperately slow. In December 1785 coal boats from Selby took sixteen days for the return journey to Ripon. The tolls charged were 1s. per ton per mile for merchandise, including lead, against the authorized 3s. By the second decade of the nineteenth century the management and condition of the canal had both fallen into disarray. The commissioners were in debt to the tune of £16,400, and there was no one left alive legally qualified to act as a commissioner. Despite this, trade along the line was doing well and there was a widespread call for improvements together with the construction of additional wharves and warehouses. In October 1819 the Mayor and Corporation of Ripon asked Parliament for authority to form The Company of Proprietors of the River Ure Navigation to Ripon. With subscriptions of £3,033 they obtained an Act in June 1820. The new company was empowered to raise £34,000 in 200 shares and £3,400 more if necessary through loans. The bulk of the new shares was reserved for creditors, who were given the choice to either accept them or keep their securities.

Bondgate Basin before 1900

Yorkshire Image/College of Ripon and York St John

The company immediately set about upgrading the line, including spending £2,730 on new warehouse facilities at Milby and Ripon Basin. The improvements were much welcomed and by 1837 Keddy & Co.'s, William Scatchard's and the 'Ripon Fly Boats' all worked to and from the city basin. By the end of the 1840s vessels carrying 70 tons and drawing 4 ft 6 in were reaching Ripon.

The Great North of England Railway from Darlington to York was enabled by an Act of 1837 and opened in 1841. The branch line which passed from the GNER at Thirsk via Ripon to Leeds was authorized in July 1845. The main line had already taken a good deal of the coal traffic to Boroughbridge and Ripon but the branch could finish the canal traffic for good. At the time, more than 26,000 tons of coal p.a. were being imported from the A&CN with three boats a week reaching Ripon Basin. The average Ure and Ripon tolls for the ten preceeding years had been £2,013 with expenses at £1,127, giving a net average profit of £886 p.a.

It appears that with the suggestion of the branch, the owners of the navigation promptly sought to sell the concern to the new railway company. The Leeds & Thirsk (later the Leeds Northern) Railway thereby agreed to purchase the Ure Navigation and the Ripon Canal in 1844. The railway's shareholders confirmed their support in January 1845. When the company sought its Act, there was little or no opposition, and in January 1846 the sale was confirmed at £34,577. Transfer took place on 1 July 1847, £16,297 being paid in cash, the rest in railway shares. The Ure & Ripon Company was then dissolved.

The railway Act compelled the new owners to keep the canal and navigation in good repair. However, with the completion of the railway to Leeds on 9 July 1849 and the opening of the York & Newcastle Railway's branch from Pilmoor to Boroughbridge on 17 June 1847, the minerals traffic was lost to rail despite a cut in canal tolls. Canal-borne A&CN coal also suffered from competition with pits from the south Durham area. In 1854 the Leeds Northern became part of the North Eastern Railway which, rather like its southern counterparts the Great Western, regarded canals as an out-of-date nuisance. The company reduced its dredging or maintenance operations to a minimum. By 1857 boat owners of Ripon–Hull craft were complaining that the navigation had not been dredged for some time with the effect that vessels were unable to pass along the route fully laden. Trade recovered slightly in the 1860s with twenty boats working the Boroughbridge coal business. Linton Lock records show that 18,000 tons of coal p.a. were moved up the line between 1854 and 1864. Coal traffic from the A&CN, however, which in the 1840s had exceeded 26,000 tons p.a., was down to 10,956 tons in 1871 and to just 1,922 tons in 1891.

By 1892 only 5,000 tons were being carried on the navigation, with a revenue of £161 against a maintenance bill of £683. Most of this traffic was wheat and coal to Boroughbridge Mill. By now there was no traffic on the Ripon Canal, which had gradually become disused; a situation exacerbated by high tolls and lack of dredging. In 1894 the NER tried to abandon the canal but had to drop the proposal after strong opposition from local people who saw the survival of the canal as a way of keeping down railway rates. Unable to close the line, the company offered the navigation as a gift to York Corporation, who declined it. The then President of the Board of Trade, A.J. Mundella, reiterated in a public enquiry in 1894 that the navigation must be kept open and must be well maintained 'for the use of all persons desirous of using the same, without any hindrance, interruption or delay'. The NER was presumably not pleased with the outcome of the enquiry and sought by all means to prevent the line being used above Boroughbridge. In 1898, 9,001 tons of cargo were carried, yielding receipts of £178. By 1905 these levels had fallen to 3,409 tons and £71 respectively. The main business at the time was in gravel, flour and sand. In June 1906 the Royal Commission reported that it was impossible to reach Ripon and that lock-gates were in a severe state of disrepair.

About 1929 Blundy, Clark & Co. Ltd, a sand and gravel merchant operating from above Milby Lock, brought an action against the London &

The former Navigation Bridge as photographed in the early twentieth century
Yorkshire Image/College of Ripon and York St John

North Eastern Railway (into which the NER had been absorbed in 1923) alleging neglect. They claimed that there was insufficient water in the navigation to allow them to move full loads and that Milby Lock had been closed for repairs unnecessarily. The case was won and the LNER was forced to pay £2,362 compensation.

The year 1931 saw a turn in the canal's fortunes. The Ripon Motor Boat Club was formed and pleasure craft started to use the navigation. The actions of these early pioneers compelled the LNER to dredge the lower reaches of the canal and to undertake repairs on both Oxclose and Bell Furrow's Locks. Although Oxclose Lock was a constant source of annoyance to boat owners, the cut became a popular winter haven for club craft and the lock was maintained using club funds. Somehow the club was also able to persuade the Royal Engineers to dredge the stretch from Renton's Bridge to the club's slipway near Nicholson's Bridge as part of an exercise.

In 1948 the canal and the Ure Navigation was nationalized (at this time it was called the Ripon & Boroughbridge Canal). In 1955 the British Transport Commission proposed to abandon the upper section of the canal to Ripon. Although at one time it was thought possible that the Motor Boat Club might take over the line, it was abandoned on the basis that there were insufficient commercial prospects to justify its retention for navigation. Rhode's Field and Bell Furrow's Locks were cascaded and Navigation Bridge culverted (in 1958). The Board of Survey for the British Transport Commission thought kindlier of the Ure Navigation, which it decided should be retained. An agreement between the BTC and the Motor Boat Club was then made in which the club (operating as Ripon Canal Company Ltd) would lease the Oxclose cut while the BTC would maintain the lock. This general situation was confirmed by the Transport Act of 1968 when the Ure was deemed suitable for development as a cruiseway. This allowed Oxclose Lock and the open section of the line to be maintained. In 1968 the club's lease of the cut expired and administration was taken over by British Waterways. In 1972 there was still some commercial traffic on the Ure Navigation in sand and gravel from above Milby Lock, as well as pleasure craft to Oxclose Lock, the limit of navigation. This situation continued until 1983 when the Ripon Canal Society was formed. The aim of the new RCS was to restore the derelict sections to navigability and to preserve any remaining features of historical interest. Society members cleared much of the accumulated debris and a Manpower Services Commission grant enabled British Waterways to restore both of the upper locks. These were reopened in 1986. Since then the society has continued to press for the full restoration of the line to Ripon Basin and it is hoped that schemes to do so will shortly be in place.

The Walk

Start and finish:	Ripon (OS ref: SE 315708)
Distance:	5 miles/8 km
Map:	OS Landranger 99 (Northallerton & Ripon)
Car park:	Signposted near cathedral
Public transport:	United and Harrogate & District buses nos. 36, 36A and X36 connect Ripon with Leeds and Harrogate BR (enquiries: (0325) 468771)

From the cathedral front, turn right to go down Bedern Bank to a small roundabout. Turn left to follow signs to Boroughbridge. This road is Bondgate Green and it passes over the River Skell which still feeds water into the canal. After just a few yards it's worth deviating for a short distance right into Canal Road. This passes the Navigation Inn and enters the area of Ripon Basin. This is the most northerly point on the English canal network! Here are a collection of buildings that include a derelict wharfinger's house and a number of warehouses with red-tiled roofs. In former times the large walled area housed a substantial wharf where coal was unloaded and stacked

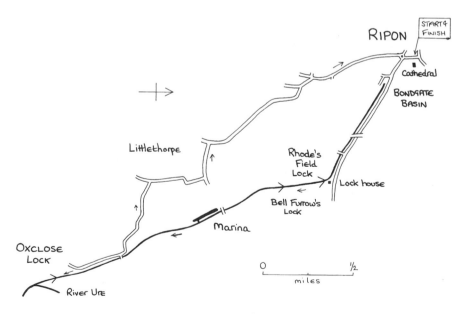

The Ripon Canal

ready for distribution. No boats have navigated the entire length of the canal since 15 May 1937 (when a flotilla turned up to celebrate the coronation of King George VI), and it's a sadly neglected spot now. The owner of the canal head has had plans drawn up with a grant from English Heritage and intends to start by restoring the cottage at the basin.

As it's not possible to cross the canal at the basin, return to Bondgate Green and turn right to start the walk along the left-hand bank with the Boroughbridge Road to left. Although we are walking alongside a busy road with industrial works to either side, the stretch along here is not unpleasant. The canal is overhung by trees, and groups of ducks stare listlessly at the passing towpather. Ripon Basin is presently not accessible to canal-borne traffic and you will shortly see why. Two pipes cross the waterway at a level low enough to preclude shipping.

The Leeds & Thirsk Railway viaduct dates from 1848 and has been redundant since 1969. It is shortly to be restored as part of the new Ripon bypass. This may seem unimportant but at one time it looked as if the new road would add a further blockage at the northern end of the canal. A little further on, however, Littlethorpe Road Bridge does effectively seal the basin from the rest of the line. The original bridge here, Navigation Bridge, was demolished in 1958. The canal is presently culverted under the road and will remain so without some not inconsiderable expenditure of cash.

Things improve shortly, for the canal bends right to pass Lock House and Rhode's Field Lock. The two upper locks along the canal were both cascaded in 1956 after the upper section of the canal was abandoned. The lock was restored and reopened in 1986; an event that was attended by a large number of interested Ripon citizens who, it is reported, viewed the occasion with great civic pride. It's just one of those little wonders of Britain that make it so hard for foreigners to understand us. Why should anyone spend all that time, money and effort restoring a lock to nowhere? It has to be said that I'm glad they did.

The walk continues along a fine straight section that improves step by step as we move away from the main road. Shortly we reach the second lock, Bell Furrow's, which was cascaded and restored at the same time as Rhode's Field. Those who own the same, rather old, copy of Nicholson's *Guides to the Waterways* as I do, will note that this lock was once known as Littlethorpe Lock and was, until 1986, the end of the navigation. From here we pass a small wooden jetty and a bridge (Nicholson's Bridge). On the far side of the bridge is Ripon Motor Boat Club's Littlethorpe Marina. The club has been the prime mover in the maintenance and restoration of the canal and are owed a huge debt of gratitude for their efforts. Formed in October 1931 and, apparently, the oldest inland cruising club in the north, it is widely recognized that without their presence, their campaigning and physical labour, the canal and the river down to Linton Lock (see below) would be mostly derelict and unnavigable.

The newly restored Rhode's Field Lock

The walk continues into ever quieter country with the gallops of Ripon racecourse just the other side of the hedge. This stretch of the canal was not built by one team of navvies but by two. One worked from the north (Bell Furrow's Lock down) and the other from the south. When the two gangs met each other, they discovered that the levelling had been so awry that the southern pound was two feet lower than the northern. This forced Smeaton to revise his plans for Oxclose Lock, which now had to take an extra rise of two feet. This necessitated the banks of the canal to be raised and strengthened to take the strain. At the next bridge (Renton's Bridge) the towpath changes to the right-hand bank where we walk along a metalled lane. When this bears right, cross a stile and carry on along the towpath to Oxclose

The Ripon Canal wends its way south towards the River Ure

Lock. From here the final stretch of cut can be seen on its way to join the River Ure, which itself becomes the Ouse en route via York to the Humber. The Ure was made navigable under an Act of 1767 and falls by two locks before joining the Ouse at Ouse Gill Beck. The line was used by commercial vessels as far up as Boroughbridge until fairly recently. A firm dredging aggregates from nearby river beds used the navigation for transport.

To return to central Ripon, you can either retrace your tracks along the towpath or you can take a slighter shorter 'inland' route via Littlethorpe. Walk back to the metalled lane and on to the bridge. Instead of crossing the canal, continue along the lane which twists and turns before reaching a minor road at a T-junction. Turn right to follow this road through the village of Littlethorpe. At a T-junction near St Michael's Church, turn left. At the next T-junction, turn right. This is Littlethorpe Lane and it leads right into Knaresborough Road with a view to Ripon Cathedral to the right. At a crossroads, turn right and walk on to Bedern Bank. If you then walk up the short hill, you will reach the cathedral and the town centre.

Further Explorations

Although you have now walked the entire length of the Ripon Canal, the Ure Navigation makes for splendid walking and is blessed with a public right of way along much of its length. One particular place of interest nearby, albeit on the Yorkshire Ouse, is Linton Lock. (This can be found on OS Landranger 100 at ref: SE 500602.)

Park the car in a small lay-by on the northern side of Newton-on-Ouse (on the road to Linton-on-Ouse). From here walk back a short distance towards Newton and a bridge which goes over the River Kyle. Just before this, turn right to go through a gate and down some steps into the field. Follow the clear path ahead which leads to the River Ouse. Turn right with the river to the left.

The Ouse has been a navigable waterway since Roman times connecting these upper reaches (to the junction of the Ure Navigation and Ouse Gill Beck) with the city of York and then on to the River Humber and the sea. Today the lower reaches between Selby and Goole are the only commercially used sections although most of the 55 miles are used regularly by holiday cruisers. The clear path passes through some pretty countryside to go over a stile and on to cross Linton Ings Outfall. From here the path joins a driveway and passes to the right of a small caravan park. After passing through a car park we reach Linton Lock with its shop and café.

Linton Lock is one of only two on the Yorkshire Ouse, and the upstream cut bypasses the splendid looking weir that is partially hidden from view behind some trees. The lock was originally designed by John Smeaton but the resident engineer John Smith had his own ideas. In the event, Smith's lock (and dam) cost £10,400 against Smeaton's estimate of just £3,000. The lock was ready in August 1769 and the first tolls taken on the 24th. By 1962 the lock had fallen into disrepair following a long period of neglect by the perpetually impecunious commissioners and was closed. Thankfully a group of local boating enthusiasts formed the Linton Lock Supporters Club. A large number of the supporters are from the Ripon Motor Boat Club whose members are required by the rules of the club to join and subscribe to the LLSC. It is they who managed to find the £3,000 necessary to restore the lock (it was reopened in 1966) and who now support the maintenance of it. This, they hope, will not be an endless task and, apparently, suggestions of a British Waterways take-over could be in the air.

The seats outside the lock-house café make a pleasant spot to sit and contemplate the scene. After a cup of tea it's worth wandering about for a while, to see the weir and to contemplate what life was like in February 1991 when the river level reached waist height. To return to the car, simply retrace your steps back along the river.

Further Information

The Ripon Canal Society was formed in 1983 with the object of restoring the line to navigability throughout. It can be contacted at:

Mr David Reasons,
The Lock House,
Boroughbridge Road,
Ripon,
North Yorkshire.

Linton Lock fans can become supporters by contacting:

Mr David Evans,
75 Kirkby Road,
Ripon,
North Yorkshire RG4 2HH.

The Ripon Motor Boat Club has published:

Jones, P.E., *A Short History of the Ure Navigation and Guide from Ripon to Naburn*. Ripon Motor Boat Club, 1986.

However, it is now long out of print. Should you wish for further information, you should therefore refer to:

Hadfield, C., *The Canals of Yorkshire and North East England*, Vols. I and II. David & Charles, 1972.

11
THE ROCHDALE CANAL
Sowerby Bridge to Littleborough

Introduction

Those who ply the M62 between Manchester and Leeds will know the Rochdale Canal. Just on the western side of Junction 20, there it is. It doesn't go under the motorway. It doesn't go over it. Instead it ploughs straight into it; the two halves of the line firmly, and on the face of it irrevocably, separated. And this is just one of the problems for those hoping to restore this wonderful waterway.

The Rochdale Canal starts its 33 mile long cross-Pennine trek at Castlefield in Manchester, where it forms a junction with the Bridgewater Canal. The first mile and a half passes through the city to a junction with the Ashton Canal at Ducie Street, just north of Piccadilly station. From here the line heads north-east to Failsworth before turning north to Chadderton and the south-eastern fringes of Rochdale. The canal completes its ascent of the Pennines just north of Littleborough. The downhill side passes along the Calderdale valley, through Todmorden, Heptonstall and Hebden Bridge before reaching its easterly terminus at Sowerby Bridge where, at one time, the Rochdale formed a junction with the Calder & Hebble Navigation.

Although never far from 'civilization', this is a fine, airy walk all the way from Yorkshire to Lancashire. It's wonderful stuff and enough to make even the most fervent canal hater put on their walking boots.

History

Following the launch of the Leeds & Liverpool Canal on 2 July 1766, it was just six weeks before Richard Townley of Belfield, near Rochdale, called a meeting to promote a rival cross-Pennine route. The aim of those who assembled at the Union Flag Inn, Rochdale, on 19 August was to take a line from the Calder & Hebble Navigation at Sowerby Bridge to the Mersey &

Irwell Navigation or the Bridgewater Canal in Manchester. The meeting stimulated enough enthusiasm to sponsor James Brindley to undertake two surveys: one that went via Bury, which Brindley estimated to cost £102,625, and one along (approximately) the present route, estimated to cost £79,180. Peculiarly, it wasn't until 1790 that a meeting was held at Hebden Bridge to take the project further. By June 1791 a committee had been appointed, a £200,000 subscription had been raised (it's said that the first £60,000 was raised in just an hour) and John Rennie had been asked to undertake a survey for a route via Todmorden. This was to include branches to Rochdale and Oldham although the precise nature of the Manchester terminus was still a matter of great debate. The key problem that faced the promoters was the objections of the mill owners whose opposition led to the defeat of the first Rochdale Canal Bill on 21 March 1792.

This defeat did not downhearten the supporters of the plan who quickly regrouped. The second bill was significant in that the committee had come to an agreement with the Duke of Bridgewater for the Rochdale to join the Bridgewater Canal en route for Manchester. The duke had originally opposed the plan but changed his mind and agreed to it if only on the promise of a transfer fee of 3s. 8d. a ton (later reduced to 1s. 2d.). The new bill also increased reservoir capacity, for what was to be a narrow boat canal, and included a plan for a 3,000 yd long summit tunnel. Despite these alterations, the bill was again lost (albeit by just one vote) forcing the promoters to regroup once more.

The third attempt at a Rochdale Canal Bill started almost immediately with considerable effort put into mollifying the mill owners and to agreeing the availability of water. William Jessop was now engaged to resurvey the more contentious parts of the line. In his scheme, the idea of a tunnel was dropped in favour of two sets of seven locks up to a higher summit pound. Extra attention to the water supply problem was rewarded when the third bill was passed on 4 April 1794; the same day, interestingly enough, as that for the line's main rival, the Huddersfield Narrow Canal.

As planned the Rochdale Canal was to run from the Bridgewater Canal at Castlefield (where there was to be a junction lock via which the duke would receive his fee) via Failsworth, Littleborough and Hebden Bridge to join the Calder & Hebble at Sowerby Bridge. There were to be branches to Castleton and Hollinwood (later dropped). The Ashton Canal was to be allowed to form a junction with the new line at Ducie Street, Manchester. From Hull to Manchester, the Huddersfield route would be 15 miles shorter but the Rochdale promoters (rightly, as it turned out) saw the advantage of building a broad canal; one that the wide Yorkshire keels that plied the C&HN would be able to use all the way to Manchester. The Act enabled the company to raise £291,900 in shares with powers to obtain a further £100,000 should it be needed. Rather optimistically, this estimate was based on one made by Rennie for his narrow canal even though the new plan was for a broad line. With establishment of the company, William Jessop was appointed engineer with William Crosley as resident engineer.

The work began almost immediately at Sowerby Bridge, at the summit and along the section in central Manchester between the Ducie Street and Castlefield Junctions. The work seems to have gone smoothly and the canal was opened on 24 August 1798 as far as Todmorden and on 21 December to Rochdale. A slight delay had occurred on the section in central Manchester although this was probably opened during 1799. A new Act was passed in 1800 giving the company powers to raise an additional £100,000 by annuities and promissory notes and this enabled the work to continue so that by September the canal had reached Hopwood. However, by the middle of 1803, funds were again running low. The Napoleonic Wars had led to a period of sustained inflation, which meant spiralling costs and an increasing inability of shareholders to raise cash when requested. In addition, the bank had refused to advance further funds until the company had reduced its overdraft. The committee was forced to seek another Act (passed in 1804) to raise a further £70,000. This was sufficient to take the line to Ducie Street and the canal was finished on 21 December 1804.

The Rochdale Canal was opened with a good deal of celebration including the ringing of bells and the cheering of enormous crowds. The new line was 33 miles long and comprised ninety-two broad locks. The size of the locks meant that it was compatible with the Bridgewater Canal but not with the C&HN, the locks of which were too short to take Rochdale barges. Cargo heading to the north-east was therefore transhipped at Sowerby Bridge. Cargo going south in Yorkshire keels (57 ft long by 14 ft 1 in wide) could, however, go the entire way to Manchester. As a consequence, the line was an almost immediate success and simply brushed aside competition from the Huddersfield when that line opened in 1811. Indeed, of the three cross-Pennine routes, the Rochdale was always to be the most heavily used.

The canal carried coal, stone, corn and lime among a wide range of goods and there were numerous carrying companies plying the route. Pickfords, for example, operated a fly-boat service from Manchester which took twelve hours to reach Todmorden. In 1812 some 199,623 tons were shipped along the Rochdale Canal. In 1819 this figure had risen to 317,050 tons. As a comparison, the Huddersfield Narrow Canal at this time was carrying just 38,899 tons. This success enabled the Rochdale to pay its first dividend in 1811 at £1 per share. An investment programme meant that this figure could not be increased until 1821 when it rose to £4, an amount that was maintained until 1835.

In 1830 a railway was proposed from Manchester to Sowerby Bridge via Littleborough and Todmorden. The response of the Rochdale was to try to unite the various navigation companies against both this proposal and any others that may come to compete with the waterway network. A meeting of interested parties was held in London but little seems to have resulted. The company successfully opposed the railway bill and set about encouraging traffic along its line. As part of this a 1½ mile branch was opened to Heywood in April 1834. A packet-boat service, operating on Monday, Wednesday and Friday, left Rochdale at 1 p.m. and arrived in Manchester at 8 p.m. From there, cargo could be transhipped on to similar services to Liverpool, Birmingham and London. These actions, coupled with the opening of the Liverpool & Manchester Railway, increased traffic levels considerably. Trade had increased to 500,559 tons in 1829 and reached 875,436 tons in 1839 with annual receipts of £62,712 and a dividend rising to £6 per share. A further boost to trade was received in 1839 when the Manchester & Salford Junction Canal was completed to link the Rochdale with the River Irwell in Manchester. And, in a most responsible way, the company still sought to further improve its canal wherever possible.

The Manchester to Sowerby Bridge railway, renamed the Manchester & Leeds, was reborn in 1835 and this time was successfully proposed to Parliament. By July 1839 the railway was open from Manchester to Littleborough and then to Leeds on 1 March 1841. The Rochdale responded by cutting tolls. The C&HN followed suit as a way of combatting what was now a full Manchester to Hull rail route. Goods carried for the entire length were given special rates. The declining tolls had a severe affect on the value of Rochdale shares which deteriorated from £150 to £40 in two years. The railways were also feeling the pinch, and by January 1843 everybody agreed that they should be talking to each other in order to 'terminate the present injurious contest'. After a meeting in February 1843 the Rochdale increased toll rates slightly and the railway company followed suit. In fact the two companies became so close that they even agreed to act jointly against potential competition from the HNC.

In March 1843 the canal committee decided to offer the M&L a lease on the canal. This was not taken any further until 1845 when the two companies once again held discussions; this time concerning possible amalgamation, purchase or leasing. At this time, no doubt encouraged by the toll-cutting, traffic along the canal had reached its peak at 979,443 tons in 1845. In January 1846 the M&L offered the Rochdale a dividend guarantee in perpetuity of £4 per £85 and both groups of shareholders agreed the deal. Unfortunately a bill to enable this move was defeated in Parliament in 1847 following opposition from the Aire & Calder Navigation which, presumably, feared potential problems with the waterway route to Manchester should a railway company take it over. What this action did result in was an agreement between the various companies on toll rates.

It wasn't until 23 July 1855 that agreement was finally reached for four railway companies (the Leicester and Yorkshire; the Manchester, Sheffield

George Stephenson's Charleston Bridge at Hebden Bridge as sketched in 1845
Ware/The Boat Museum Archive

& Lincolnshire; the North Eastern and the London & North Western) to lease the Rochdale for twenty-one years at £37,652 p.a.; an amount sufficient to pay a dividend of £4 per share and yet still allow £15,000 p.a. to be spent on maintenance. Dividends during the 1840s and early 1850s had, in fact, been less than this (£2 10s. from 1848–53) with toll income standing at £25,000–£30,000 for the whole of the previous decade, despite the amount of cargo being carried remaining at 750,000–850,000 tons. Thus the shareholders must have seen the deal as a good one and were obviously eager to sign the agreement. Traffic and toll income remained at a similar level during the course of the period of the lease (i.e. until 31 August 1876) and it was extended until 1890. Such was the success of the arrangement that the canal committee had turned down an offer from the new Bridgewater Navigation Canal (from which nearly 60 per cent of the Rochdale's traffic came) of a dividend of £4 in perpetuity together with an option to buy the £85 Rochdale shares for £90.

In the late 1880s the company formed its own carrying capacity, issuing £48,000 of debentures to fund it. In 1887 it bought out several bye-traders. By 1892 the company owned fifteen steam cargo craft, thirty-eight narrow boats and fifteen short wide boats to work the C&HN. By this stage, however, most of the traffic was being carried within a few miles of each end of

Horse-drawn narrow boats, breasted together, pass Clegg Hall, Milnrow, *c.* 1920
Ware/The Boat Museum Archive

the line with relatively few craft passing along the entire length. Trade remained good in the early 1900s but the First World War hit the line hard. The carrying business was closed on 9 July 1921 and trade declined rapidly. In 1922 there were only 8,335 boat passages along the canal compared with 25,130 just nine years before. This decline was even greater at the summit where 3,223 passages in 1913 was down to 639 by 1922. A virtual admission of defeat came in 1923 when a number of the company's reservoirs were sold to the Oldham and Rochdale Corporations for £396,667. Some of these funds had to be paid to the Manchester Ship Canal Company in respect of the MSC's loss of rights to take water from the Rochdale. The remaining £298,333 was used to reduce the stock issue.

The decline in traffic during the 1930s was such that the last working trips over the summit of the canal came in 1939: a ladened narrow boat reached Sowerby Bridge on 6 April and an empty boat, *May Queen*, reached Manchester on 12 June. On the 1 January 1948 the British Transport Commission took over most of the nation's waterways but did not incorporate the Rochdale, which was already considered to be semi-derelict. By 1952 an Act enabled the closure of the line from the junction with the Ashton Canal northwards. The westernmost section of 1¹/₄ miles through the city remained open although, on 29 May 1958, the last boat from Bloom Street power station passed down the canal to the Bridgewater. The remaining section between the Ashton and the Bridgewater Canals in central Manchester remained (barely) open despite an Act in 1965 which allowed it to be closed if (or perhaps when) the Ashton Canal were ever closed. David Owen records how it took twelve days to navigate the 1¹/₄ miles to the junction with the Ashton Canal in the early 1960s.

Although much effort had already been expended by canal restoration enthusiasts, sadly, by the early 1970s, the prospects for the canal were still poor. The inherent dangers of an open watery rubbish tip running through the suburbs of Manchester led the city council to infill 2¹/₂ miles of the canal from Great Ancoats Street northwards. However, by this time there was an increasing interest in leisure use. The Rochdale Canal Society was formed in the 1970s with the aim of promoting full restoration between Manchester and Sowerby Bridge. This has resulted in the formation in 1984 of the Rochdale Canal Trust Ltd with the view to raising funds for this purpose. And, slowly but surely, life is being breathed back into the old canal. The Calderdale section between Sowerby Bridge and Todmorden is now fully restored although there is presently no access to the C&HN. The Trust hopes to be able to open the full line by the early part of the twenty-first century. Full restoration will, according to an estimate in early 1991, cost £15.9 million. Yet the benefits in reopening the waterway will be substantial. We can only hope that it will be done.

The Walk

Start:	Sowerby Bridge railway station (OS ref: SE 062234)
Finish:	Littleborough railway station (OS ref: SD 938162)
Distance:	15¹/₂ miles/25 km (or shorter, see text)
Map:	OS Outdoor Leisure 21 (South Pennines)
Return:	British Rail Littleborough to Sowerby Bridge (enquiries: (0422) 364467)
Car park:	Sowerby Bridge station
Public transport:	British Rail at either end

Thanks to the proximity of the railway line, this walk can be divided into three shorter lengths. Sowerby Bridge to Hebden Bridge is approximately

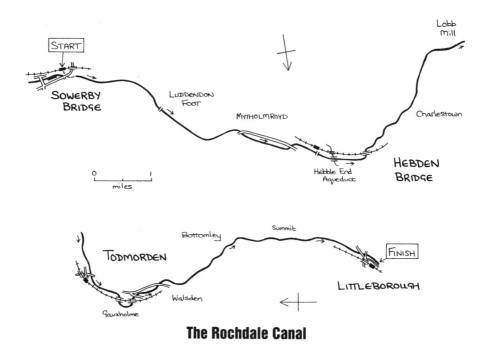

The Rochdale Canal

6 miles (9¹/₂ km), Hebden Bridge to Todmorden is 5 miles (8 km) and Todmorden to Littleborough is 6 miles (9¹/₂ km). (These figures include the distance to and from the relevant railway stations.)

Sowerby Bridge to Hebden Bridge

From the town side of Sowerby Bridge station, go down the steps to a lane. Turn left and then right along an alley, following signs to the town centre. This path crosses the River Calder with its riverside warehouses to the left. Continue along an alley between houses to the main road (Wharf Street).

Before starting the walk to Hebden Bridge, it is well worth having a look at the Sowerby Bridge Basin. To do this, turn right to walk along Wharf Street. After passing some traffic lights near a church, the road goes over the stub end of the Rochdale Canal. Turn right to walk with the rather sad looking Rochdale Canal to left. Our way passes two locks to reach the Sowerby Bridge Canal Basin. The basin was originally the westernmost terminus of the Calder & Hebble Navigation; a line engineered by John Smeaton and completed to Sowerby Bridge in 1774. The first warehouses were built in

The Rochdale at
Sowerby Bridge

the 1780s and over the following years more warehousing, offices and stabling were added. At its peak there were 3,673 square yd of covered warehousing and 1,800 square yd of outside storage here. When the first section of the Rochdale was opened from Sowerby in 1798, the basin became a major transhipment point for those Rochdale barges which couldn't work the C&HN short locks. The shorter (but equally wide) Yorkshire keels, however, were able to work both navigations and would not have needed to have stopped here. Some of the original basin buildings are still standing but the features are now hard to pick out. Many of the structures have large arches to enable the boats to go into the building. The last barge to use these facilities left in September 1955.

The walk continues round the towpath to reach the C&HN. Walk on to go under a bridge and up the other side to the road. Cross the canal and then return to Sowerby Bridge along the wharf side with the basin still to the left. Here we can see the new function of the basin: in part home to Shire Cruisers; in part, converted into The Moorings Bar and Restaurant.

To start the walk proper return to the station alleyway. Cross Wharf Street via the pedestrian crossing and walk into the car park opposite with the public conveniences to your right. Bear left in the car park to reach the truncated end of the Rochdale Canal which starts near the Kwiksave supermarket. To rejoin the Rochdale Canal with the C&HN will be no easy matter; the cost of doing so was estimated in early 1991 as £2.5 million. The walk starts along the left bank of the canal with a series of old canal buildings to the left. The line soon bends right and goes under a footbridge. Within a relatively short distance the canal seems to leave the town behind with the River Calder down to the left and an embankment right which appears to have been hewn out of solid rock.

Further on, we go through the 43 yd (and hence rather misnamed) Sowerby Long Tunnel, originally the only one on the canal. After the tunnel we pass a steep cliff up to the right to a suburb of Sowerby Bridge called Friendly. A large, rather tentatively perched, house (called, appropriately enough, Canal View) peers down on the canal. The river is now quite close to the left. We cross an overflow weir and pass a sewage farm. After High Royd Bridge the scene becomes much quieter. Although the river below is swathed in rubbish, the view is typical south Pennine countryside: green and lush fields separated by angular stone walls. The next bridge is Longbottom Bridge, which for a long time was culverted. It has now (since 30 September 1987) been restored by Calderdale Metropolitan Borough Council as part of the Manpower Services Commission Community Programme. The towpath has also received considerable attention in recent years under the Rochdale Canal Restoration Scheme partly paid for by the European Commission Regional Redevelopment Fund. As a result, the towpath all the way along is wide, dry and firm under foot.

We pass some mills to the left and go under a bridge to reach Luddenden Foot where a steep bank rises to the A646 to the right. After passing some large mill buildings to the right there is a picnic spot on the left and a Rochdale Canal Company post. The car park just before Station Road Bridge marks the site of the old Luddendon Foot Wharf. After the bridge the route goes over a small aqueduct which crosses the River Ludd, a tributary of the Calder. If you are thirsty, go up to the bridge and cross the main road to the Coach and Horses pub. After the next bridge the towpath goes over a weir via stepping stones and on to two locks in quick succession: Brearly Lower and Upper Locks (Nos. 5 and 6). The canal now makes a sweep to the left with a road high on the bank to the right. The next bridge is a new one and, rather unsympathetically, has no towpath so we have to go over the top. We enter Mytholmroyd by going under the A646, the bridge of which has a building on top of it. Immediately on the left just after the bridge is the White Lion pub. Shortly thereafter there is a newish road bridge followed by some industrial works.

After passing the much-restored Redacre Mill Hotel and going under a minor road bridge (note the well-worn rope marks on the sides of the arch), the canal heads back into open country and on to Broad Bottom Lock (No. 7) with an old mill building on the left. Among other things, the old mill houses Walkley Clogs; apparently the only mill in the UK still making traditional clogs using age-old methods. It is open for visitors, many of whom arrive from Hebden Bridge via Calder Valley Cruising's water bus which stops just here near a winding hole.

The second tunnel on the route is met shortly after the mill. It is not blessed with a towpath. Walkers are therefore forced to cross the road and to go down the slope behind the crash barrier. The canal continues under a bridge and on to Mayroyd Mill Lock (No. 8). By now, the outskirts of Hebden Bridge are beginning to make themselves felt and we reach bridge 128 which offers access to Hebden railway station to the left.

Hebden Bridge to Todmorden

If coming from Hebden Bridge railway station, leave the station and follow the road round over the River Calder to the canal bridge. Turn right, walk down the slope to the canal and turn left to walk along the left-hand bank and under bridge 128.

The path starts with a park to the left and the backs of a row of houses to the right. Shortly we reach central Hebden Bridge with Hebden Bridge Marina on the opposite bank. This is the base of Calder Valley Cruising which runs regular trips along the canal using both powered and horse-drawn boats (enquiries to: (0422) 844833). After going under a footbridge the towpath rises up to Blackpit Lock (No. 9). The lock itself is a bit of a split personality as it has two sets of recesses in the lock chamber walls. At one time, as a water conservation measure, it was intended to install two sets of gates so that when a short boat from the Calder & Hebble used the lock, it wasn't necessary to use the full chamber.

The towpath changes sides here but if you walk on along the left bank for a short distance, you will cross Hebble End Aqueduct which goes over the River Calder. A good view of the rather heavy, Brindley-esque, aqueduct with its four low arches, can be had by walking briefly up a path to the left. Return to Blackpit Lock and continue the walk along the right-hand bank. The route continues by crossing the aqueduct and passing a craft centre before going under a bridge at Hebble End. We now pass two locks (Stubbings Locks) and go under a minor road bridge. The river is now close to the right and when I was here it was stained a rather peculiar purple colour. To the left is a steep, thickly-wooded slope.

After going under the railway and passing the buttress of an old bridge, we walk on to Rawden Mill Lock (No. 12). The countryside is now opening out again although we are still relatively close to the main (A646) road, which is on the right. The rest is hills, trees and willow-herbs. Go under another minor road bridge, Callis Bridge where the Pennine Way crosses the canal, to Callis Lock (No. 13) and then on for what must be the closest most people would want to come to a sewage works. After an accommodation bridge and a minor road, we reach Holmcoat Lock (No. 14). We now wend our way through some peaceful country under a small road bridge and then past Calder Bank House which serves lunches and teas. After the next small road bridge, we pass Shawplains Lock (No. 15) and Lob Mill Lock (No. 16) in which the lock bridge appears to be upstream rather than down.

We reach the outskirts of Todmorden at the large, and now sadly dilapidated, Thomas Binns clogmaker mill building. This fine structure was built in 1832 and has a series of canalside loading bays. Immediately after this is Woodhouse Bridge and lock 17 which, when I was here, was doubling as a swimming pool for a large group of local kids. Factories and old mill buildings now begin to surround the line and, after a couple of small road bridges, the canal heads into the centre of Todmorden. We pass a wharf area to the right (now a car park) and then Shop Lock (No. 18). Go under the main road bridge (dated 1864) via the horse tunnel and rise up to Todmorden Lock (No. 19). If continuing here, cross the tail of the lock and walk on along the left-hand bank. For Todmorden station, turn right to reach the road and then left.

Todmorden to Littleborough

From Todmordon BR station go down the road ahead and turn right to reach a bridge over the canal near the Golden Lion pub. The walk starts along the left-hand bank at Todmorden Lock (No. 19).

The canal now bends left with a high railway embankment to the right. Go under a bridge and pass the cenotaph-like abutments of an old tramway. There are now two locks in quick succession. This is Gauxholme and in front of us is the extravagant, castellated viaduct which carries the Calderdale railway line from Todmorden to Littleborough. The canal goes under the viaduct and past the two Gauxholme Locks. Here the waterway seems to head straight towards the moorland but, in fact, it winds around some houses and under another road bridge to reach Gauxholme Upper Lock. Here is an old warehouse, now home to a building supply company, with its own canalside entrance.

Gauxholme Viaduct near Todmorden

The railway recrosses the canal, this time by a somewhat less flamboyant structure and the towpath then continues on to a new bridge which doesn't have a towpath. As the towpath changes sides here, cross both the road and the canal to continue along the right-hand bank. Walk on to the next lock, Smithholme Lock (No. 25), the gates of which were donated by the Halifax Building Society in April 1990. The canal bends right then left to Pinnel Lock (No. 26) near Walsden station. After a side weir and a small road bridge the canal bends right to another bridge and Travis Mill Lock (No. 28), with Birkshall Wharf on the left. The canal is now winding itself up to the summit and the locks are coming thick and fast. There now follows: Nip Square Lock (No. 29), Winterbutlee Lock (No. 30), White Bank Lock (No. 31), Sands Lock (No. 32), Bottomley Lock (No. 33), a lock known to me only as No. 34, Warland Lock (No. 35) and, finally, East Summit or Longlees Lock (No. 36) with its white-painted lock-keeper's cottage (Longlees).

The summit pound was just a trickle when I passed this way and the mud at the bottom was littered with miscellaneous debris including the bones of a narrow boat and a tub-boat. There is also a motley collection of bedsteads, children's toys, bicycle wheels, old wheelbarrows and, yes, the ubiquitous supermarket trolley. All this but little sign of the chemical works which once, apparently, straddled the canal here. It must have been a bleak place for a factory in winter, and it's said that boats were often frozen up for weeks on end during really harsh weather.

After about half a mile we reach a feeder from nearby Chelburn reservoir. This is shortly followed by West Summit Lock (No. 37) and the start of the descent into Manchester. The canal itself is now in water again and pretty soon we reach another lock (No. 38) with the outskirts of Littleborough to the right and a huge quarry to the left. Six locks now follow in quick succession. Some of the upper locks are so close together that the intervening pounds have been widened in order to ensure that sufficient water is available to keep the system in operation. On the way down this descent we pass the substantial Courtaulds Mill which dominates the canal. If this isn't dark and satanic then what is? While passing the mill keep an eye open to the right for a fine cast-iron aqueduct which carries a small stream over the railway line, which has just reappeared from its sojourn into the summit tunnel. There are now only three further locks before we reach Littleborough Bottom Lock (No. 48). The canal now bends right to reach a road (Canal Street) which runs parallel with the waterway. After 100 yd turn right to leave the canal and go under the railway to reach Littleborough station and bus-stops.

Further Explorations

The canal in Manchester has received considerable attention of late from the Manchester Development Corporation. It is no longer the eyesore it once was and, although some industrial archaeologists hate such restorations, it is now a fine place for a stroll of a little over 2 miles. Start from the Metrolink station at G-Mex. Walk along Liverpool Road towards the Museum of Science and Industry. When you reach the Castlefield Hotel, turn left to go down some steps to the Castlefield Basin and Potato Wharf (where, despite the name, various kinds of produce were unloaded). Continue along the left-hand bank and under a splendid series of railway viaducts which provide the passing towpather with the feeling of walking into a high cave. Continue across a new cast-iron bridge and out into the open. Now bear left across a bridge.

To the left here is the end of the Rochdale Canal and to the right is the Manchester terminus of the Bridgewater Canal. The lock is the ninety-second on the Rochdale (although it was actually built by the Bridgewater) and the start of what is known as the Rochdale Nine through central Manchester. To the left of the lock is one of the original lock-keeper's cottages, reputedly built on the skew in order to enable the keeper to look up and down the line. Just across the canal is Dukes 92, a pub-cum-wine bar which has been built in an old stable. If you look closely at the lock you'll notice that the arms of the downstream gates are so close to the bridge that they're shorter than would otherwise be expected. A chain and wheel system has been employed to help open them.

Our route goes left here along the Rochdale but, before proceeding, its worth just walking on to the right of the pub to peer over the wall towards a loop that runs towards the River Medlock. This arm, originally part of the Bridgewater Canal, turned a full half-circle to disappear into a tunnel that we shall see later. On the far wharf is the superb 1830 middle warehouse that has been converted into offices and apartments and renamed Castle Quay. The massive loading and unloading bays have been cleverly incorporated into the design of what is now a splendid looking building.

Return to Dukes 92 and turn right to start the walk. Almost immediately, the canal passes through a section that has been cut through solid sandstone. On the left are a series of derelict warehouses after which is another railway bridge. We then pass an arm that seems to disappear into a low cave. This was formerly the underground part of the Bridgewater that was mentioned earlier. It's low height is explained by the fact that the Rochdale has risen a lock's depth compared with the Bridgewater. In its operational days (which probably began in 1765), boats from the Worsley coal-mine were driven into the tunnel and under a shaft which led up to street level. Coal was loaded into a container which was hoisted upwards by a swivel crane. The container was then unloaded and returned down the shaft to the boat.

A little further on is an old coal wharf where the wooden jetties are, remarkably, still in position. After passing under Deansgate Tunnel we reach the next two locks (Nos. 90 and 91). Albion Street Bridge and then lock 89 follow. Shortly after the lock, on the opposite bank, is the site of the former Manchester & Salford Junction Canal. This short line, of about a mile, ran almost due west to link the Rochdale with the Irwell Navigation and the Manchester, Bolton & Bury Canal. It was opened in 1839 and had a 500 yd long tunnel and four locks. Although closed in the 1870s, some parts of the line are still extant. The tunnel was apparently used in the Second World War as an air-raid shelter.

The line now bends left to approach Oxford Street Bridge with a view of the Refuge Building tower and a glimpse of the Palace Theatre. We are followed under the bridge by a pipe which was formerly part of a steam-heating scheme operated by a nearby power station. The extravagant and intriguing loops that occur frequently along the way are expansion bends. On the other side of the bridge we pass lock 88 to reach Princess Street. Go up to the road and cross the canal to walk along the left bank or, more correctly, along Canal Street. We pass lock 87 and Sackville Street to reach

Chorlton Street where a lock-keeper's cottage straddles the canal. Lock 86 follows almost immediately and we recross the line at Minshull Street to resume along the towpath on the right bank. Shortly we enter into the chasm that is known as the Undercroft. The canal here passes directly under some buildings and there is even a lock buried in the cave.

We see light again at the final lock of the flight at Dale Street. Here we can see the old line of the Rochdale ahead. If you turn right to reach the road and right again, you can walk up to the former entrance to the Rochdale company's Dale Street Basin. In 1822 the company built this rather fancy gateway to their wharves and it still serves this purpose today. Sadly, however, the entrance is not to a canal basin but to the company's modern mainstay, a car park. If you peer through the gates, the canal can be seen heading west. To the right in the distance is Ducie Street Junction and the bridged entrance to the Ashton Canal, a line to both the Peak Forest and the Huddersfield Canals.

Continue along Dale Street to reach the centre of town and Manchester Victoria station. To reach Piccadilly station, return back over the canal and turn right along Ducie Street and then turn left.

Further Information

The Rochdale Canal Society is the leader in the task of reforging the cross-Pennine route via Hebden Bridge:

The Rochdale Canal Society Ltd,
3 The Broad Ing,
Passmonds,
Rochdale,
Lancs. OL12 7AR.

The history of the line is described within the pages of:

Hadfield, C. and Biddle, G., *The Canals of North West England*, Vols. I and II. David & Charles, 1970.
Owen, David, *Canals to Manchester*. Manchester University Press, 1977.

12
THE SANKEY BROOK NAVIGATION

Sankey Bridges to Widnes

Introduction

The accolade of being the first English canal is one that is busily fought over. The Sankey Brook Navigation, or the St Helens Canal, is, if we are absolutely correct, the first navigable waterway that was built as an artificial cut separate from a river. This differentiates it from the Fossdyke or the Exeter Canal, both of which are based on natural water courses. It would be worthy of inclusion here for that fact alone but there is far more to the Sankey than just that.

The 12½ mile long navigation begins in central St Helens. The current terminus, near Safeways car park, is not the original. In its working days the line was about half a mile longer, running along the course of the present Canal Street. From St Helens, the canal heads north-east and then east to Newton Common on the outskirts of Newton-le-Willows. Now the line gradually bends south to go under the M62 and past the outskirts of Warrington to Sankey Bridges. Originally the canal terminated here but later additions mean that the line now turns westwards to run parallel with the Mersey for an almost straight 5 miles into Widnes. At Spike Island the canal ends with a lock down to the river for a route on to Liverpool.

It has to be said that the thought of walking between Warrington and Widnes didn't appeal very much. I put it off and put it off and it ended up as the last canal that I walked for this series of books. I was wrong! The Sankey has a lot going for it. It's not pretty but there's plenty to see and lots of, almost, fresh air.

History

In the seventeenth century the salt producers of Cheshire made the first move to use coal rather than wood to heat their salt pans. The heavy demand for coal which this switch produced was one that was difficult to meet using the then primitive road system of the area, but with the opening of the Weaver Navigation during the 1730s, coal from Lancashire was able to reach this important market. At about the same time, Liverpool was beginning to become an important port and a significant industrial centre. With its growth came another major potential market for Lancashire coal.

Although access to the coalfields around St Helens had been improved with the opening of various turnpikes into the area, in June 1754 the high level of the tolls persuaded Liverpool Council to survey the Sankey Brook from the Mersey to the coalfields of Parr and St Helens. The brook was already navigable for 1¼ miles from the river to the quay at Sankey Bridges where there were a number of private wharves established by 1745. By the time of the survey there were also a group of warehouses, a coalyard and a public house. The survey was undertaken by Henry Berry, a dock engineer at Liverpool. He reported to the council that the brook could be rendered navigable and it was agreed that a bill should be sought to construct the new navigation. In an invitation to investors, the council claimed that the line would supply Liverpool with coals 'which of late years are become scarce and dear and the measure greatly lessened to the great imposition and oppression of the trades, manufacturers and inhabitants'.

The two key subscribers to the venture were John Ashton, a Liverpool merchant who owned the Dungeon salt-works, and John Blackburne, who owned salt-works in Liverpool and Northwich. Ashton took up 51 of the 120 available shares in the project. The bill was widely supported by local landowners and colliery proprietors, and was passed on 20 March 1755. The Act authorized the making navigable of the Sankey Brook and the construction of three small branches. In practice, Sankey Brook was far too small to render navigable and the new line was to be built as a canal rather than an upgraded river. This point wasn't stressed in the Act and the proprietors essentially built a canal with a navigation Act. Permission to construct any artificial cuts at all was only included as a rather bland clause which allowed them to be built where deemed (by the proprietors) to be necessary. Thus the proprietors avoided stimulating opposition from landowners which may otherwise have led to the defeat of the bill. Apart from detailing the course of the line, the Act determined a maximum toll rate of 10d. a ton with limestone, paving stones, granite, soapers waste, manures and road-building materials to be carried free. The Act also stated that the work had to be completed before 29 September 1766. There were no limits set to the company's powers to raise capital.

The work began on 5 September 1753. It appears to have gone smoothly for an advertisement in November 1757 announced that the line was operational from Sankey Lock, in Sankey Bridge, as far as Haydock and Parr collieries. There were eight single locks and a staircase pair, the Old Double Lock. The construction to this point included the north (Penny Bridge or Blackbrook) branch (5/8 mile), and the west (Gerard's Bridge) branch (1¹/2 miles). The cost so far was £18,600. By February 1758 Sankey coal was being advertised in Liverpool and a triangular trade had grown in which coal was loaded on the navigation, taken to Liverpool and unloaded. From there the goods were taken for delivery to Bank Quay. The vessels then ran empty back to the Sankey. By spring 1759 the navigation was built to Gerard's Bridge. The Blackbrook branch was probably completed by 1762 (it was extended in 1770).

It wasn't until 1772 that the canal was extended beyond the Old Double Lock past another staircase pair, the New Double Lock, to Boardman's Bridge to complete the original scheme. The extension to the St Helens terminus that we see today (and on a further ¹/2 mile to the site of the Ravenhead Copper Works and St Helens Crown Glass Works) wasn't completed until later in the 1770s. As soon as the full line was open, the company set about encouraging trade, and in April 1761 toll rates were reduced to 7d. a ton. Boats that plied the line carried cargo loads of 40–8 tons, drawing up to 5 ft 4 in.

When first opened the southern terminus of the canal was a lock into the Sankey Brook at Sankey Bridges. When tides were favourable, boats passed along the brook to the Mersey. Having to rely on the tides was soon found to be inconvenient, and in 1762 the company authorized an extension of nearly 1¹/2 miles to Fiddlers Ferry where traffic could lock directly into the Mersey. Craft were charged an extra toll of 2d. for using the new cut. Fully loaded boats were still held up at neap tides for three days a month but this was far better than hitherto. The original lock was still sometimes used thereafter when the line was busy and the tides favourable.

From opening, the navigation was a success. In 1771, 89,721 tons of goods were carried along the line: 45,569 for Liverpool and 44,152 tons for Warrington, the Weaver and elsewhere. (It should be noted that a ton was measured as 63 cubic feet, the weight of which will vary according to the quality of the coal. Estimates suggest that a ton here may well mean 27–8 cwt and not 20 cwt as would be expected.) From October 1774 the canal faced competition for the Liverpool coal business from the newly opened Leeds & Liverpool Canal which ran from Wigan via the Douglas Navigation. To counter this, the SBN appointed its own coal agents in Liverpool (to prevent malpractice) and it sought lower prices. The proprietors also entered the carrying business. The moves must have been a success for the company paid its first dividend in April 1761 and by 1785 its £155 shares were valued at £300 each; suggesting dividends of £15 a share.

The SBN was also stimulating a considerable growth of industry along its route. The British Cast Plate Glass Manufacturers established a plant in St Helens in 1773. It was the first plate glass works in Britain. By 1780 the Ravenhead copper smelting works had also been built and new collieries were being opened. Coal and raw materials were now being shipped into St Helens as well as away from it and finished goods were being exported. As a consequence, the beginning of the nineteenth century saw the profitability of the SBN still increasing. Dividends between 1805 and 1816 averaged between £44 and £58 p.a. and in 1815 the volume of traffic carried was up to 181,863 tons.

Despite continuing success and dividends of 33 per cent, the proprietors seemed unwilling to develop the line and several suggestions for links with the Mersey & Irwell Navigation were rebuffed. It wasn't until the dawn of the railway age that the SBN finally moved to improve the navigation. That threat first came in October 1824 when the promoters of the Liverpool & Manchester Railway made it plain that they viewed the shipment of coal from the St Helens field to Liverpool as a major target. If this wasn't bad enough, the colliery owners themselves were actively promoting a line of their own from St Helens to a dock at Runcorn Gap, Widnes. This move resulted in an authorizing Act for the St Helens & Runcorn Gap Railway that was passed on 29 May 1830. As if to add insult, the new line was to cross the SBN twice: at Hardshaw Mill and Burtonwood.

The SBN's proprietors met the threat by extending the canal for a further 3½ miles along a line parallel with the Mersey to the Runcorn Gap where there was to be a dock and river entrance. This scheme had been originally discussed in 1819 and was intended to avoid Mersey shallows which impeded traffic at low tide. The extension was authorized by an Act in 1830 with the entrance lock at Fiddlers Ferry remaining operational until 1846. The Act enabled the company to borrow £30,000 to finance the new works which were engineered by Francis Giles and opened in July 1833. Twin locks, 79 ft by 20 ft, were built at the Mersey entrance. The 1830 Act also enabled the SBN to form a new incorporated Company of Proprietors which could issue four new £200 shares for each of the 120 old £155 shares. The nominal capital therefore became £96,000. Toll charges, however, were frozen at the original level.

The St Helens & Runcorn Gap Railway opened on 21 February 1833 and was blessed with its own dock at Widnes; reputedly the first purpose-built railway dock in the world. The SH&RGR had cost £200,000; a considerable overspend which stopped the company from building branch lines to as many collieries as it had hoped. Passenger traffic was not forthcoming

The United Alkali Co. Ltd's jigger flat *Santa Rosa* after having been launched from the Clare & Rideway's yard at Sankey Bridges in 1906

Ware/The Boat Museum Archive

and movement along the line was hindered by two inclined planes. By July 1834 the railway was only barely solvent while the canal was still flourishing. In fact the competition between the canal and the railway had stimulated a round of toll reductions and, in turn, this had increased traffic levels. In 1836 the SBN moved 170,000 tons of coal whereas the SH&RGR shipped 130,000. By 1845 the equivalent figures were 440,784 and 252,877.

Although trade was flourishing, a meeting of SH&RGR shareholders in February 1838 considered amalgamation with the SBN. The SBN agreed but the SH&RGR's financial position postponed the deal. It wasn't until January 1844 that talks were reopened and the SBN agreed to sell to the railway for £144,000 (together with the transfer of its debt of £29,450). The new company, the St Helens Canal & Railway Company was authorized by an Act passed on 21 July 1845. This Act obliged the new company to maintain the canal although, should any part of the line prove uneconomic, closure would be allowed with the agreement of the Board of Trade. At the time, the income on the canal was £21,373 p.a. with a surplus of £13,581. The most important traffic was coal with 693,000 tons being shipped from the St Helens field in 1846. Most of this was moved to Liverpool although there was still a substantial trade to the Cheshire salt-works. In 1847 the company also started to sell some of its canal water to neighbouring factories.

The future of the amalgamated company was, however, definitely with track rather than water. A series of new railway construction projects were keenly supported whereas the old canal line was essentially allowed to stay as it was. By 1853, mostly as a result of the opening of a railway line to a new Garston Dock, the company's railway tonnage exceeded that of the canal for the first time: 613,805 tons by rail against 510,668 by canal. Despite this, canal traffic continued to increase; tonnage was up 17 per cent between 1853 and 1856.

In 1860 the company leased the Warrington–Garston line to the London & North Western Railway. By 1864 the LNWR had absorbed the whole of the St Helens railway and canal and the old company was dissolved. Under the Act of 1845 powers to close parts of the canal were removed and the LNWR was committed to keeping it open to traffic, dredged and in water to a depth of 6 ft 3 in. This condition must have been aggravating to the LNWR as by this time the canal was said to have been in 'miserable order'. LNWR was forced to spend £23,000 to get it into a navigable state. The Act continued the toll exemptions of the original 1755 Act and ensured that if any locks were rebuilt they were no smaller than that at Newton Common. Canal tolls were also revised to a maximum of 8d. a ton on coal, and 10d. on other commodities.

There were more canal related problems for the LNWR in the 1870s when the line became seriously affected by chemical pollution which undermined the mortar of its lock walls. The water also occasionally overflowed into the Sankey Brook and on to nearby fields. The LNWR obtained an injunction against the polluting company but was forced to buy the meadows as the quickest way of settling compensation issues. Mining subsidence also became progressively more serious after 1877. In one case the line dropped 18 ft in a year. In 1892 the LNWR sued the coal owners for damages and only won on appeal the following year. It was then agreed that the colliery company should maintain the canal banks.

Despite the attentions of the LNWR, towards the end of the century traffic levels were beginning to decline. In 1888 the line had moved 503,970 tons, but by 1898 the level was already down to 381,863, and by 1905 it had dropped to 292,985. Tolls were dropping with similar rapidity: £6,275 (1888), £4,275 (1898) and £3,010 (1905). The most important feature of the new century traffic was the disintegration of the coal trade. The canal's cargo was now primarily alkali, soap, silicate, river sand, acid, sugar, oil, tallow, manure, copper ore, silver sand, salt and copper.

By the First World War the navigation was in terminal decline and although the SBN carried 211,167 tons of cargo in 1913, it was reported that not more than twenty boats had passed the New Double Locks to St Helens since the turn of the century. In 1919 just seven boats passed Newton Common Lock at Newton-le-Willows. By the 1920s traffic had

Newton Common Lock and the Stephenson railway viaduct from a print published in January 1831

The Boat Museum Archive

stabilized at approximately 130,000–150,000 tons p.a. with income at around £5,000–£6,000. In 1923 the LNWR was absorbed by the London, Midland & Scottish Railway and it, with the support of the St Helens Corporation, closed the, now largely unused, 5 miles of canal north of Newton Common Lock in 1931 (2¹/₈ miles of main line plus the Gerard's Bridge, Blackbrook and Boardman's Bridge branches). Bridges in St Helens were then fixed at Raven Street in 1932, at Redgate and Old Fold Double Lock in 1934 and at Pocket Nook in 1934. The main line channel itself was retained as a water feeder. Below Newton Common, diesel barges continued to work to the Sankey Sugar Company's wharf (near the railway line just south of Newton-le-Willows).

By the Second World War 94,016 tons of cargo were still being moved and the company had an income of £4,289. But by 1946 the figures were 20,638 tons and £2,558. The extant portions of the line were nationalized in 1948 although with no obvious enthusiasm. Traffic continued for a while. In 1957 about 35,000 tons a year were being carried including raw sugar to Sankey, lead to Sankey Bridge and chemicals from the Widnes area. But all traffic ceased in 1959 when bulk transport of sugar was introduced. The SBN was finally abandoned in 1963.

During the course of the 1960s and 1970s, the canal was totally neglected and much of it was infilled. Some cosmetic improvements were made in the late 1970s and early 1980s but nobody, it seems, thought the line worth protecting. However, in 1985 the Sankey Canal Restoration Society was formed with the aim of stopping the decline and eventually restoring the route for navigation. The society is encouraging local authorities to become involved and successfully persuading British Waterways not to dispose of bits of it on the quiet. St Helens Council have obviously been convinced and, with the aid of a Derelict Land Grant, have regated the New Double Locks in St Helens. Other councils are also showing interest. The society's efforts have culminated in the publication of a strategy document, *Sailing the Sankey*, which hopefully should forward the restoration of an historic waterway.

The Walk

Start:	Sankey Bridges (OS ref: SJ 585876)
Finish:	Spike Island, Widnes (OS ref: SJ 514843)
Distance:	5¹/₂ miles/11 km
Map:	OS Landranger 108 (Liverpool)
Outward:	North Western buses nos. T9 or T10 (enquiries: (0925) 30571)
Car park:	Central Widnes, at Spike Island or on-road at Sankey Bridges
Public transport:	British Rail Widnes

There is a bus-stop (for buses from Widnes) close to the canal on the Liverpool Road near a pub called The Sloop. The walk starts on the pub side of the road and is along the left-hand bank of the canal. The bridge here is actually quite recent, being built in 1972 to replace an old bascule bridge (a type of lifting bridge that uses counter-balancing weights to raise it) and, before that, a swing bridge. The rusting metal framework of a reserve bridge is still in position just beyond. The third of the trio of Sankey Bridges follows with the railway crossing. How a restored waterway will get past the two used bridges is uncertain. The road will need another opening bridge. Perhaps the only hope for the railway bridge is if Fiddlers Ferry power station stops using the line for coal deliveries. In its operational days the railway bridge could swing out of the way; a system that once resulted in an almost brand new locomotive crashing into the canal.

The line now turns abruptly right. Here, barely noticeable on the towpath side, is the pre-1762 route that went straight on via a lock into Sankey Brook

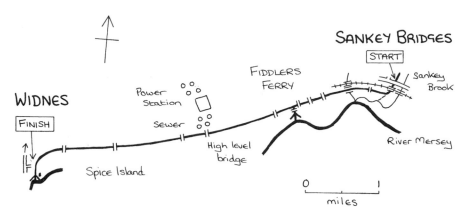

The Sankey Brook Navigation

and then the Mersey which is about a third of a mile distant. The new 1½ mile long section of canal was built to avoid having to wait for appropriate tides, making passage both easier and more convenient.

The line passes the now-fixed Mayers Swing Bridge and over Whittle Brook before reaching Penketh Bridge and Fiddlers Ferry Swing Bridge. The large concrete monolith on the left takes water from the Mersey and supplies to the Fiddlers Ferry power station that now looms over the canal. Interestingly, it is the power station's waste which is used to keep the canal in water along this stretch. Fiddlers Ferry has become quite a spot for pleasure boats with the Ferry Inn and Fiddlers Ferry Yacht Haven. This stretch of the canal isn't the original 1762 course as it was altered slightly when the railway was built in 1847. After passing another swing bridge and the boatyard, the towpath crosses the upper side of the Fiddlers Ferry Lock on a small swing bridge. From 1762 two locks here allowed vessels to pass to and from the River Mersey which is now visible to the south. The extension of the Sankey to Widnes was completed in 1833 but the two locks stayed in use until 1846. One lock was removed when the line was moved as part of the railway realignment. By the 1950s the lock had fallen into disrepair but it has been restored and is used as a quick route to and from the Mersey.

The canal now passes the no-longer-swinging Marsh House Swing Bridge before entering into the environs of the Powergen power station. The area to the left, beyond the fence and not visible to towpathers, is firstly where the waste ash is dumped and later comprises a series of settling lagoons. To the right, the massive power station dominates everything. If this sounds ecologically unpromising, you may wish to know that I had the highly improbable view just here of a heron actually perched on top of the barbed wire fence to the left. The canal is blocked along this section by a series of obstructions, the first of which contains the pipe connecting the power

Fiddlers Ferry power station

station with the settling lagoons. Between this and the high connecting bridge, there were two swing bridges - now both missing. A little further on there has even been some infilling of the line following a suggestion that the settling lagoons were about to settle into the canal. After crossing an open sewer (which isn't as awful as it sounds), the canal starts its final stretch into Widnes. An area of scrubby marshland known as Cuerdley Marsh dominates the view to the left down to the Mersey. On the right is a series of buildings that comprise part of the Widnes chemical industry. As you would expect from a canal of the 1830s, the line is almost uncompromisingly straight with the railway line hugging the navigation to the right.

After Carter House Bridge (near Carter House signal-box), the last stretch of the canal begins. Cross the line at the small wooden footbridge (the first after the signal-box) and continue along the right-hand bank of the

Fiddlers Ferry Lock to the River Mersey

canal. The land to the left is known as Spike Island. The Countryside Commission claim that Spike Island was the birthplace of the British chemical industry. In the nineteenth century this small patch of land was home to two large chemical factories: Hutchinson's and Gossage's. The Sankey was an important factor in the development of the area and enabled transport out to the important markets not just along the Mersey but around the world. The tall tower that overlooks the island is all that remains of Gossage's soap and chemical factory and it is now used as a science museum. By the turn of the twentieth century these factories were already outdated and the area declined. Gossage's itself closed in 1932. By 1975 the area was widely viewed as a disgrace. Halton Borough Council and Cheshire County

Council then started a reclamation scheme. Just seven years later Spike Island was transformed into a popular green space; abhorred by industrial archaeologists but preferred by most.

Just beyond the information centre, the Sankey Brook Navigation comes to an end at the twin locks which take the canal down to the Mersey. Cross the lock-gates and you'll be able to investigate some of the remnants of the chemical works. There are the last earthly remains of the pyrites kilns and the bases of some acid absorption towers. The large wet-dock was once connected to the Mersey via a lock. It had been built to serve the St Helens & Runcorn Gap Railway whose wagons tippled their coal into waiting boats. From the wet-dock it is possible to wander all the way round the island and back to the canal. To complete the walk, leave Spike Island by the drive near the car park to a road junction near the Swan Inn. Turn right along Upper Mersey Road and then go right along Waterloo Road. Go under the railway bridge and straight on at the traffic lights. At a large roundabout bear right for central Widnes.

Further Explorations

Although not always visible, it is possible to walk along virtually the entire 12¹/₂ miles of the Sankey Brook Navigation in a great arc from St Helens to Widnes. If you wish to do this, it is well worth buying the restoration society's guide to the canal which is available from the address below.

For those with less ambition and who are merely seeking a gentle amble, make your way to the Sankey Valley Park which is off the A572 between Newton-le-Willows and St Helens. You can park near Penkford School which is on OS Landranger 108 at ref: SJ 563949. There is a SVP car park a little further along the road. From the school return to the A572 and turn left to reach the canal bridge (Penkford Road Bridge). This was originally a swing bridge but was replaced in 1935 after the canal was abandoned. Before crossing the bridge, turn left to walk along the course of the SBN.

Pretty soon the canal line is obliterated but the clear path meanders around what remains of the infilled bed. Just after the aforementioned car park, avoid the obvious lane that bears right towards Sankey Brook and instead keep to the slightly higher ground to the left. Near here was Newton Common Lock although I'm unable to say precisely where. Interestingly, the canal restoration society reports that beneath the infill the upper lock-gates are still in position. The location of Sankey Viaduct isn't so hard to ascertain. This splendid structure was built as part of Stephenson's Liverpool & Manchester Railway, the first passenger-carrying railway in the world. Work started here in January 1827 and the line was fully opened, with great ceremony, on 15 September 1830. It is said that when that first train trundled over the viaduct, a sea of faces peered up in wonder from the canal and towpath below. The viaduct rapidly became a popular scene for artists and engravers. And it is still a splendid structure worthy of anyone's photographic genius.

If you continue on under the viaduct, the path becomes progressively more rural although for a while there are some industrial workings up on the bank to the left. This was formerly the site of the Sankey Sugar Company, an important customer for the canal and, indeed the last. The final barge shipped cargo from here in 1959. A little further on, a broad flattened expanse marks the site of a winding hole where the sugar boats turned. Just beyond is a section of waterway that was restored in the 1970s and the remains of Bradley Lock with its stone walls and upstream gates still in place. Once here you can wander gently down, past the infilled Hey Lock, to a point where the restored canal disappears and the Sankey Brook crosses the line in a concrete channel. To return to the car simply about face and enjoy your second view of Stephenson's viaduct on the way back.

Further Information

The Sankey Canal Restoration Society (SCARS) was formed in 1985 with the principal aim of publicizing the canal with a view to its restoration. The society has produced a strategy document: *Sailing the Sankey – the case of restoration*. There is a strong possibility that a major programme of restoration could begin over the next couple of years. To find out more, contact:

SCARS,
c/o The Groundwork Trust,
27 Shaw Street,
St Helens WA10 1DN.

The society has published a fine towpath guide:

Greenall, C. and Keen, P.G., *The Sankey Canal*. 1991.

For historical information, however, the details are enclosed within the pages of:

Hadfield, C. and Biddle, G., *The Canals of North West England*, Vols. I and II. 1970.

APPENDICES

A: General Reading

This book can, of course, only provide you with a brief glimpse of the history and workings of the waterway network. Other authors are far more qualified than I to fill the gaps and the following reading matter may help those who wish to know more.

Magazines

There are two monthly canal magazines that are available in most newsagents: *Canal & Riverboat* and *Waterways World*. Both have canal walks columns.

Books

There is a wide range of canal books available, varying between guides for specific waterways to learned historical texts. There should be something for everyone's level of interest, taste and ability to pay. Libraries also carry a good stock of the more expensive works and are well worth a visit.

For a good introduction to canals that won't stretch the intellect, or the pocket, too far:

Smith, P.L., *Discovering Canals in Britain*. Shire Books, 1984.
Burton, A. and Platt, D., *Canal*. David & Charles, 1980.
Hadfield, C., *Waterways Sights to See*. David & Charles, 1976.
Rolt, L.T.C., *Narrow Boat*. Methuen, 1944.

This can be taken a few steps further with the more learned:

Hadfield, C., *British Canals*. David & Charles, 1984. New edition: Alan Sutton, 1993.

There are a number of books that are predominantly collections of archive photographs. Examples include:
Ware, M., *Canals and Waterways. History in Camera* series, Shire Books, 1987.
Ware, M., *Narrow Boats at Work*. Moorland Publishing Co., 1980.

Gladwin, D., *Building Britain's Canals*. K.A.F. Brewin Books, 1988.
Gladwin, D., *Victorian and Edwardian Canals from Photographs*. Batsford, 1976.

At least three companies publish boating guides:

Nicholson's *Guides to the Waterways*. Three Volumes.
Pearson's *Canal & River Companions*. Eight Volumes (so far)
Waterways World. Eight volumes (so far).

Of the three, Pearson's guides are the most useful for towpathers. However, the only one which covers any of the northern canals is:

Canal Companion: Cheshire Ring Companion. J.M. Pearson & Son, 1990.

An essential read for many will be:

Rowland, C. and Simpson, J. *The Best Waterside Pubs*. Alma Books, 1992
    (a CAMRA publication).

Readers seeking further walking books should look no further than:

Quinlan, Ray, *Canal Walks: Midlands*. Alan Sutton, 1992.
Quinlan, Ray, *Canal Walks: South*. Alan Sutton, 1992.

B: Useful Addresses

British Waterways

BW are the guardians of the vast majority of the canal network and deserve our support. There are offices all over the country but their customer services department can be found at:
    British Waterways,
    Willow Grange,
    Church Road,
    Watford WD1 3QA.
    Telephone: (0923) 226422

Inland Waterways Association

The IWA was the first, and is still the premier, society that campaigns for Britain's waterways. They publish a member's magazine, *Waterways*, and provide various services. There are numerous local groups which each hold meetings, outings, rallies etc. Head office is at:

Inland Waterways Association,
114 Regent's Park Road,
London NW1 8UQ.
Telephone: (071) 586 2556

Towpath Action Group

The Towpath Action Group campaigns for access to and maintenance of the towpaths of Britain and publish a regular newsletter. They are thus the natural home for all keen towpathers.

Towpath Action Group,
23 Hague Bar Road,
New Mills,
Stockport SK12 3AT.

C: Museums

A number of canal museums are springing up all over the country. The following are within reach of the area covered within this book and are wholly devoted to canals or have sections of interest:

THE CANAL MUSEUM,
Canal Street,
Nottingham.
Telephone: (0602) 598835

THE CANAL MUSEUM,
Stoke Bruerne,
Towcester,
Northamptonshire NN12 7SE.
Telephone: (0604) 862229

THE BOAT MUSEUM,
Dockyard Road,
Ellesmere Port,
Liverpool LL65 4EF.
Telephone: (051) 355 5017

INDEX

CANAL WALKS

MIDLANDS

CONTENTS

ACKNOWLEDGEMENTS

This book would have been impossible without the splendid resources of various libraries: communal ownership in practice. Despite chronic underfunding, the information and help received was substantial.

Help and advice on the routes came from: Neil Bough and members of the BCN Society; Roger Cook of the Grantham Canal Restoration Society; Mary Awcock of the Shropshire Union Canal Society; Malcolm Sadler and members of the Stratford-upon-Avon Canal Society; and various employees of British Waterways.

Assistance with the archive photographs came from Roy Jamieson of BW at Gloucester, Lynn Doylerush of the Boat Museum, Ellesmere Port and C. Wilkins-Jones of Norfolk County Council Library & Information Service. Assistance with, though no responsibility for, the author's pictures came from Mr Ilford and Mr Fuji and two, old and increasingly battered, Olympuses fitted with 28 mm and 75–150 mm lenses.

For on the ground assistance during walking trips, thanks to Howard and Jan, Ruth and Rebecca (and Simon) and Paul and Geraldine. Thanks also to Taffy for much of the transportation.

Thanks to Jaqueline Mitchell for helpful comments on the contents and the text.

Many thanks, of course, to Mary – for her patience more than anything else.

Finally, to Humphrey Bogart, whom I've always wanted to be able to thank for something.

LOCATION OF WALKS

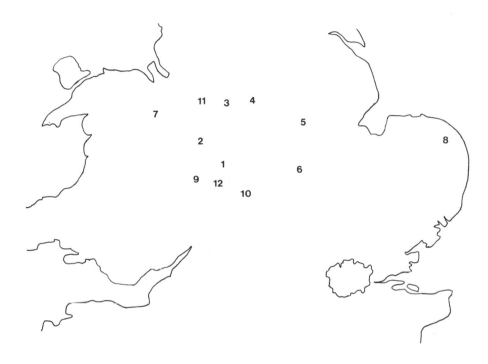

KEY TO MAPS

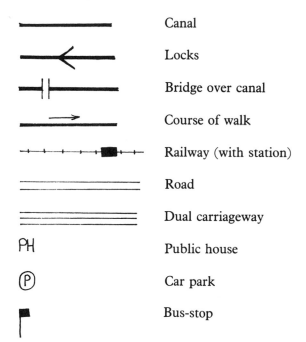

Canal

Locks

Bridge over canal

Course of walk

Railway (with station)

Road

Dual carriageway

PH   Public house

Ⓟ   Car park

Bus-stop

INTRODUCTION

Some two hundred years after the first canal age, Britain has entered a second. Having allowed its waterways to fall into a serious state of decay or even, in many cases, to disappear entirely, the country is now seemingly desperate to clean up its navigable act. The once popular final resting places for old bedsteads, broken pallets and the ubiquitous supermarket trolleys are now being dredged and made into linear water parks, nature trails or public amenity areas. Suddenly, what local authorities once called 'eyesores in need of filling in' are now being described as 'green-fingers into the urban environment'. Councils are putting up signposts and information notices. Canal societies and trusts are opening museums and running boat trips. Enthusiasts, keen enough to forsake an annual sun-drench on the beach in Majorca, are up to their knees in mud restoring lock systems and repairing canal beds. Everywhere it seems, water and boats are being seen where they haven't been seen for years.

Perhaps one of the most interesting things about this renaissance of Britain's canals is that it is only partially for the benefit of those who wish to be waterborne. British Waterways, the guardian of the vast majority of the nation's canal network, admit that although one million people use their lines for boating, some six million use them for walking. Indeed, some canals, such as parts of the Cromford, have been restored with no prospect of any significant boating traffic. The towpaths of the Cromford, and of all the other canals around the country, have become a magnet for walkers seeking to combine a pleasant spot for an afternoon stroll with a place rich in interest and industrial heritage.

What a nineteenth-century boatman would have thought about all this can, of course, only be guessed at. Towpaths were built and maintained not for the benefit of the tourist taking in the scenery but as a part of a means of communication. They were the industrial 'motorways' of the eighteenth and nineteenth centuries. The towpath gave the canal a special advantage. It provided the towing horse with a firm, wide footing. River navigations often didn't possess such luxury and boats were either sailed or bow-hauled by groups of tough, and often rough, men. The towpaths were, therefore, usually as well maintained as the course of the canal itself. After all, if there was no towpath, there was no motive power for the narrowboat on the canal.

Prospects for the towpath changed with the advent of the self-powered narrowboat. From then on it wasn't needed and, in some instances, may even have been removed in order to widen the navigation. Standards of maintenance began to fall and the increased wash from speeding boats caused many sections to erode. Large stretches of many of the country's canals, in this area the northern section of the Stratford Canal, now have no towpath to speak of and are not accessible by foot. And British Waterways aren't required to replace them. Indeed they're not even legally obliged to provide them.

Contrary to popular belief, canal towpaths are mostly not public rights of way. They are technically privately owned land and we are only able to use them through the open-access policy provided by British Waterways. Those canals that weren't nationalized under the 1948 Act, and hence aren't under the control of BW, such as the North Walsham & Dilham Canal, are not open at all and walkers cannot gain access to those lines other than on public rights of way. The permission granted by BW is, of course, welcome by all of us but it raises many a spectre for the future. With increasing economic constraint, BW will, naturally, be forced to investigate ways of maximizing the return from its resources and the popularity of the towpaths will undoubtedly attract attention. It could well be argued that whereas boaters and fisherman have to pay for the privilege of using the waterways, walkers do not. Worse still is the possible position of a privatized BW in which the new owners could remove or restrict access entirely. This would be totally legal given that the routes are not public rights of way.

In the past, the canal companies went to great lengths to ensure that their towpaths were kept off the definitive rights-of-way maps. This was done despite the fact that in many cases the paths had become locally important thoroughfares. This was achieved by periodically closing the path for the supposed reason of maintenance or safety so that the standard definition of a public right of way – that the route has been in uninterrupted use for a period of twenty years – did not apply. Alternatively, notices were erected clearly stating that passage along the path was under sufferance or that access was granted for a particular purpose only. In both cases, it was plain to those using the towpath that it was private land and so could not be declared a public right of way.

Even today, towpaths can be closed without either warning, explanation or the provision of a diversion. The Towpath Action Group has campaigned for some time, for example, over the closure of the towpath on the Ashton Canal in Manchester which has been shut for more than two years while Piccadilly Village has been built. Both the Inland Waterways Association and the Towpath Action Group are seeking to ensure that, where possible, towpaths are declared public rights of way and that the genuine rights of access are maintained. All walkers, especially towpathers, should be alert to these possibilities. Support moves to make your local lines into public rights of way and do report to British Waterways any obstructions and problems along the towpath. In time, it may well be a case of use it or lose it.

The Midland Canals

In so many ways the Midlands are the heart of the English canal network. Many of the earliest and the most successful as well as the most used and battle-scarred navigations are located here. Many of the lines aren't pretty but then they weren't built to be so. They were industrial arteries that fed and thrived on the collieries, quarries and factories of central England. The Trent & Mersey Canal, built under the watchful eye of Josiah Wedgwood, supplied the burgeoning pottery industry around Stoke. Much of Birmingham's industry was deliberately sited next to the city's ever expanding canal network. The Staffordshire & Worcestershire Canal was so busy that it was still able to pay dividends to its shareholders right up to nationalization in 1948, even though trade on most canals had dried up more than fifty years earlier.

It is, however, possible to find peace and solitude on the Midland lines. The once busy, quarry-led, Caldon and Cromford Canals are now splendid rural retreats and the North Walsham and Grantham Canals always were. The Birmingham & Liverpool Junction Canal (or Shropshire Union as it became) passes through some of the remotest parts of the West Midlands while the Llangollen Canal is a stunning construction through some simply wonderful countryside. And who would believe that the Stratford and Worcester & Birmingham Canals are so close to civilization; both make for fine country walking.

Here is just a selection of the canals of the Midlands and, although by no means arbitary, it is strictly a personal choice. Walkers should not ignore the Ashby Canal (near Hinckley and Nuneaton), the surprisingly rural Coventry Canal, the Warwick & Napton Canal (near Leamington Spa), the River Soar Navigation (near Loughborough), the Erewash Canal (near Nottingham) or the various remains of the long-lost lines such as the Leominster, the Oakham, the Shrewsbury, the Derby or the Nottingham Canals. All these are worthy of further investigation and, with the aid of an Ordnance Survey map, can be traced both on paper and on the ground.

As with all personal choices, some people's favourite lines or stretches of waterway may have been omitted. There is also the problem as to what constitutes a Midland canal. The Oxford, for example, runs further north than the Stratford and yet will be included in *Canal Walks: South*; the Trent & Mersey goes as far north as the Macclesfield, even though the latter has been designated as northern. In all these matters, only the author is to blame. I have not tried to be the sage on these matters merely a stimulus. Walking the canals of the Midlands should be an adventure with plenty to see and to discover for yourselves. And it is quite likely that you will find even more than I did and enjoy them every bit as much.

Walking the Towpaths

The walks in this book are all straightforward and require no special feats of strength or navigation. Towpath walks have two great virtues: they are mostly on the flat and they have a ready made, unmistakable course to follow. Getting lost should therefore, in theory at least, be relatively hard. The key problem with towpath walks is that if you want to spend most of the day by the canal, circular routes to and from a vehicle or a particular station or bus-stop become difficult to organize. Many of the walks described within this volume involve walking one way and returning by public transport. This means that you must check the availability of the bus or train before travelling. Telephone numbers are provided for your use and your local library should have the main British Rail timetable.

Walkers should generally plan for 2 to 2½ miles an hour so that stops can be made for sightseeing or a break. Head-down speedsters should easily manage three miles an hour on a good track. You should, of course, add a little time for stoppages for refreshment and add a lot of time if you are accompanied by photographers or bird watchers.

No special equipment or provisions are needed to walk the towpaths of Britain. On a good day and on a good path, any comfortable footwear and clothing will do, and you'll be able to leave the laden rucksack at home. However, for longer walks through more remote country you should be more prudent. Even in a drought, towpaths can be extremely muddy and, from experience, it can not only rain virtually anytime but usually does. Boots and a raincoat of some sort are therefore advisable. Similarly, although pubs and small shops are often fairly common along the way, it may be useful to carry some kind of snack and drink with you.

This book includes sketch maps that show the route to be taken. However, the local Ordnance Survey map will always be useful and the appropriate map numbers and references are provided in each chapter. Again your local library may well have them for loan.

Finally, the dangers inherent in walking along a waterway are often not fully appreciated. Over the 1990 Christmas holiday, three children died after falling into a lock on the Kennet & Avon Canal at Burghfield. Locks are deep, often have silt-laden bottoms, and are very difficult to get out of. Everybody, especially children, should be made aware of this. If somebody does fall in, you should not go into the water except as a last resort. You should LIE on the bank and use something like a coat for the person to grab so that you can then pull them in. Better still, keep children away from the edge.

Otherwise, please enjoy.

1
THE BIRMINGHAM CANAL NAVIGATIONS

The Smethwick Round

Introduction

Ask virtually anybody to say what they know about Birmingham and they will quite possibly tell you that it has more canals than Venice. While this is not the place for a discussion about quantity and quality, it is a remarkable fact which has produced one of the most convoluted waterway systems in the country. To the uninitiated a map of the canals that run around the city, and from there into the Black Country, looks like a dropped plate of spaghetti, randomly arranged and impossibly tangled. To the initiated, however, the Birmingham Canal Navigations, as the canals are collectively known, are an intimate and historic network of waterways with a kind of grubby charm and a lasting appeal.

In the middle of the nineteenth century there were over 160 miles of canals in the BCN. Over the years this has gradually dwindled to about 100 miles of interconnecting waterway made up of a series of individual lines. The earliest of those lines and the main core of the BCN is the Birmingham Canal. This main line runs from Aldersley on the Staffordshire & Worcestershire Canal through Wolverhampton to Birmingham centre at Newhall and Gas Street. Off this line run a series of later canals. From the north-west the first is the Wyrley & Essington (the Curley Wyrley) which runs from Horseley Fields Junction to Brownhills. At Tipton, the Dudley No. 1 Canal goes south-east through the 3,172 yd long Dudley Tunnel to Primrose Hill and on to join the Stourbridge Canal. A little further on, the Netherton branch goes through the 3,027 yd long Netherton Tunnel to the Dudley No. 2 Canal which at one time joined the Worcester & Birmingham at Selly Oak. At Pudding Green, the Walsall Canal winds its way north to reach the W & E at Birchills Junction. From the Walsall at Ocker Hill, the Tame Valley Canal sets off east to Gravelly Hill. Further along, the Birmingham & Fazeley Canal runs from the main line at Farmer's

Bridge to meet with the Tame Valley Canal at Gravelly Hill and then on to a junction with the Coventry Canal at Fazeley Junction. From this, the Digbeth branch joins with the northernmost section of the Grand Union Canal (formerly the Birmingham & Warwick Junction Canal). Although there are numerous other lines and branches, the only remaining section of any length is the Rushall Canal: a north–south line which goes from the Rushall Junction with the Tame Valley Canal, through Aldridge to meet with the W & E at Brownhills.

A walk along any of these canals, through the often derelict industrial parts of an inner city, isn't to everybody's taste but here you will see the last remains of a once busy canal system. With an awareness of the remnants comes an intensity of interest. These aren't pretty lines that look like rivers flowing through attractive countryside. These are canals that have worked for their living and which more often than not show their scars. Towpathers should, therefore, enjoy them for what they are and leave their pretty-view photography for the Llangollen.

History

The thirty thousand people of the small town of Birmingham must have looked eagerly at their Aris's *Birmingham Gazette* in January 1767. Within those pages, a correspondent (a 'well wisher to the town') had written glowingly of the potential for a canal to run from the recently authorized Staffordshire & Worcestershire Canal at Wolverhampton into Birmingham. The letter, promising cheaper coal and other commodities, obviously fell on eager ears for in the following week (28 January) a meeting was held at the Swan Inn to discuss the letter and its suggestion. This meeting decided to ask James Brindley, the engineer of the Bridgewater Canal, to carry out a survey and to make recommendations. To this end, some 165 individuals contributed a guinea each towards the cost. At a follow-up meeting on 4 June Brindley duly reported on a line that went from Newhall in central Birmingham (near Paradise Circus) via Smethwick, Oldbury, Tipton Green and Bilston to Aldersley. It would cost, he said, £50,000, a sum that was easily raised by the following August.

The Birmingham Canal Company was authorized by an Act on 24 February 1768 to build a waterway along Brindley's suggested route with powers to raise £55,000 with an extra £15,000 should it be needed. The new company appointed Brindley as engineer and the work began virtually immediately with Robert Whitworth and Samuel Simcock acting as the great man's assistants.

The first ten miles from Birmingham to the Wednesbury collieries were opened on 6 November 1769. The entire 22½ mile route was completed in 1772 and opened for traffic on 21 September. It was a typical Brindley contour-following canal with twelve locks lifting it up the hill to Smethwick and another twenty taking it down to Wolverhampton and Aldersley. The final cost was over estimate at £112,000 but this included the funds needed to build branches to Wednesbury and Ocker Hill.

Despite the overspending, the line was an immediate success, benefiting enormously from the fact that it passed directly through an area of great mineral wealth and rapidly developing industry. The canal was, virtually from the start, supplying coal, pig-iron, limestone and raw materials to industry, and stone, brick, slate, timber and other cargoes to the town. The Birmingham Canal Company was, however, soon seen to be exploiting the position that they had developed and they came to have a dreadful reputation. They were widely regarded as holding a monopoly on coal supplies and to be 'creaming' the market. This reputation was to be maintained for many years and the company's aggressive, high-handed dismissal of such charges carried through to their future dealings with other canal companies.

With the high profile held by the Birmingham Canal Company and the evident profits to be made, others soon sought similar fortune. The Birmingham & Fazeley Canal was first discussed at a meeting in Warwick in August 1781. The B & F was originally going to run east from Wednesbury via Fazeley to join the Coventry Canal at Atherstone. This was later amended to a line that started in central Birmingham and stopped at Fazeley, where a new stretch of the Coventry Canal would meet it. Another new line from Fazeley to the Trent & Mersey at Fradley would also be built to open a further route to the north. This proposal was warmly supported by the Oxford, Coventry and Trent & Mersey Canal companies all of whom saw potential for business to the east of Birmingham. The Birmingham Company, however, was furious and decided upon 'all possible opposition' to this potential competitor. An unseemly row blundered on for a year and a half before the heavily contested Birmingham & Fazeley Act was passed in June 1783. In 1784, the still antagonistic Birmingham Company solved the issue by simply flexing its financial muscle. It bought its potential competitor and then amalgamated with it to form the Birmingham and Birmingham & Fazeley Company, later renamed the Birmingham Canal Navigations. The new company soon built the Broadwaters Canals (via Ryders Green) as well as the new Fazeley line, which was opened in August 1789.

This growing network of waterways was soon generating considerable business and traffic jams were frequent along the main line. The company felt compelled to make improvements and the engineer John Smeaton was called in to advise. During 1789–90 the summit level at Smethwick was lowered by 18 ft in order to remove six locks. Other locks were duplicated to permit two-way traffic. By 1793, the congestion had eased somewhat and a hundred boats a day were passing along this section of the canal.

Again this success attracted others to enter the navigation business. The Dudley and Stourbridge Canal Companies had, with Thomas Dadford as engineer, built a line from the mines around Dudley to join the Staffordshire &

Worcestershire at Stourton Junction. This had been authorized on 2 April 1776, much to the annoyance of the BCN who feared that its more northerly route to the Staffordshire & Worcestershire Canal was being subjected to a southerly short cut. By 1779, this line from Stourton was open to a point just below the present Blower's Green Locks about a mile south-west of Dudley town centre and the two companies planned to extend their line through a tunnel to join the BCN at Tipton. After several setbacks this extension was started following an Act passed in July 1785. The biggest engineering work on the new line was the 3,172 yd long Dudley Tunnel, a venture that was to take the two companies almost to breaking point. At the beginning of 1787, work had to be stopped when it was realized that the tunnel was out of line. The contractor, John Pinkerton, was relieved of the contract and the company set about the work itself, with Isaac Pratt as engineer. Later he also had to be replaced, this time by Josiah Clowes. It was Clowes who finished the tunnel and continued the new line into Tipton. The new route from Birmingham to the Severn was opened on 6 March 1792 and it did indeed fulfill the BCN's worst fears by providing a much quicker route to Worcester and Bristol. As a result, the BCN proved to be difficult working partners at Tipton Junction, demanding high tolls and being generally uncooperative.

The Dudley company, tired of the pedantic nature of the BCN, sought alternative routes into central Birmingham. With the passing of the Worcester & Birmingham Canal Act in 1791, they decided to build a new line to join the W & B at Selly Oak. This move would allow coal from the Netherton collieries to be taken to markets to the south without entering into BCN waters at all. The W & B were clearly delighted as they too had found dealings with the BCN unrewarding – having been refused access to BCN water at Gas Street. The new Dudley No. 2 Canal, 11 miles long, was opened in 1798. It boasted two tunnels including one at Lappal which was the fifth longest in Britain. It was cut with enormous difficulty, suffered continuously from subsidence and roof falls and was frequently closed for repair. The financial strain of building and maintaining the canal crippled the Dudley company which was eventually absorbed by the BCN in 1846. Lappal Tunnel finally succumbed to its inherent instability and was closed in 1917.

In the north, it was the Wyrley & Essington Company which dared challenge the BCN's supremacy. The W & E from Wolverhampton to Wyrley was opened in 1795 under the direction of William Pitt. This line also proved to be a prosperous one and was extended firstly to Brownhills and then to the Coventry Canal at Huddlesford. This important through route to the Trent & Mersey at Fradley was sadly abandoned in 1954. Several branches were added to the W & E to service the collieries in Cannock and Brownhills. These provided the W & E with a good trade especially when the Black Country pits began to decline.

While the W & E was expanding to the north-west, the BCN was spreading northwards to Walsall. To the outsider, it was an obvious step that an additional link between the two lines should be made. But, typically, the intense rivalries between the companies meant that a junction wasn't built until 1840 when the Walsall Canal was extended via eight locks to Birchills Junction.

Traffic on the BCN during the early nineteenth century continued to increase and with it so did the wealth of the company. By the 1820s, Brindley's main line at Smethwick was again becoming congested and there were continuing water supply problems to the very short summit pound. In a major upgrading of the line, steam pumps were installed to restore lost water, and Thomas Telford was appointed to advise on the condition of the main line. Telford, never a man to do things by halves, modernized the canal in a dramatic fashion between 1825 and 1838. He constructed massive cuttings and embankments to produce a straight, virtually lock-free line from Deepfields to Birmingham that chopped about a third of the distance off the route from Aldersley to Newhall. As the old line remained in use, the improvements also increased the amount of waterway available.

The improvements had a significant impact on both the ease of navigation and on the BCN's finances. They had cost £700,000 and had increased the company's debts to over half a million. This was partly paid off immediately by shareholders or later recovered from them by reducing dividends. But overall, the company was still in a very healthy state with revenue throughout the 1830s and '40s being over £100,000 p.a.

A zinc engraving by L. Haghe, from *c.* 1830, of Telford's fine Galton Bridge which at the time it was built was the largest bridge span over the biggest earthwork anywhere in the world. The scene includes a Pickford's Joey boat and horse underneath the bridge

British Waterways

By this time, however, there was growing concern over the incursion of the railways and in particular the proposed line from Birmingham to Liverpool. The Grand Junction Railway was opened into the city in January 1838 and, in the following November, it was followed by the London to Birmingham line. Perhaps it was this threat, together with the need to build the Tame Valley Canal, that led, in 1840, to the merger of the BCN and W & E. The Tame Valley Canal was built as a bypass for the increasingly congested Farmer's Bridge Locks. Even though the locks were open for twenty-four hours a day and seven days a week, the line along this part of the Birmingham & Fazeley was causing severe traffic jams. The new Tame Valley line was opened in 1844 and ran from the Walsall Canal to the B & F near Gravelly Hill. At roughly the same time the Bentley Canal (which joined the W & E with the Walsall) and the Rushall Canal (which joined the W & E with the Tame Valley) were also built.

By this time, the BCN Company was considering its future in relation to the railways with increasing seriousness. In 1845 the company was adhering to its normal aggressive stance by becoming involved in the construction of a railway between Birmingham and Wolverhampton (the Stour Valley line). But by 1846, when the Dudley Canal Company had been amalgamated into the BCN, the situation had crystallized into one in which the recently formed London and North Western Railway agreed to lease the BCN in exchange for a 4 per cent guaranteed dividend on the shares. The ultimate control of the waterway was thus handed over to railway interests in a rather uncharacteristically timid fashion and the hub of the nation's waterway network had lost its formerly fiercely defended independence forever.

Perhaps contrarily, railway control of the BCN led initially to an expansion in the use of the system. New lines were built, for example the Cannock Extension Canal and Netherton Tunnel (opened on 20 August 1858), and a large number of railway/canal interchange basins were constructed to promote inter-transport traffic. The use of the canals by the railway company resulted largely from the way that Birmingham had developed. Many of the city's factories were built around its canals, did not have access to railway lines and were unlikely to gain access in the future. The importance of the canal infrastructure was therefore considerable and would remain so even when most of the other canals around the country were in rapid decline. In contrast, trade on the BCN continued to grow and by the end of the nineteenth century topped eight and a half million tons p.a., most of it local. But these figures hide the fact that the company was finding it harder to cover its costs and in 1867 it could not pay its dividend from income. In 1868, the railway guarantee was called on and the influence of the LNWR increased still further.

After the turn of the century, the BCN's reliance on local business meant that when the Black Country collieries and mines became exhausted so the trade along the canal declined substantially. Perhaps more seriously, the newer

Joe Taylor at the Island Toll House, Smethwick, in 1940. The toll man assessed the load aboard the passing narrowboat and applied the tolls which were either collected immediately or logged. Sadly, these interesting octagonal toll booths were demolished during the 1950s

British Waterways

factories in Birmingham were built alongside the railways and roads and out of reach of the canal system. Yet even as late as 1949 there was still over a million tons of trade on the line, half of it coal traffic. The BCN continued operating right into the 1960s when nearly a third of a million tons of cargo were moved annually. The canal age only came to an end in Birmingham in 1966 when the coal trade finally stopped. Remarkably, there remained a small amount of commercial traffic into the 1970s and, perhaps even more remarkably, some of it was still horse-drawn. Michael Pearson reports that the last commercial load was of chemical waste taken from Oldbury to Dudley Port in 1974.

As trade declined so some of the less important parts of the network fell out of use and were closed. By 1968, a third of the BCN canals had either been officially abandoned or allowed to fall derelict. In the Transport Act of that year, only the main line and the Birmingham & Fazeley Canal were classified as cruising waterways and were given the official protection of the Act. All the rest were classified as 'remainder' and thus, even today, only survive through goodwill rather than by right. But the West Midlands County Council have recognized the amenity value of the BCN and many areas have been steadily cleared and renovated. There is no greater sign of this than the areas around Gas Street where a number of modern developments have been focussed on the waterway rather than shunning it as they would have done just a decade or two ago. The bulk of the BCN has therefore made it through the worst of times and may, perhaps, be heading towards the best.

The Walk

Start and finish:	Smethwick Rolfe Street BR station (OS ref: SP 022887)
Distance:	4½ miles/7 km
Map:	OS Landranger 139 (Birmingham)
Car park:	There are two signposted car parks near the Smethwick shopping centre off Stony Lane and Church Hill Street
Public transport:	Smethwick Rolfe Street is on the BR main line from Birmingham New Street

Warning: my feet got wetter on this walk than on any other in this book.

Smethwick Rolfe Street station is one stop west of Birmingham New Street. The railway runs parallel to the BCN main line for virtually the whole distance so, if you come that way, it's a good idea to sit on the left-hand side of the train in order to get a good view. Car travellers will not be so blessed.

From the station, cross the road to turn right and then left down North Western Road. This bends left to run for about 150 yd before turning right to go over Brasshouse Lane Bridge. From the bridge, or from the footbridge over the railway nearby, there are excellent views up and down the canals. To the right they head off towards central Birmingham, to the left they go towards Wolverhampton.

Two canals and the remnants of a third can be seen from this spot. The higher waterway is the Old Main Line: Smeaton's adaption of Brindley's original Birmingham Canal. The stretch to the left is the summit level of this line. When asked to improve Brindley's canal in 1790, John Smeaton made a cutting slightly to the south of the then summit level so that the canal could be lowered slightly. It now runs at what is known as the Wolverhampton level of 473 ft above sea level. This new line enabled Smeaton to remove three locks at either end and thus to save both time and water loss. The line of Brindley's original canal is still visible as a ledge which runs about 15 to 20 ft above the present upper waterway.

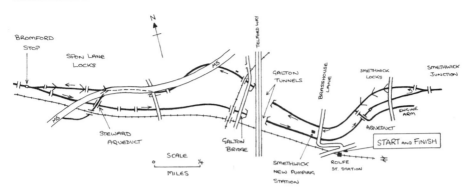

The Birmingham Canal Navigations

When Telford was asked to improve the navigation in the mid-1820s, building technology had moved on considerably. The great man wasn't very flattering about Brindley's canal or Smeaton's renovation of it. 'I found . . . a canal little better than a crooked ditch with scarcely the appearance of a haling-path [towpath], the horses frequently sliding and staggering in the water.' He remedied the situation by building a much straighter line that as far as possible avoided the use of locks altogether. Telford's New Main Line (the lower of the two in front of us) is deeper, with much bolder cuttings, enabling him to take an even lower level than Smeaton; this is the Birmingham level of 453 ft – the same as that in central Birmingham. Telford's line was opened in December 1829 and reduced the travelling distance from Gas Street to Aldersley from 22½ miles to 15 miles and with fewer locks.

Half way across Brasshouse Lane Bridge, pass down a gap in the wall on the left to join the towpath of the Smeaton level. Here immediately on the left is the Smethwick New Pumping Station. Water supply to the short Old Main Line summit was always a problem and, as a consequence, a number of steam pumping stations were built in and around the area to return water lost down the locks. When this engine house was opened in 1892, it contained two steam engines coupled to pumps which lifted water from the New Main Line up to the Old. It continued in use into the 1920s but has since become derelict.

The towpath soon reaches a patch of woodland to the left as the canal bends right. Deep motorbike tyre ruts in the path suggest that some users haven't read BW's carefully considered towpath regulations, nor I suspect have those who have deposited the small mountain of assorted car seats, settees, lumps of metal and planks of wood into the canal itself. Despite this, the line now feels peculiarly rural even if the air smells metallic and fume ridden, and the dull rumble of motor traffic constantly seeks to remind the towpather of the location.

In a couple of hundred yards, the canal reaches Galton Tunnel, a misnomer as the structure is more like an earth bridge. The tunnel dates from 1975 when a new road, The Telford Way, was built to carry traffic from the M5 into Smethwick. The prefabricated concrete cylinder was put into place and then the whole thing covered in soil to make the artificial hill on which the road was built. As far as I know it didn't win any prizes for style but it does have a good towpath all the way through and it eventually opens to reveal the Summit Bridge, built in 1791 and now a scheduled ancient monument. This is closely followed by a blue brick railway bridge.

The canal is now in a modest cutting which still provides a peculiar feeling of isolation from the hubbub that surrounds it. One hundred yards further on to the right is a large bunker-like structure which contains a series of shutes used to load narrowboats with coal. It was built about 1930 and received coal from Sandwell Park and Jubilee collieries via a cable tramway. It was last used in 1970.

The Old Main Line now gradually bends to the left where it suddenly becomes engulfed by the massive concrete pillars that hold the M5 high above the canal. The stark contrast between the gently meandering waterway of the eighteenth century and the harsh, brash motorway of the twentieth is almost shocking in its impact. It's a curiously interesting, if not wholly pleasant, stretch of waterway and not without its hazards. Towpathers will have to stumble over the rubble and mud through here. The situation isn't improved by the constant attention that the flyover seems to need in order to keep it upright.

After surfacing briefly, with the motorway now over to the right, the line passes the high clock tower of Archibald Kendrick & Sons to reach Spon Lane Bridge, an original turnover bridge sadly out of place in this particular location. Cross the canal here and take the right-hand bank which soon winds around to reach Spon Lane Junction. Here a signpost supplied by the BCN Society points off left along the Old Main Line to Wolverhampton via Oldbury. By forking right, we leave the Old Main Line and the M5 and take the line to Bromford. This right turn was formerly a branch off the main line to Wednesbury called the Wednesbury Old Canal.

The first of Spon Lane Locks is immediately on the left. These locks have a single top gate and a doubled bottom. The associated lock bridge is a cantilevered type seen most frequently on the much more rural Stratford Canal. The design allows tow ropes to pass easily through without the need to uncouple the horse. The three Spon Lane Locks take the line down from the Wolverhampton level back to the Birmingham level and are thought to be the oldest working locks in the country. The scene in the early nineteenth century was described by Telford: '. . . at the locks at each end of the short summit crowds of boatmen were always quarrelling, or offering premiums for a preference of passage'. Telford estimated that it was not uncommon for two hundred boats to pass through these locks in a day. Following the completion of the New Main

Line, this situation had apparently changed. John Rickman wrote that the boatmen then met and passed 'with good humour and with mutual salutations'.

After passing a massive car crusher and two more locks, we reach Bromford Junction. Cross the Wednesbury Old Canal via the fine cast-iron bridge, produced at the famous Horseley iron works in Tipton in 1829, to reach Telford's line which now comes in from the left. The two lines join here only to split again about a $\frac{1}{2}$ mile further on when the New Main Line continues on to Wolverhampton while the Wednesbury Old Canal turns northwards towards Walsall.

The island in the middle of the canal to the right is the Bromford Stop. Originally a small toll-house stood on the island from where keepers collected tolls from passing vessels: those going one way passing on one side, those going the other on the opposite. A barrier was held across the waterway to stop the boat. The tonnage carried was then assessed by measuring the depth of the vessel in the water. After the exchange of cash or a chit, the boat was allowed to proceed onwards. Sadly the little hut was demolished in the 1950s and the island is now just another awkward obstacle for passing traffic.

The M5 crosses Telford's
New Main Line on
the Birmingham Canal
Navigations

Turn back almost 180° to continue the walk along Telford's New Main Line. Almost immediately, the straightness and greater width of the newer line is obvious. Telford was particular in avoiding pinch-points at the bridges, a common hinderance in older canals. There are also two towpaths on the Telford line although walkers should stay on this left-hand bank to pass another Horseley bridge.

A ¼ mile further on, the canal again passes under the M5, fortunately this time only briefly, and then Steward Aqueduct. This latter structure, built by Telford in 1828, carries the Old Main Line as it swings round from Spon Lane Junction to go south. In building his new line, Telford dug this staggering cutting underneath the old canal thereby enforcing the construction of the aqueduct which was named after a member of the company's committee. The new line is now wedged between the high walls of Chance's glassworks to the left and the railway to the right. For those who might think the area bereft of wildlife, a heron lept into the air and disappeared over the railway wall as I passed here. Two more bridges are passed before we reach a large blue brick bridge, Spon Lane South Bridge.

The line now opens out slightly as Telford's cutting proceeds onwards for about a ½ mile before reaching a dull grey railway bridge, underneath which is the place where shopping trolleys go when they die. Immediately after, however, is the magnificent Galton Bridge. This is Telford at his best. Built in 1829 to carry Roebuck Lane over the new canal, Galton Bridge is 150 ft wide and 71 ft above the canal. When it was built it was the largest single-span bridge over the largest earthwork in the world. Like the lesser footbridges that abound up and down the BCN, it was cast at Horseley works and has recently been sympathetically painted to enhance its fine structure. Its importance has been acknowledged and it is now a Grade I listed building.

The same cannot be said of Galton Tunnel, the mirror image of the one met on the Old Main Line. The only good thing is that we are soon through it and out on to a resurfaced towpath which leads around to the Smethwick New Pumping Station passed earlier. This time we go under Brasshouse Lane Bridge which was built in 1826. From the low level of the Telford canal, it is now possible to see the high retaining wall that holds North Western Road in place.

The line bends slowly to the left to reach another splendid cast-iron structure: the Engine Arm Aqueduct which carries the Engine Arm Canal from the Old Main Line across the New and over to Lime Wharf to the south. The prime purpose of the Engine Arm was to act as a feeder canal from the Rotton Park reservoir (about 1½ miles to the south-east) to the Smethwick summit. The Smethwick engine which did the pumping was built by Boulton and Watt and installed in 1779. It was one of their first and still survives, albeit in the Birmingham Museum of Industry and Science. It is reputed to be the oldest working steam engine in existence. It was Telford who made the arm navigable (so that coal could be delivered directly to the engine site) and who designed the aqueduct. The engine (and another that had been installed later) was removed in 1897. The aqueduct luckily remains and, in its own small way, is every bit as fine as Galton Bridge. Again the paintwork is sympathetic with the structure which has the superficial appearance of the inside of a church. The cast-iron trough which runs across the length of the aqueduct is clearly visible but the design shows how a fairly routine structure can be made beautiful with a little careful effort. If only Telford had designed the M5 flyover!

After the aqueduct, the canal narrows to reach a gauging station similar in theory to the one at Bromford. The only difference is that at this station there are two islands dividing the canal into three. Here the toll clerk measured the height between the water level and the upper side of the deck of the narrowboat (a distance known as the freeboard). From this he was able to work out the weight of cargo being carried and hence the toll that was due to the company. This was done by having previously recorded the unladen freeboard, measured when the boat was first built, and knowing how far the boat sank into the water at different cargo weights.

Further on, eagle-eyed towpathers will have noticed that the embankment to the left has gradually declined. Eventually as we pass under Rolfe Bridge, it disappears altogether and we reach Smethwick Junction where the Old Main Line and the New Main Line merge to head off towards central Birmingham. The two Horseley cast-iron bridges date from 1828 and, like the others on this walk, were assembled on-site on brick abutments. A BCN Society signpost again marks the junction. The installation of these very useful signposts (there are thirty of them altogether) was started in 1983 and has been undertaken by the society with the occasional contribution from a sponsor. The posts were made by the Royal Label Factory at Chipping Norton. It is a sad reflection that these fine posts, purchased by a voluntary society, have been regularly vandalized. Incredibly some have even been sawn off at the base!

We turn left to return to the Old Main Line. This soon reaches Pope's Bridge which carries Bridge Street from Smethwick towards Handsworth. When Smeaton revised the height of the Old Main Line summit, he not only removed three locks at each end but he also built a duplicate set for the three remaining locks at this end. Brindley's set remained in use into the 1960s when they were filled in. Just as you pass under Pope's Bridge, you can see the remnants of Brindley's original line on the right where the entrance to the bottom lock still exists with a reed-ridden side pound and a bricked-up arch in the bridge. On the other side of the bridge, we pass the first of Smeaton's locks and then go on to the next two which take us up to the Wolverhampton level. At the end of the third lock, it is again possible to make out the entrance to the Brindley lock on the other side of the canal. Roughly opposite this lock on the left-hand side is the Engine Arm Aqueduct. If you have time, you can take the left hand bank of the arm round for about a $\frac{1}{2}$ mile to Lime Wharf and a building which was formerly a malt-house.

The final stretch of the walk up to Brasshouse Lane Bridge is now in view. On the right is the site of Smethwick Brasshouse which was established in 1790 and had its own canal wharf. It was later converted into an iron and steel works but was closed in the early 1980s.

Just before passing under the bridge, take the slope to the left which goes up to the road and turn left to reach the railway station and the end of the walk.

Further Explorations

With about one hundred miles of waterway still open for navigation, the BCN has potential for a seemingly endless combination of walks. It has to be said that not all of it makes for particularly attractive scenery but it is all a matter of taste and there is an amazing variety. You can walk through the inner city scrapyard of the Walsall Canal or the open countryside around the eastern end of the Birmingham & Fazeley. You can go through the open tunnel formed by Spaghetti Junction as it rides high above the Tame Valley Canal or the real 3,027 yd long pitch blackness of the Netherton Tunnel on the Dudley Canals. And not only is the choice of view tremendous but from the BCN you can walk to virtually anywhere on the canal system. Go along the Birmingham Canal to Wolverhampton and you can join the Staffordshire & Worcestershire and the Shropshire Union Canals. A walk from Gas Street in the centre of the city will take you immediately on to the Worcester & Birmingham Canal and later on to the Stratford. Take the Birmingham & Fazeley eastwards and you will join the Coventry Canal and then the Trent & Mersey. Go along the Digbeth branch from Aston and before long you'll be able to start a trek to London along the Grand Union.

Despite this huge choice, it's virtually impossible to visit the BCN without visiting the Gas Street Basin and the central Birmingham & Fazeley. For a walk of approximately 2 miles start at New Street station in the centre of the main shopping area. Leave the station to reach New Street where the first of a series of useful signposts points towards the Convention Centre. Follow these signs to Paradise Circus and through Paradise Forum and out towards the Repertory Theatre and the Convention Centre. Turn left here to cross Broad Street and go down Bridge Street. After passing in front of the Hyatt Regency Hotel, you will reach the James Brindley pub and the entrance to the Gas Street Basin.

Originally the main area of wharfage was on the left-hand side of Bridge Street, stretching right over to the BCN's offices – a spot now occupied by Central TV. Gas Street is still the place where the Worcester & Birmingham Canal joins the BCN at a point known as the Worcester Bar. For many years the BCN refused the W & B access to their line and cargo had to be manhandled across the bar which prevented boats from moving between the two canals. Eventually common sense prevailed and the bar was replaced by a lock where passage was possible but where tolls were exacted. Looking from the pub, the W & B is on the left. To see what remains of the bar take the left-hand pathway which leads round the side of the pub and over to the small Horseley iron bridge. The clearly defined remnants of the lock can still be seen from the bridge. This is also a good vantage point for views of the basin as a whole. The extent of recent 'improvements' can be seen here with the old and new sitting uneasily with each other. I, for one, only regret that I didn't see it before. Old timers speak fondly of the atmosphere of the old basin and harshly of the yuppie pub and the brash hotel. It's not exactly unpleasant now. It's simply that it has less authenticity.

Retrace your steps to the pub and take the path which goes left over the small bridge across an ex-factory arm and under Broad Street Tunnel. Shortly the sight of the new Convention Centre hits you. Is there any more extraordinary sight in the British canal network? This architect's view of heaven is juxtaposed with the former working waterway and, across the line, a stretch of scrappy dereliction. Best to walk on quickly to reach Farmer's Bridge Junction. Here a series of lines run off at angles. The first left, under a roving bridge, is the Oozells Street Loop, part of Brindley's original winding main line. Next right is Telford's much straighter New Main Line. We, however, turn right to follow the Birmingham & Fazeley Canal towards Fazeley. Incidentally, after crossing the bridge, it's worth pausing a while to read the interesting and informative notices put up by the Birmingham Inner City Partnership as part of their canal improvements programme.

The city scenery is now more in keeping as you pass under Tindal Bridge to reach Cambrian Wharf, heavily renovated but not unpleasantly so. This was formerly the Newhall Branch Canal that went off right for a $\frac{1}{2}$ mile to Newhall Wharf. The wharf fell into disuse in 1901 and has since been filled in and built on. Another short section called Gibson's Arm also offered wharfage. This area has now been converted into gardens between the new Convention Centre and Paradise Circus.

Bear left to pass a small toll office to reach the first of the highly atmospheric Farmer's Bridge Locks. By means of this flight of thirteen locks, the canal gradually slides down the hill towards Aston Junction. On the way, the intimacy produced by the close crowding of the buildings is awe inspiring. Many of them have their own wharves including covered loading areas. By lock 8, a joint council–company project has produced the Canning Walk on the opposite bank but the most peculiar section is that after lock 9. Here a high-rise office block completely engulfs the canal and lock 10 only to release them into a canyon made up of office blocks on either side of the line.

After passing lock 12, the canal goes through the massive cave that is formed by the railway bridge which carries lines into Snow Hill station. This is followed

An office block engulfs the canal and two of the Farmer's Bridge Locks on the Birmingham Canal Navigations

by the final Farmer's Bridge Lock and Snow Hill Bridge. Here ascend to the road and turn right to follow signs to Snow Hill station and then New Street station to complete the walk.

It is possible to continue for a further 2 miles by going on to Aston Junction and turning right to go along the Digbeth branch. This eventually leads round to the Warwick Bar and the Grand Union Canal. As the GU turns off right continue over the roving bridge to reach the road. Turn right to return to the city centre.

Further Information

The Birmingham Canal Navigations are blessed with a vigorous and enthusiastic society which holds regular meetings and publishes a newsletter. If you wish to join or help or both, you should contact:

Neil Bough,
BCN Society,
60 Tresham Road,
Great Barr,
Birmingham,
B44 9UD.

There are a number of locally available leaflets that describe various parts of the BCN. One set has been produced by the Birmingham Inner City Partnership and the other by the Sandwell Metropolitan Borough Council. Copies of both sets are available from local libraries.

The best guide to the BCN as a whole is:
Pearson, Michael, *Canal Companion Birmingham Canal Navigations*. J.M. Pearson & Associates, 1989.

For historical detail, however, the following are recommended:
Bainbridge, S.R., *The Birmingham Canal Navigations*, vol. 1 (1768–1846). David & Charles, 1974. (NB there is no vol. 2.)
Hadfield, Charles, *The Canals of the West Midlands*. David & Charles, 1969.
Rolt, L.T.C., *Thomas Telford*. Penguin Books, 1958.

2
THE BIRMINGHAM & LIVERPOOL JUNCTION CANAL

Gnosall Heath to Grub Street

Introduction

The Birmingham & Liverpool Junction Canal, or the Shropshire Union as it became, is a thoroughly modern waterway. As an antithesis to Brindley's contour followers that called in at all the local villages, Thomas Telford made the B & LJ as straight and lockless as possible. The aim was simply to produce the fastest practicable route from the city of Birmingham to the River Mersey. As a result, the Shroppie is littered with dramatic embankments and cavernous cuttings, built using techniques that were unknown in Brindley's time and barely achievable in Telford's. Indeed, it may be said that the B & LJ's cut and fill engineering makes it more akin to a modern motorway than to the Oxford Canal or the Staffordshire & Worcestershire.

From Autherley, on the Staffordshire & Worcestershire Canal near Wolverhampton, the B & LJ runs north-west, taking a direct line through the Staffordshire countryside past Brewood and Wheaton Aston. After Gnosall Heath, the canal passes over the Shelmore embankment to Norbury Junction, from where a line formerly ran west to join the Shrewsbury and Shropshire Canals at Wappenshall. The B & LJ, meanwhile, continues briefly into Shropshire and through the 100 ft deep Woodseaves cutting to Tyrley. Here, unusually, are five locks that take the line down to pass Market Drayton and on to Adderley and Audlem. The fifteen locks that constitute 'the Audlem Thick' take the line down 90 ft into the Cheshire countryside. After crossing an aqueduct over the River Weaver, the canal takes a straight course to the outskirts of Nantwich where, near Nantwich Basin, it crosses an aqueduct over the A51 and joins the old Chester Canal. The Shropshire Union now runs past Hurleston Junction (for the Llangollen Canal) and Barbridge Junction (for the Middlewich branch to the Trent & Mersey Canal) and then on for another

16 miles to Chester. From there the Wirral line of the old Ellesmere Canal continues for a further $8\frac{1}{2}$ miles to Ellesmere Port.

The Shroppie is a tricky canal for towpathers. Its straightness and its route through some remote rural countryside puts circular walks or those that use public transport for a return journey virtually out of the question. But the B & LJ makes an excellent long-distance path and, anyway, Telford's massive earthworks are well worth seeing twice.

History

The story of the B & LJ starts in 1824 when Thomas Telford, in making recommendations for the renovation of the Birmingham Canal Navigations' main line, suggested that an improved waterway to the Mersey would increase that company's business to the north-west. With the threat of a railway line between Birmingham and Liverpool hanging over them, the BCN Company grabbed the idea as a way of competing with the new fangled, and as yet largely unproven, mode of transport.

Virtually from the start, the line was established as running from Autherley, on the Staffordshire & Worcestershire Canal near Wolverhampton, to the Ellesmere & Chester Canal near Nantwich. The Chester Canal, a line of nearly 20 miles from Nantwich to Chester, was opened in 1779. The Ellesmere had been started as a much grander venture designed to link the Mersey with the Severn. Such splendid hopes, however, had petered out into a Welsh branch from Hurleston, just north of Nantwich (see the Llangollen Canal), and a northern stretch of $8\frac{1}{2}$ miles from Chester to Ellesmere that became known as the Wirral line. This meant that by 1796 a canal from Nantwich to the Mersey was open but with no southern connection. After the amalgamation of these two companies in 1813, the new Ellesmere & Chester Canal Company actively sought to extend their waterway south and welcomed the promotion of the B & LJ. Naturally, the company hit hardest, the Trent & Mersey, weren't so enthusiastic and responded to the potential newcomer by appointing Telford to upgrade their line with doubled locks and a new Harecastle Tunnel (see the Trent & Mersey Canal).

The new project was widely promoted during the course of 1825 and, in May 1826, an Act was passed to authorize the raising of £400,000 capital with £100,000 in reserve. Thomas Telford was appointed as the canal's principal engineer in what was to be his last major project. A later Act, in 1827, added the Newport branch: a line that went west from Norbury to join the Shrewsbury and Shropshire Canals at Wappenshall.

Even at the time, many people felt that to build a canal from Birmingham to

Nantwich was desperately old fashioned. Indeed, other than the Manchester Ship Canal, which was altogether a different kind of venture, the B & LJ was the last major canal project to be started in Great Britain. Even before the canal was open, the country was well into a new age and was developing a more modern form of long-distance transport. As plans for a railway along a similar route had already been proposed, the lobbying for the B & LJ must have been both intense and highly efficient. We can only imagine the excitement, or maybe it was relief, that was felt by the promoters when so many pledged their support for the new canal in direct conflict with the prevailing mood.

The construction contracts were divided into three areas: Autherley to Church Eaton, Church Eaton to High Offley (just north of Norbury) and High Offley to Nantwich. W.A. Provis, who had worked with Telford on the Holyhead Road, took the middle section and the Newport branch. John Wilson, who had also worked with Telford, this time on the Llangollen Canal, was appointed contractor for the two ends. Alexander Easton was appointed as resident engineer based at Market Drayton.

The route taken by Telford was a bold one. The intention was to shorten the Birmingham to Mersey route both in distance and time and so the course ignored previous canal building concepts of following either the contours or the easiest line. The B & LJ drove across the country on high embankments and through deep cuttings. It was thoroughly modern and inordinately expensive. Work started in the northern section in January 1827. By late 1831, with building work under way all along the line and the northern section nearly ready, the company had spent £442,000 and were forced to apply to the Exchequer Bill Loan Commissioners (an early job creation scheme) for a loan of £160,000 which was to be repaid in three years. Part of the problem was the stiff opposition put up by the local landowners. The company expended some £96,119 in compensation, more than they could possibly have expected.

In 1832, a further £24,600 was obtained from the Exchequer with the hope that this would carry the work through to completion in 1833. But the embankment at Shelmore (near Gnosall) was causing both technical and financial problems. Continuing subsidence meant that a single mile long stretch took some six years to build at a time when pressure to complete the work was tremendous. By 1833, Telford's health was failing fast and William Cubitt was called in to deputize. More and more money was needed to make fast the Shelmore embankment and the Grub Street cutting and when the final stretches were opened on 2 March 1835, the total cost of the canal had risen to £800,000, equivalent to £16,000 per mile, not including any interest arrears. Sadly, Telford, who had died on 2 September 1834 at the age of seventy-seven, did not live to see the completion of his last great work.

As built, the B & LJ was $39\frac{1}{2}$ miles long with twenty-nine locks including fifteen at Audlem. It shortened the route from Birmingham to the Mersey by nearly 20 miles and thirty locks and to Manchester by $5\frac{1}{4}$ miles and thirty locks. Cargo from Birmingham could now reach Liverpool in just forty-five hours.

The Staffordshire & Worcestershire Canal viewed the new line as a major threat to its fortunes and saw an opportunity to obstruct the new line at Autherley Junction. Carriers from Birmingham destined for the B & LJ took the BCN main line to Aldersley, then passed along a ½ mile of the S & W to Autherley where they entered the B & LJ. The S & W complained bitterly that it couldn't afford the excessive water loss that this traffic involved and imposed a crippling toll for every lockful that flowed into the B & LJ. In addition, whenever the S & W ran low, permission to withdraw water was suspended and the B & LJ were forced to buy extra supplies from the Wyrley & Essington Canal (the northernmost of the Birmingham Canal Navigations). Not surprisingly, the B & LJ soon grew tired of these tactics and, in a novel manoeuvre, it released a plan for a mile long aqueduct 'fly-over' from the Birmingham Canal to the B & LJ. The S & W was so shaken that it promptly agreed to reduce the tolls. This successful trump card was used twice more on different issues with similar effect in 1842 and 1867.

But the company was not as successful in many of its other ventures. As would be expected with such a late canal, the B & LJ faced immediate railway

The scene at Norbury Junction in 1905 shows that fishing was as popular then as now although it appears that much less equipment was necessary. The entrance to the Newport Branch is on the right. The craft at the company wharf is said to be the narrowboat *Penguin*

The Boat Museum archive

competition and tolls had to be reduced practically from the start. The Grand Junction Railway, which opened for freight traffic in January 1838, could carry goods considerably faster for roughly the same cost and the B & LJ had to respond in order to gain any traffic at all. This meant that with the high building costs, financial problems plagued the B & LJ from birth. By October 1839, the canal was in dire straits, owing well over £350,000. The situation was made worse by the need to expend even more money to enlarge the Belvide reservoir (on the western side of the canal about 1½ miles north of Brewood). Without it there was not enough water to guarantee traffic flow along the line. A further £20,000 had to be spent before it was open in 1842.

The canal was, however, relatively busy carrying a range of cargo. Along the Newport branch, for example, the canal was conveying iron from the Coalbrookdale area on to the main line and then around the country. From Ellesmere Port, there was a large trade in iron ore, wheat and raw materials. The Llangollen branch brought significant quantities of limestone and lime. Limestone was used extensively in the iron industry in Wolverhampton and lime was becoming an essential 'modern' fertilizer. The canal was also carrying some 140,000 tons of coal p.a. Goods on fly-boats were shipped from the Mersey to Birmingham in just twenty-nine hours, a time achieved by giving them priority over all other traffic and by having teams of fresh horses at roughly 20 mile intervals. By 1842, the profits being made by the carrying companies attracted the B & LJ who obtained powers to start their own carrying company for both cargo and passengers.

But the financial situation remained extremely serious. Revenue along the line reached a peak in 1840 when income was just £30,859. No dividends were being paid and the results clearly demonstrated to all those who said that it was a canal 'white elephant' that they were probably right.

It came as no surprise therefore when, in May 1845, the B & LJ announced that it was to merge with the Ellesmere & Chester Canal and that the resultant Shropshire Union Railway and Canal Company (as the company was eventually renamed in 1846) would, in deference to the inevitable, convert their canals into railways. The company had realized that the innovation started by the Ellesmere & Chester of using steam tugs to pull boat trains couldn't be used on its Welsh lines and would never be cheaper to run than a railway. The decision must have been cemented when their engineer, W.A. Provis, assured them that the cost of converting a canal to a railway was half that of building a railway from scratch. The company would therefore be able to compete with the many lines that were springing up in direct competition with its own canals throughout the north-west. This decision was not met with much enthusiasm by the rest of the canal world. The Staffordshire & Worcestershire Canal, vehemently anti-railway at the time, insisted that a stoppage to any part of the inland waterway would 'cripple the whole system'.

The new SU was unmoved. In the company's prospectus, published in

October 1845, four major railways were proposed: a line that went from Crewe to Nantwich and then along the Llangollen and Montgomery Canals to Newtown and Aberystwyth; a line from Wolverhampton to Nantwich; a line from the Trent Valley to Shrewsbury; and a line from Shrewsbury to Worcester. The proposed capital was £5,400,000 in £20 shares and William Cubitt, Robert Stephenson and W.A. Provis were named as engineers. To extend its activities, the SU incorporated the Shrewsbury Canal and, later, the Montgomery Canal (in February 1847). It also took a lease on the Shropshire Canal (on 1 November 1849). The first railway to be built by the company was not, interestingly enough, on a canal bed but on a line between Shrewsbury and Stafford. This joint venture with the Shrewsbury & Birmingham Railway was chosen as it didn't interfere with any current canal activity and would thus give them a 'feel' for the new business before finally committing them to the proposed conversion.

Despite this apparently vigorous start, the period of renaissance of the SU as a railway company was short-lived. By the autumn of 1846, the newly formed London and North Western Railway, seeing that the SU was about to become a serious threat to its own activities and yet could become an important conduit into non-LNWR areas, offered a perpetual lease to the SU and the SU committee agreed. The canal company was to receive 4 per cent on its existing canal stock until its railways were built and then a rent equal to half the ordinary dividends of the LNWR stock. The canal's debt was to be serviced out of canal operating profits. The Act authorizing this move was passed in June 1847.

By 1 July 1849 (by which time the Shrewsbury to Stafford line had opened), the LNWR had persuaded the SU to give up all thoughts of further railway building in return for having its debt serviced. With this, the company's railway ambitions effectively evaporated for ever.

The SU committee may well have made this decision after having seen that the levels of canal traffic had increased substantially. In 1850, the revenue on the line was over £180,000 with a useful profit and a very vigorous carrying business. It must be said that the interest on the company's debts was still £34,473 p.a. and that the SU never earned enough to cover this figure in any year after 1850, the balance being paid by the LNWR. But dividends were paid and, with the LNWR lease, prospects must have appeared reasonable. Only later did the increasing railway competition (from lines which at one time the B & LJ itself had intended to build) make itself felt. From an operating surplus of £45,000 p.a. in the late 1840s, profits dropped to around £11,000 by the late 1860s even though the level of receipts was similar. The SU had been forced to accept lower tolls in the face of railway competition. An agreement with the Great Western Railway for example forced canal rates to below those of the railway for traffic to Llangollen, Ruabon and Chirk. At this time, most of the company's income was, in fact, coming from its canal carrying business rather than from tolls; in 1870 the company operated some 213 narrowboats (of which about a third were employed carrying iron ore or manufactured products) and by 1889 it operated 395. By the end of the century, the company's own carrier was handling over 70 per cent of the trade along the line.

The gradual decline in trade was perhaps inevitable although in the 1890s, with the construction of the Manchester Ship Canal, a brief respite in the downward trend was felt. At that time, the company built new quays, improved facilities and considered the possibility of widening the SU main line to Autherley. The cost was estimated at almost £900,000. Although some minor improvements to a number of wharves were undertaken, the line was not substantially altered and the widening scheme was quietly forgotten.

The new century started with some modest operating profits but, by the time of the First World War, the canal as a whole was running at a loss and the LNWR appeared to lose interest. After the war, the canal began to sustain large losses and the LNWR insisted on economies. On 1 June 1920, the SU

Cadbury's had a fleet of seventeen boats that brought milk from farmers along the canal to Knighton where it was processed and then shipped on to the chocolate factory at Bourneville. This photograph of horse-drawn boats arriving at Knighton was taken in the 1920s
The Boat Museum archive

announced that it was to give up the carrying business that still ran some 202 boats up and down the line. The company also reduced its maintenance effort and closed the locks at weekends. This action reduced the losses from £98,384 in 1921 to £26,473 in 1922 but it was clearly a retrograde step for a working canal. At the end of 1922, the SU company was absorbed by the LNWR which in turn was itself absorbed into the London Midland & Scottish Railway the following year.

In 1929, the SU carried 433,000 tons of cargo but by 1940 this was down to 151,000 tons. The LMS did nothing to encourage traffic along the SU and the lack of maintenance must have acted as a positive discouragement. Road transport was by then becoming the main threat, being both easier and cheaper. Although a widening scheme for the main line was again considered in 1943, it was inevitable that in an Act of 1944 the entire lengths of what today are known as the Montgomery and Llangollen Canals were abandoned (although the Llangollen line was maintained as a water feeder for a reservoir at Hurleston). The last commercial traffic to run along the SU main line came in the form of oil and tar boats from Stanlow near Ellesmere Port to Oldbury in the Black Country. At one time, this trade alone took one thousand boats a year and they were horse-drawn into the 1950s. But by 1957–8, the trade had all but stopped and, although a small amount of traffic continued into the 1960s, the commercial life of the canal was effectively over.

The SU main line was never in the same kind of danger of imminent closure as many of the nation's canals. Even in the Report of the Board of Survey to the British Transport Commission in 1955, the line was deemed worthy of retention. But true safety only came in 1968, when the main line, together with the Llangollen Canal, was classified as a cruiseway under the Transport Act. Since then the line has been gradually improved into the popular cruising route that it is today.

The Walk

Start:	Gnosall Heath (OS ref: SJ 820203)
Finish:	The Anchor Inn, near Grub Street (OS ref: SJ 774256)
Distance:	5 miles/8 km each way
Map:	OS Landranger 127 (Stafford & Telford)
Return:	Along canal
Car park:	On street at Gnosall Heath near The Navigation pub
Public transport:	Happy Days buses go to Gnosall Heath from Stafford (which is about 6 miles to the east) where there is a BR station. For bus enquiries telephone: (0785) 74231

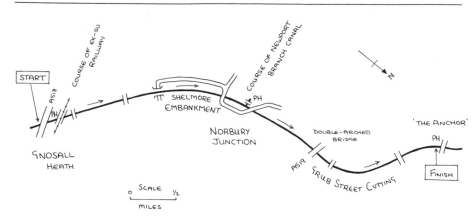

The Birmingham & Liverpool Junction Canal

There is no public transport from Gnosall to The Anchor Inn. Towpathers therefore have to find somebody nice enough to pick them up from The Anchor Inn or befriend a fellow towpather so that they can perform a two car trick. The only other alternative is simply to turn round and walk back. Luckily, this stretch of canal is so pleasant that this latter option is no hardship. In addition, if you start out in mid-morning, then the pub can serve its wonderful Wadworth's 6X to you so that you won't notice that you've seen all the scenery before.

The walk starts by The Navigation which is on the western outskirts of Gnosall Heath on the A518 Newport to Stafford road. Take the path which runs down the right-hand side of the pub to walk on the left-hand bank of the canal. You may not wish to know that Gnosall Bridge is reputedly haunted by the ghost of a man who drowned in the canal in the mid-nineteenth century.

The towpath shortly passes a map of the village and some mooring spaces before bending gently left to go under a former railway line. This line is significant because it was the SU's one and only venture into railways before being bought off by the London and North-Western Railway. The line ran from Stafford to Shrewsbury and opened on 1 June 1849. After the rail bridge, the canal enters a small cutting and becomes delightfully peaceful and relaxing. At the next bridge (No. 36), the deep rope cuts in the bridge metals confirm just how busy the line must have been at one time.

At the end of the cutting, the scenery ahead opens out and the canal stretches directly across the landscape, more like a Dutch dyke than an English waterway. The line is held high above the farmland on either side and shows how far Telford had moved on in terms of engineering expertise and ideas from Brindley just fifty years before. Brindley would presumably have locked down here by, say, two locks, possibly adding half an hour or so to the journey time. However, such exposed embankments can be a problem in the face of a gale. As

Tom Rolt found, forward progress can be virtually impossible in some conditions when it is easier just to switch off the engine and sit it out.

After passing an old B & LJ milepost (not a SU one), the canal goes under bridge 37 and on to the start of the Shelmore embankment, announced by the positioning of a massive stop gate on the canal near a small fishing hut. Within a few yards, it becomes clear that the land to the left has fallen away and the canal is now riding high above it on a massive embankment. Indeed, almost unnoticed, the line runs over a minor road by means of a small aqueduct. The canal channel is now concrete lined, a move which only hints at the problems that the B & LJ had with the canal during the construction of this section.

It was never Telford's intention to build such a vast embankment here at Shelmore. Originally, the canal was to run on the level along a line to the east of its present route to take a more direct course from Grub Street to Gnosall. However, Lord Anson refused to allow the company to make a cut through his game reserves in Norbury Park. He claimed that the building of the line and its operation would disturb his fowl and the company was forced to move the canal westwards. To retain height (and hence avoid a sequence of locks), Telford designed an embankment, a mile long and 60 ft high: the Shelmore Great Bank. If you look at the local OS map, the preferred course can be easily traced as the current line makes a sudden and rather awkward westward 'lean' away from Norbury Hall.

The construction of the embankment, under contract to W.A. Provis, began in June 1829 and at both ends. Four hundred men and seventy horses were in action piling up the spoil that was being extracted from the Grub Street and Gnosall cuttings. By July 1830, some half a million cubic yards of soil had been moved to the site but, by the following winter, the embankment was still only a quarter of the desired height. More importantly, it was subsiding. Telford promptly concluded that the problem was with the material they were using and ordered that only sandy soils could be used to build the embankment. This, he said, would be more reliable and he confidently predicted that the works would be ready by the autumn of 1832. By July 1832 he had changed his mind. The embankment, although nearly at full height, was still slowly slipping into the valley below. In August, the gradual slide stopped and a rapid one began. Eight hundred yards of earthworks, half the total length, collapsed. It was clear that the Shelmore embankment was about to become the Shelmore embarrassment.

The situation was not helped by the seventy-five-year-old Telford's failing health. The great man was often simply unable to provide the help and guidance that was necessary. Indeed, it was Alexander Easton who, in the following winter, had to report that, although 300 yd of embankment had been virtually finished, the rest was still slipping. In February 1833, William Cubitt was asked to deputize for the ailing Telford. Cubitt approached the problem with a good degree of self-confidence and recommended that hardcore be used to form a firm base on the bank. By May, the company was adding some 40,000 cubic yd of

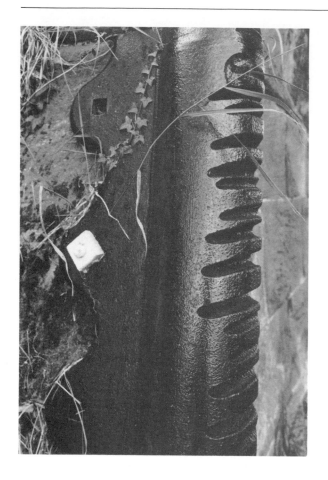

Rope-wear marks on
Machins Barn Bridge near
the Shelmore embankment

hardcore a month to its moving hillside. This they did until January 1834 when Cubitt was, finally, able to report that the line was nearly ready and that the puddled bed was virtually complete. Although he added that there were still signs of shifting along some stretches, there must have been a feeling that the trauma was over at last. History does not recall a meeting with Lord Anson when Cubitt and Telford visited the embankment in March 1834, but presumably some wrath must have been vented on the man who had forced them to construct an otherwise totally unnecessary earthwork simply in order to keep the game happy.

In May, the embankment slipped again and when Telford died on 2 September, the Shelmore embankment was still not passable; his final great engineering work remained as two unconnected halves. Cubitt was forced to breach the slide and to continue adding hardcore. Finally, on Monday 2 March 1835, the first traffic went along the line. Although only a single boat's width, it was passable.

By July Cubitt was able to report that the line had been completed to full width. It had taken six years to build.

This story of near disaster for the B & LJ company is hard to credit when one looks at the scene today. Shelmore Wood now covers both banks and one would be excused the thought that the canal is one of Brindley's running around a perfectly natural hillside, such is the air of permanence and solidity that it exudes. Only the concrete channel and the history tell us otherwise.

After another milepost, the canal bends around to the right, straightens to pass some moored vessels and arrives at Norbury Junction. Norbury is now a busy little canal centre where British Waterways have an office (responsible for the maintenance of the canal from Autherley to Audlem) and there is a boatyard, a wharf with an old crane, a pub (Junction Inn), a shop and a café. The small chimney stack is all that remains of the steam engine which used to drive the workshop machinery. It's a pleasant spot and suitable for a brief rest before continuing onwards.

Norbury became a boating centre because at one time the Newport branch canal ran from here via a set of locks called the 'Seventeen Steps' to Newport and then on to Wappenshall. There it joined the Shrewsbury Canal and thus the Shropshire and Donnington Wood Canals. This, of course, gave the B & LJ access to the industry around Coalbrookdale and the line was an important conduit for the iron from the area to the rest of the country. The branch was just over 10 miles long and was opened in March 1835. The line was still carrying a small amount of traffic in 1943 but was abandoned as part of the infamous Act of 1944 which put paid to a number of canals owned by the LMS (including the much lamented and soon to be restored Montgomery). The line has since been filled in, built upon and generally neglected. The remains can be seen by walking up to the roving bridge which passes over the former entrance and looking left. The scene may appear to be beyond redemption but the Shropshire Union Canal Society haven't given up all hope and are reportedly undertaking a preliminary feasibility study on the restoration of the branch.

After passing over the bridge and by the Junction Inn, would-be shoppers can cross the next bridge to the small store. Those who wish to can continue on under the bridge (past the stop planks to protect this side of the Shelmore embankment) and into a straight length of canal that appears to be heading directly for a wooded hillside.

Almost imperceptibly, the trees on either side of the canal get taller and it's not until you're some way in that you realize that you've entered another cutting. This is the Grub Street cutting and the site of more headaches for the contractor W.A. Provis. The hill is composed of a friable rock that alternates with beds of clay which crumble when dry and act as a lubricant when wet. Almost from the start of building in 1829, men were employed virtually continuously to clear up the earth slips that plagued the cutting. The worst slip was in May 1834 when 10,000 cubic yd of marl and rock came down to block the

High Bridge in Grub Street cutting on the Birmingham & Liverpool Junction Canal

canal completely for about 60 yd. The only answer was to take the sides further back from the waterway. Similar problems were had at the Woodseaves cutting which is about 8 miles to the north near Tyrley. Grub Street cutting is approximately 80 ft deep and 2 miles long. The depth isn't fully appreciated until the canal reaches the A519 road bridge (No. 39 'High Bridge') which is a curious double-arched affair with an old-fashioned telegraph pole stuck half-way up. It's a strange landmark but oddly memorable. It is also said to be haunted by a black monkey, all that remains of another long-drowned boatman.

Bridge 40 (a footbridge) is nowhere near as interesting but by now the canal is deep into the cutting. The Anchor Inn is about a mile and a quarter away and we walk through a waterway chasm before reaching bridge 41. From here our walls begin to subside and a notice brings our attention to the proximity of the pub.

Just before bridge 42, a gate leads off the towpath and into the beer garden. The Anchor Inn has been known as the New Inn and the Sebastopol but has always been a peculiarly canal-locked hostelry. At one time the pub had its own stables so that hospitality could be offered to the horses as well. Thomas Pellow and Paul Bowen describe evenings in the pub as well as the extraordinary story of Lily Pascall who ran the place from the turn of the century into the 1960s.

Probably nothing too extraordinary will happen to you while you're here but you may be lucky enough to meet the kindly soul who is picking you up or you will have a chance for some R and R before returning to Gnosall.

Further Explorations

The lack of public transport along the B & LJ suggests that one of the best ways for towpathers to see the line is to make it into a long-distance path. With a full length from Autherley to Nantwich of 39 miles, three days and two nights are sufficient for a steady amble along the line.

Day one starts at Autherley Junction (OS Landranger 127, ref: SJ 902020). From Wolverhampton central bus station, take the West Midlands Transport bus 504 or 505 to the corner of Oxley Moor Road and Blaydon Road. Take a lane which runs past the waterworks to the Water Travel boatyard. This provides access to the canal and the towpath at bridge 1 close to the stop lock which controlled water flow from the Staffordshire & Worcestershire Canal.

Having swallowed deeply, towpathers can now put their best foot forward with the hope of reaching the Hartley Arms at Wheaton Aston for lunch (about 7½ miles). The morning will have taken in the splendid Avenue Bridge near Brewood (pronounced brood) as well as the Stretton aqueduct which takes the canal over the A5. The pub is close to the canal on the left-hand side not long after Wheaton Aston lock, the first after the Autherley stop lock. After 13 miles, you reach Gnosall Heath (turn off at bridge 35 when you reach The Navigation) which is a convenient place to aim for the first night. You will however have to take a bus into Newport for accommodation. Check with the bus company for times (0785 74231). Accommodation in Newport includes the Bridge Inn (0952 811785) and Norwood House (0952 825896).

Day two follows the course of the main walk and then goes beyond to the Wharf Inn at Shebdon. The afternoon takes in the peculiarly misplaced factory at Knighton (an ex-Cadbury's milk plant) and then the chasm-like Woodseaves cutting: 100 ft deep and yet another crumbly problem for the hapless Provis and the long-suffering British Waterways. After the five locks at Tyrley, the sizeable Market Drayton Wharf follows within 1 mile. Turn left at the bridge which carries a major road into the town centre (about a ½ mile). Market Drayton accommodation includes the Corbet Arms (0630 2037). You have now done 26½ miles.

The morning of day three is a time for locks. A steady down hill walk should pass seventeen of them before reaching the Bridge Inn next to the canal at Audlem (32½ miles). The final afternoon crosses the Weaver Aqueduct on through the Cheshire countryside to an aqueduct which goes over the A51. Here turn right to take the main road into Nantwich. The town centre is about a ¾ mile from the canal and proffers trains to Crewe and Shrewsbury (enquiries to 0270 255245) as well as a range of buses (0270 212256).

Good luck.

For those with less ambition, a short (4½ mile) walk from Tyrley is worthy of a warm summer's day. Tyrley is signposted along a minor road off the A529 just to the south of Market Drayton. There is a small canalside car park next to the bridge at Tyrley Wharf. Here are some fine Telford-designed buildings (including a lock-keeper's house, a warehouse and other cottages), all of which date from the late 1830s. Here also is a series of five locks (four to the Market Drayton/northern side), the first going north for about 17 miles. Turn right to take the right hand bank past the top lock. Within a short distance, the canal starts to become enclosed by the magnificent Woodseaves cutting. A mile long and, in places, nearly 100 ft deep, Woodseaves typifies Telford's modern approach to canal building. As at Grub Street, landslides were the plague of contractor William Provis' life and still are a problem for British Waterways. A report to BW, made in the autumn of 1990, confirmed that there were continuing stability problems caused by water draining from the sandstone bedrock into the alternating layers of mudstone. This leads to the mudstone slipping and causes the walls of the cutting to sag and collapse. At the time of writing BW are contemplating a major (and expensive) project to secure the walls and prevent any further landslips. From a purely selfish point of view, the towpath through the cutting can also suffer and it's certainly the only one in this book that sports a healthy growth of duckweed.

As at Grub Street, part of the magnificence of the cutting are the bridges and the third bridge from Tyrley, known as High Bridge for fairly obvious reasons, is the best of all. This must be the most outstanding accommodation bridge (a bridge built simply to enable a farmer to pass from one part of his land to the other) in the country. After passing under the next and last bridge in the cutting, the view opens out and the towpather reaches Goldstone Bridge and Wharf where the Wharf Tavern offers a canalside garden, bar foods and drinks: all to be enjoyed before the gentle stroll back to Tyrley.

Further Information

The Shropshire Union together with the Llangollen Canal and the Montgomery Canal are all very ably supported by members of the Shropshire Union Canal Society Ltd. Their address is:

Membership Secretary,
'Oak Haven',
Longdon-on-Tern,
Telford,
TF6 6LJ.

The society is currently engaged in restoring the Montgomery Canal and is investigating the possibility of reopening the Newport branch. All these issues are discussed and news updated in the society's magazine *Cuttings*. New members are, of course, always made welcome.

An excellent history and guide to the Shropshire Union Canal as a whole is:
Pellow, Thomas and Bowen, Paul, *The Shroppie*. The Landscape Press, 1985.

The history of the B & LJ, along with the rest of the canals that make up the Shropshire Union, can be traced through the chapters of:
Hadfield, Charles, *The Canals of the West Midlands*. David & Charles, 1969.

If a fan of Tom Rolt's highly readable style, then a good account of the trials and tribulations of Thomas Telford's last great venture can be found in:
Rolt, L.T.C., *Thomas Telford*. Penguin Books, 1958.

Those considering the long-distance route might find the following of interest:
Morris, Jonathon, *The Shropshire Union. A towpath guide to the Birmingham & Liverpool Junction Canal from Autherley to Nantwich*. Management Update Ltd, 1991.

3
THE CALDON CANAL

Leek to Cheddleton

Introduction

It's hard to believe that the quiet woodlands around Cheddleton are so close to the heaving civilization of the Potteries. You can almost hear the boat crews, who have struggled up from Etruria, sigh with relief as they reach Hazlehurst Junction. There they enter an altogether different world. It's not that the area is without its industrial history, there are remnants of a busy past all the way from Cheddleton to Froghall, but, from Hazelhurst on, nature has regained its rightful place. This is a countryside of rolling hills and abundant wildlife that belies its proximity to the five towns.

The Caldon's $17\frac{1}{2}$ miles start at Etruria where the branch leaves the summit pound of the Trent & Mersey Canal from a spot close to Etruscan Mill and Stoke Top Lock. After winding around the town of Hanley, with its working and historic potteries, the canal reaches the villages of Foxley and Milton and then Engine Lock, so-called because it was formerly the site of a beam engine used to drain the local colliery. A little further on, the feeder from Knypersley reservoir enters and a mile or so further on boats pass through the five Stockton Brook Locks into the Caldon summit pound. At the top level the line passes Endon Basin, now used as a mooring place by the Stoke-on-Trent Boat Club, and skirts around the village of Endon to reach Park Lane Wharf and the Hazlehurst Junction. Here the Leek branch turns right to cross the main line a little further on by means of Denford (Hazlehurst) Aqueduct. The branch then takes off to the north-east to ride high on the hillside and through Leek Tunnel, only to come to an end about a $\frac{1}{2}$ mile outside Leek. Meanwhile, the main line descends by three locks to go under Denford Aqueduct and on to Cheddleton with its fascinating Flint Mill Museum. From Cheddleton the canal makes its way into increasingly remote country. After Oakmeadow Ford Lock, the line joins the River Churnet for about 1 mile. River and canal separate again at Consall Forge, once a busy industrial centre, and move on to Froghall. The canal goes through a short tunnel and past the former entrance of the Uttoxeter Canal to come to a complete stop at Froghall Basin. This once frenetic wharf with its historic lime kilns is now, perhaps rather ignominiously, a picnic site.

Although the Caldon has its origins as a highway for the quarries of the area, it now has a wonderfully tranquil and relaxing atmosphere. Use it as the natural antidote to the Trent & Mersey.

History

Even before the Trent & Mersey Canal had fully opened, the possibility of a branch line from Stoke to Leek was being considered. Indeed it was the Caldon Canal that finally laid James Brindley to rest. In September 1772 Brindley had been surveying the route of the proposed canal at Ipstones, between Leek and Froghall, when he had been soaked to the skin during a thorough downpour. The oft-told story now relates that he then slept in those still wet clothes, caught a chill and finally succumbed on 27 September 1772. With him, for a while at least, went any thoughts about the proposed branch. However, one plan that was circulating at the time was for a very narrow canal that used five ton tub boats in place of narrowboats and inclined planes instead of locks.

It wasn't until November 1775 that Josiah Wedgwood revived the idea and then it wasn't Leek that was his target. This time, it was the potentially lucrative Caldon Low limestone quarries that attracted the entrepreneurial eye. The quarries, owned by the Earl of Shrewsbury, were a few miles east of Froghall (OS Landranger map 119, ref: SK 080485) and, according to John Sparrow, clerk to the Trent & Mersey company at the time, they had an 'inexhaustible fund of Limestone'. Hugh Henshall, Brindley's brother-in-law, surveyed the line and prepared an estimate. The branch was to run 19¼ miles from Etruria via Hanley and Norton to Cheddleton from where 'a railed way' was to be made for the carriage of coal, stone and other goods from the canal to a place called 'Sharpecliffe'. Another length of canal then ran from Sharpecliffe to the Caldon Low quarries. The proposed line would cost exactly £23,126.

Although the quarries offered the potential for more traffic along the line, from some accounts, the T & M's main interest in developing the Caldon Canal was in gaining access to additional water supplies for the main-line summit pound at Etruria. The need for water, described as insatiable by some authors, was stressed by John Sparrow, who clearly saw the Caldon as being able to produce 'a plentiful supply'.

On 12 February 1776, the T & M submitted a petition to Parliament to build the Caldon branch. There were, of course, objections. The millowners at Cheddleton feared loss of water to their works and filed a counter petition

Froghall Wharf was the terminus of the tramways from the Caldon Low limestone quarries and where, as this 1905 photograph shows, narrowboats were loaded for subsequent transport along the canal

Dr J.R. Hollick, The Boat Museum archive

against the proposed line. The proprietors of the T & M were clearly not the kind that were ruffled by challenges of this nature. Their response, on 18 April 1776, was to expand their original proposal with a continuation of the line through to Froghall via Consall. From there, they said, rail lines would go up to the quarries at Caldon Low. As the vast majority of the landowners were in favour of the line, the Act was passed without too much fuss on 13 May 1776. Funds of £25,000 were authorized, a sum which was raised through various loans taken on the credit of the tolls. Three colliery owners from Froghall alone contributed some £5,000. The tolls on the new branch were to be 1½d. per ton per mile for coal, stone, timber and other goods.

In order to guarantee the use of the canal, the company signed contracts with the quarry owners many of whom were already involved with the T & M. The contracts stipulated the amount of stone, gave the right to the T & M to operate the quarries itself if insufficient was delivered, detailed the tolls, and committed the T & M to building branch railways from the canal to the quarries.

There appears to be some confusion about when the Caldon branch opened to traffic but tolls are recorded as being received from Christmas 1778 onwards. The completed canal was just over 17 miles long and cost approximately £23,560. Originally there were fifteen locks with eight uphill and seven down.

This number was later increased to seventeen. The first horse-drawn railway to the quarries from Froghall was opened in about 1779. It started from Froghall terminus (on what is now the Cheddleton side of Froghall Tunnel) by the original kilns on the north side and ran up Shirley Hollow along the Shirley Brook to Shirley Common. This tramroad clearly wasn't the most reliable available and was described as being 'laid and placed in a very inconvenient course and direction' and had 'not answered all the good ends and purposes thereby intended'. As a result additional funds were raised in 1783 to improve matters. As part of these works, the canal at Froghall was extended by 530 yd through a 76 yd tunnel to a new basin. The new railway ran about 3 miles up Harston Wood and then turned right beyond the Harston Rock and over Cotton Common to the quarries. This new work cost £2,671.

By March 1785, the T & M had further developed its Caldon branch by building warehouses, reservoirs for yet more water (Stanley, Knypersley and Bagnall) and new railway lines at a cost of £6,000. The route of the railway was further revised in 1802, this time with John Rennie as engineer. The new line was a double track of flanged plateway that was spiked to stone blocks. It rose up the 649 ft of hillside on five inclined planes using horse power: one horse to twelve wagons.

In March 1797, the company decided to realize its original ambition and obtained an Act to make a branch from Endon to Leek. The prime rationale was not in the likely traffic that this would generate but was related to the additional supply of water from a proposed feeder reservoir at Rudyard Lake (2 miles north-west of Leek). The plan was initially obstructed by the Chester Canal Company who feared, apparently, that the T & M was aiming to 'lay hold of all the Supplies of Water in the Country'. The Leek branch, which had been surveyed by Hugh Henshall but was engineered by John Rennie, was 2¾ miles long and forced an alteration to the summit of the Caldon branch. The top pound was originally between the Stockton Locks and Endon (Park Lane). In the new layout, the locks at Endon were removed and the summit extended along a wholly new cut to Hazlehurst where the junction with the Leek branch was constructed. The new branch was fully open in 1802.

It was also in 1797 that the company obtained an Act to continue its line beyond Froghall to Uttoxeter, a distance of 13¼ miles. It is said that the main purpose for building the extension was to defeat the projected Commercial Canal, a proposal for a broad navigation from the Mersey along the Grand Junction to London. This direct line would have been a major competitor to the T & M and the company was keen to undermine the new venture wherever possible. Although there may have been this covert *raison d'être* behind the proposal, the Uttoxeter Canal was still seen as having plenty of scope for new business with the main trade coming from the collieries at Cheadle and Kingsley Moor and the copper and brass works at Oakamoor and Alton.

The project met immediate problems when the Earl of Shrewsbury and the

Cheadle Brass Company objected to possible loss of water to their wire mill. Because of this the work was delayed while an answer to the problem was sought. After toying with the idea of building a railway instead, the company decided to start on the new canal while the negotiations continued. This it did in 1807 with John Rennie as engineer. By August 1808, the line reached Oakamoor where work stopped while the agreement on how to proceed past Alton Wire Mill was finalized. With a decision to build a new weir and pond for the mill, work was restarted and the line finally completed on 3 September 1811. The opening of the canal was received in Uttoxeter with a 'great demonstration of joy by the inhabitants'. Guests received a 'cold collation' as well as speeches, toasts and songs, including one specially written for the occasion. The 'poor' received the benefits of two roast sheep and a quantity of bread and ale.

The Uttoxeter extension had seventeen locks, a 40 yd long tunnel at Alton (near the now (in)famous towers) and an iron aqueduct over the River Tean. Such was the enthusiasm engendered by the new line, that imaginative plans were brought forward including one that continued the canal to Burton-on-Trent.

In 1836, the company obtained an Act which allowed it to widen and straighten a section of canal between Oakmeadow Lock and Flint Mill Lock near Consall Forge. Further improvements to both the canal and the tramways were made during the next five years, including a major rebuilding of the Hazlehurst Junction which improved traffic flow considerably. But in 1846 the North Staffordshire Railway (NSR) took over the Trent & Mersey Canal Company, a move that would normally be considered to be the middle of the end. In one way it was. The Uttoxeter extension, which had fulfilled its purpose of defeating the Commercial Canal, was losing about £1,000 p.a. and was promptly closed. The Churnet Valley Railway line (seen as one section of a route from Manchester to London) was built partly on its bed. However, the commitment of the railway company to the rest of the Caldon Canal was as positive as ever. In 1849 the NSR installed a new $3\frac{1}{2}$ ft gauge cable-drawn railway from Froghall to the Caldon limestone quarries and, by the 1860s, a recently discovered band of haematite was producing 400,000 tons of iron ore p.a., about half of which went down the canal.

Such activity was not to last forever. Within a further twenty years, the original quarries were beginning to be worked out and new areas were being opened away from the tramroad and towards Waterhouses (half way between Leek and Ashbourne). It was now far more convenient to move the main quarry terminus further towards Stoke. Endon Basin, a $\frac{3}{4}$ mile from Stockton Locks, was therefore built as a transfer point for limestone brought over from the quarries by rail. By July 1905 a rail link from the new Leek-Waterhouses line was opened to serve the new workings and there was a substantial reduction in canal traffic. In 1920 a series of events effectively finished the line off: the Brunner Mond works at Sandbach, one of the principal users of canal-borne limestone, was closed; a serious landslip occurred on the canal near Froghall; and, on 25 March 1920, the cable line was shut. Transhipment at Endon Basin finally stopped about 1930 and from then on the Caldon Canal was hardly ever used.

The last of the coal trade to Leek ended in 1934 and, in 1939, the once active tar business between Leek and Milton also stopped. In 1944 the Leek branch was officially abandoned as it was deemed 'unnecessary'. After the Second World War, most of the other traffic subsided and in 1951, when the shipment of coal from Park Lane to Brittain's Paper Mills at Cheddleton stopped, the canal had no commercial traffic at all.

Following nationalization on 1 January 1948, boats continued to use the section to Hazlehurst, although the lines to Froghall and Leek were unnavigable. The canal was dredged in 1955 in an attempt to restart commercial coal-carrying to Froghall. When this failed, the future of the line looked bleak and a closure notice was duly posted at Etruria.

The Caldon Canal seemed doomed in 1961 but, in July, it was visited by the Inland Waterways Redevelopment Committee, a body set up to advise the Minister of Transport on disused canals. The scene of dereliction on the Caldon must have presented a challenge to the committee members who tried to walk along the overgrown towpath. Fortunately they recognized that the canal was fulfilling a vital function in supplying the T & M with water from its three active reservoirs and felt that it would be better if the line could be restored for navigation. A group of devotees met in 1963 and formed a committee to work for the restoration of the Caldon Canal. They brought together interested bodies, including the Staffordshire County Council and the Stoke-on-Trent City Council, and, on 4 April 1963, the Caldon Canal Committee became the Caldon Canal Society.

The rate of progress of restoration was inevitably slow, a situation that wasn't helped when the line was categorized as 'remainder' and not 'cruiseway' in the Transport Act of 1968. This categorisation gave the canal no permanent status in the eyes of the government and once again the entire future of the line depended on its role as a feeder for the T & M. But, through the actions of the Caldon Canal Society, on 22 August 1972 a two-year restoration programme was instigated when a new gate was installed at Engine Lock. This was followed by substantial renovation and dredging efforts from British Waterways. Cheddleton Top Lock and Waterworks Lock (at Stockton Brook) were completely rebuilt and major repairs were carried out to many of the other fourteen. A piped section near Froghall, that had been built in the early 1960s after some major landslips, was reopened for navigation as a concrete-lined channel.

All these efforts were rewarded on 28 September 1974 when the canal (including the Leek branch) was reopened for navigation as far as Froghall. The opening ceremony was performed at Cheddleton Top Lock. In 1983, the line was upgraded to cruiseway status and is today a highly popular and relaxing boating route.

Sir Frank Price lowers the lock gate at Engine Lock to mark the start of the restoration work on the Caldon Canal in 1972. The canal was fully restored and reopened on 28 September 1974

British Waterways

The Walk

Start:	Leek (OS ref: SJ 986564)
Finish:	Cheddleton (OS ref: SJ 973526)
Distance:	6¼ miles/10 km
Maps:	OS Landranger 118 (Stoke-on-Trent & Macclesfield) and 119 (Buxton, Matlock & Dovedale)
Return:	PMT buses run frequently between Cheddleton and Leek but there is only one service on Sundays (No. 76). Telephone: (0782) 223344
Car park:	Several signposted in town centre
Public transport:	PMT buses to Leek from Stoke which is itself on the BR main line. Telephone: (0782) 223344

From the centre of Leek, take the Stoke road (A53) for about a ¼ mile to the Safeways superstore (situated on the site of the old railway station) and The

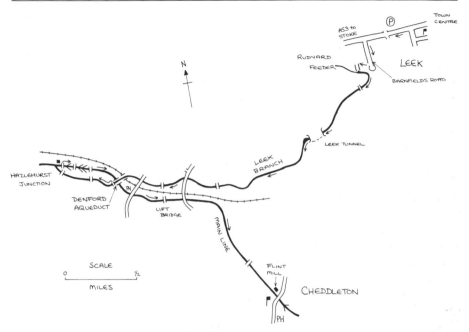

The Caldon Canal

Churnet Valley pub. Just after the pub is the site of the former Leek Wharf which, following the closure of the Leek branch of the canal, was purchased by the Leek Urban District Council in 1957 and converted into a trading estate. To reach the canal, turn left into Barnfields Road. Walk to the end of the road (not as far as it seems) to turn right along a narrow passageway between a wooden and a wire fence.

The path soon reaches a bridge that was in fact once the aqueduct which carried the line over the River Churnet and into Leek. Coming in from the right is what appears to be a small stream but is in fact the artificial feeder coming from the Rudyard reservoir some 3 miles to the north. The reservoir was an important source of water for the Trent & Mersey Canal and one of the major reasons for the construction of the Leek branch. It is said that Mr and Mrs Kipling named their son after a visit to the reservoir.

Turn left to follow the towpath around a bend to a winding hole (a last opportunity for boats to turn round) and a fine sandstone bridge (West Bridge). Already the scenery has forgotten the grubby trading estate and there are splendid views to the left and fields full of sheep to the right. The path follows a course around the contours to reach Leek Tunnel, 130 yd through solid sandstone. As there is no towpath through the tunnel, the path goes up over the hill. The effort involved is rewarded with a good view down to a small lagoon on the other side and over to a tall Gothic tower on the hill to the left, part of a hospital.

Continue down to the canal, which immediately bends left to go along a broad grassy track through a wooded stretch with mature trees and sandstone outcrops on the right-hand bank and fine views to the left. About a ¾ mile after the tunnel, the path goes under Waterworks Bridge and over a small wooden bridge which crosses a weir. In drier times the bridge doubles as a pleasant seat. A feeder on the far side of the weir appears miraculously to produce water from nowhere.

The canal now bends right with the Churnet Valley far down to the left. The view includes a railway line (now freight only) and, on the far hill, the Caldon Canal main line on its way to Cheddleton. Just before the next bridge (Horse Bridge), there is a stand full of stop planks: lengths of wood which slide down a groove cut into the sides of the canal to form a kind of temporary dam. Stop planks are used to isolate a stretch of the canal when there has either been a breach in the banks or when some maintenance is needed on the canal bed. The grooves for this set of planks can be seen just on the other side of the bridge.

The canal now bends left with a winding hole to the right. On the left is a splendid weir 'shoot' taking overflow water dramatically down to the River Churnet. The canal now winds past a pleasant group of houses and under two bridges to pass a British Waterways' yard. There are now two aqueducts in quick succession. The first, an innocuous affair which announces itself with a narrowing of the channel, passes over the railway line. There is then a short stretch before the second, and finer, structure – the Denford or Hazlehurst Aqueduct. This offers the fascinating sight of a canal over canal 'flyover', as the Leek branch goes over the Cheddleton main line. Here is the first part of the Hazlehurst Junction.

The Leek branch, and hence this junction, was added eighteen years after the main line to Froghall was built and it necessitated a major revision of this part of the canal. The original line, as it existed in 1797, had a 1½ mile top pound between Stockton Brook Locks and Park Lane (about a mile and a half west from here). It then fell by three locks to Cheddleton. This original route can still be seen today because it roughly follows the course now taken by the railway. To accommodate the branch and, more importantly, the feeder water from the Rudyard reservoir, the company had to extend the summit pound and move the locks closer to Cheddleton so that the water was fed into the top level. This was done by building a new, higher cut to the south of the old line thereby allowing the summit pound to stretch all the way from Stockton to here at Denford. The Leek branch joined the main line at a point just on the Stockton side of a staircase of three locks. That point is just here where the canal turns abruptly right immediately after Denford Aqueduct. At that time, this stretch with the two aqueducts was simply an embankment which took the Leek branch across the valley to join with the main line. To go to Cheddleton or Froghall, boats turned left at this point to go through the staircase of three locks and down to the lower canal level.

This arrangement proved to be unsatisfactory, presumably because the staircase was causing too many delays. So, in 1841, the present design was devised. Part of the old line was reopened by cutting through the embankment to form the Denford Aqueduct. This new cut then continued to join the Leek branch at Hazlehurst Junction. The staircase locks were closed and that cut filled in. Three new locks were then built along the reopened stretch. What was a junction became a sudden right turn and what was an embankment became an aqueduct.

It is possible to go down the steps alongside the aqueduct to join the main line directly. However, it is more enjoyable to cross the canal at the footbridge and to continue along the Leek branch with the towpath now on the right-hand bank. There are good views of the main line, with its three locks, to the right. The canal bends round left to go under a bridge and then back right to another. Here is Hazlehurst Junction with its lock-keeper's cottage and fine cast-iron footbridge. The line goes on for a further 9½ miles to Etruria where the canal has its junction with the Trent & Mersey. Our walk, however, turns nearly 180° right to pass the first Hazlehurst Lock (lock 10) with its cantilever bridge. The next two locks follow in quick succession, lock 11 having a neat brick-built hut.

To the left is the railway line which follows the original course of the canal

The first of the Hazlehurst Locks on the Caldon Canal

before the Leek branch was added. The point where the new route deviates can be seen where the canal bends suddenly right to pass under Denford (Hazlehurst) Aqueduct, a pleasant, unpretentious structure. Shortly after, the canal again bends suddenly, this time to the left. In the 1797 layout, the staircase locks were situated here to the right. We, meanwhile, are finally on the original course and pass the Hollybush at Denford pub.

The canal now borders the Deep Hayes Country Park to the right and passes under a bridge and then alongside Denford Lift Bridge. To the left, the Leek branch can be seen up on the side of the hill. After a fairly new road bridge (Wall Grange Bridge) and an old railway branch, a milepost reminds us that the canal once went to Uttoxeter. The towpath now passes over a wooden footbridge across a weir, similar to the one met on the Leek branch, and then proceeds gently to Cheddleton Wharf.

Here on the left is the site of the Cheddleton flint-grinding mill. Powdered flint was once used to whiten pottery. The first stage of the process involves 'calcining', a process in which the flints are heated to make them more brittle. With splendid efficiency, the kilns to do this have been built as part of the wharf. The flints were unloaded from the canal (the crane is still in position) into the tops of the kilns and then removed, after heating, at the bottom. The mills were operational by 1815 although there is evidence that the process was established here before then. Following their demise, the mills were restored by the Cheddleton Flint Mill Preservation Trust in the 1960s. It is said that James Brindley, who started as a millwright, may well have designed parts of these mills which are powered by undershot wheels. There is a small display here and the mill is open at weekends.

The towpath continues around some buildings to Cheddleton Bridge and a main road. Just the other side of the bridge are the two Cheddleton Locks which continue the descent of the canal to Froghall. At Cheddleton Top Lock, there is a plaque which commemorates the reopening of the canal in 1974.

Cheddleton has a small shop, a post office and some pubs: Old Flintlock, The Red Lion and The Black Lion. Buses for the return journey to Leek leave from the bus stop to the right on the mill side of the road.

Further Explorations

The entire length of the Caldon Canal is open to walkers: $17\frac{1}{2}$ miles from Etruria to Froghall plus $3\frac{3}{4}$ miles from Hazlehurst to Leek.

A fine walk can be taken from Cheddleton to the end of the line at Froghall (OS Landranger map 119, ref: SK 027477): a distance of 6 miles. The only problem is that there is no ready transport between the two villages. However,

A typical Trent & Mersey Canal Company milepost on the Caldon Canal near Cheddington (note that this one was actually erected by the canal society in 1981)

those willing to walk both ways, or who have access to two cars, can enjoy some superb scenery. Cars can be parked at the picnic site at Froghall Wharf, a spot which is worth visiting even if you don't intend to walk anywhere. There are some fascinating old limekilns, the point where the Uttoxeter Canal formerly connected, a tunnel and, on summer Sunday afternoons, a horse-drawn boat which rides up and down the canal. The kilns obtained their supply of limestone via a horse tramroad and inclined plane from Caldon Low quarries 3 miles away. The course of the tramway runs straight on from the canal, running to the right if looking from the lime kilns. Lime burning continued here into the 1930s, although by then the limestone was transported down from Caldon Low by lorry.

The Uttoxeter Canal was opened in 1811 and closed in 1847. It ran for 13½ miles. If you wish to trace it from Froghall, go from the wharf car park across the small road to the Caldon. A clear expansion of the waterway on the left-hand bank marks where the line started. Although long, long gone, it is still possible with the aid of an OS map to follow the course of the canal: through Oakamoor, past Alton Towers, through Denstone and into Uttoxeter.

It is 12¼ miles from Etruria to Leek and return journeys courtesy of PMT buses (albeit from Stoke) are frequent. From Etruria Junction (OS Land-ranger map 118, ref: SJ 872469), the route twists and turns through the outskirts of Hanley (with its numerous bottle kilns) and Hanley Park. The line soon fights its way to Foxley where the boats have to go through a kind of canal chicane before heading off towards Stockton Brook with its five locks. After Endon, you reach Hazlehurst Junction where you should take the right-hand course over the aqueduct for Leek.

At the Leek end of the canal, there is a popular walk of 3 miles each way along the feeder up to Rudyard reservoir. To reach the feeder follow the initial instructions for the main Leek to Cheddleton walk. The lake is very pleasant and highly frequented by fishermen.

Further Information

Although normally considered to be a branch of the Trent & Mersey, the Caldon has its own society:

Caldon Canal Society,
c/o W.G. Myatt,
Long Barrow,
Butterton,
Newcastle,
Staffs,
ST5 4EB.

The history of the Caldon Canal can be found within the pages of:
Lindsay, Jean, *The Trent & Mersey Canal*. David & Charles, 1979.
Hadfield, Charles, *The Canals of the West Midlands*. David & Charles, 1969.

However, most readers will probably find all they need in:
Lead, P., *The Caldon Canal and Tramroads*. The Oakwood Press, 1990.

4
THE CROMFORD CANAL
Ambergate to Cromford

Introduction

The village of Cromford, situated on the edge of the Peak District National Park just south of Matlock, is primarily known for its association with Sir Richard Arkwright and the birth of the factory system. Arkwright, who had developed the first successful water-powered cotton spinning mill, built his factory in the village away from the machine breakers and in an area where water power was both abundant and accessible. His only problem was transportation into and out of what was, at the time, a highly remote part of the country. The only way available to him was a hard and costly packhorse or wagon journey from the Lancashire ports. It comes as no surprise therefore to find Arkwright as one of the key protagonists in the construction of a canal to link his growing industrial village with the outside world. Although the collieries and quarries of the area produced more cargo for the line and although the great man died before the canal was opened, Arkwright's name will always be associated with this fine waterway.

The Cromford Canal is now only partly in water and large stretches are barely detectable on the ground. In its operational days, it started at a junction with the Erewash Canal, a line that runs from the River Trent near Long Eaton, through Ilkeston to Langley Mill near Eastwood. From Langley Mill, the Cromford Canal runs up the valley to Ironville, from where the Pinxton branch, of just under 2½ miles, once ran to the north. Here the main line turns west to enter Butterley Tunnel at Golden Valley. It re-emerges, 3,063 yd later, near Bullbridge, where the navigation once crossed both the River Amber and the railway line via an aqueduct. At Ambergate, the line follows Derwent Valley through Crich Chase and Whatstandwell to Lea Wood, where a short, private branch once ran to the north. The final 1½ miles of the canal begin where the line passes over the Derwent on Wigwell Aqueduct before continuing on to Cromford Wharf.

The northern section of the Cromford passes through some fine countryside which would have made good walking territory even if the canal had never been built. If one adds some fascinating industrial archaeology (apart from the canal there is a steam pumping station, the Cromford & High Peak Railway and, of course, Arkwright's Mill), then this walk makes for a splendid afternoon out.

History

The Erewash Canal runs for 11¾ miles from the River Trent to Langley Mill on the border between Derbyshire and Nottinghamshire near Heanor. Authorized by an Act of Parliament on 30 April 1777, the line was open for navigation just two years later, engineered by John Varley and built by John and James Pinkerton. The Erewash Canal was largely sponsored by local colliery owners and the line was an instant hit, with Derbyshire coal being shipped from the collieries around Eastwood, Ilkeston and Langley Mill to the River Trent at Long Eaton. From there, coal was moved both along the Trent and along the Soar Navigation into Leicestershire.

The success of the line soon prompted landowners further up Erewash Valley to consider an extension of the waterway to Pinxton, where a number of potential seams were said to be unworked because of the lack of transport. There was also considerable interest from the owners of the iron furnaces at Butterfly and Somercotes, of the limestone quarries at Crich, of the lead-works at Alderwasley and from Sir Richard Arkwright who had built his revolutionary cotton mills at Cromford. After some initial considerations, including the input of Arkwright and a meeting in Matlock, a line running from Langley Mill to Cromford, with a branch to Pinxton Mill, was proposed and widely agreed. At a meeting in Alfreton in December 1788, William Jessop presented his plan for the Cromford Canal together with an estimated cost of £42,697. Interest was such that half the necessary sum was raised at the meeting with the other half coming during the following two weeks. Peculiarly, the Erewash Canal Company contested the Cromford Act of 31 July 1789. Its objection was not to the canal *per se* but concerned its claimed exclusive rights to the water supplies from the River Erewash.

The line authorized was just over 14½ miles long and had fourteen locks; all of which were south of Butterley Tunnel. The Act enabled the company to raise capital of £46,000. With Arkwright as company chairman, William Jessop was appointed engineer and Benjamin Outram and Thomas Dadford his assistants. Thomas Kearsley of Tamworth and Thomas Roundford of Coleridge, Staffordshire, were given the contracts for the work.

Not long after the construction work had started, it was realized that the project was going to run over budget, a situation that was not helped when the contractors absconded forcing the engineers to take direct control. By September 1791, all the initial monies had been spent and additional funds were raised by calls and loans. The year 1791 was also marked by the decision to proceed with the Nottingham Canal which would run from the Cromford, at a point just above Langley Mill, through Nottingham to the Trent. The supply of water along the line was a particular concern to the Cromford Company. This problem was originally overcome for the Cromford itself when Jessop had decided to dig an especially deep summit pound to act as a kind of *in situ* reservoir for water coming from the Cromford sough. If the Nottingham Canal was to take additional water then the problem was sure to become more acute. As part of the agreement between the two companies the Cromford only guaranteed water to the Nottingham if that company built extra reservoir capacity for both.

In February 1792 the line below Butterfly Tunnel was opened for traffic and

Boatmen 'legging' their narrowboat through Butterley Tunnel in about 1895. All tunnels that lacked a towpath had to be passed in this way while the horse took a more scenic route over the hill. 'Legged boats' took nearly three hours to go through the 3,063 yd long tunnel; just over a $\frac{1}{2}$ mile per hour

British Waterways

receiving tolls. However, the 200 yd long and 50 ft high Amber Aqueduct near Ambergate was already in need of partial rebuilding. Jessop, who appears to have had a reputation for indifferent masonry work, accepted responsibility for the faults and paid the £650 needed to undertake the repairs. The situation repeated itself in 1792 when Wigwell Aqueduct over the Derwent was closed for some time as it too needed partial rebuilding. Again Jessop accepted responsibility, admitting that the failure was due to the lack of strength in the front walls. With the aqueducts repaired and the rest of the line, including the tunnel, ready, the line was fully opened in August 1794. The final cost was £78,880: twice the estimate.

This gross over-expenditure was soon forgotten as the canal was an almost instant success. In 1802–3, the line carried over 150,000 tons of cargo, of which some 110,000 tons were coal and coke and nearly 30,000 tons lime and limestone. The stone for St George's Hall in Liverpool, for example, was carried along this route. In 1802, the privately owned Lea Wood branch was opened providing canal access to a number of works and quarries in the Lea Wood area (south-east of Cromford). At about the same time plans were produced for a similar branch to Bakewell, although this line was never built.

By 1814 the company was doing well enough to be able to pay a dividend of 10 per cent, a sum that reached 18 per cent in 1830 and 28 per cent in 1841. By that time traffic along the Cromford Canal as a whole had almost doubled from the 1802–3 levels to nearly 290,000 tons. Almost twice as much coal and coke was now being carried and there had been substantial increases in the carriage of farm produce and other quarried materials such as ironstone and gritstone. Iron from the Butterley works also increased considerably during this period. Much of the coal traffic passed down the Cromford Canal and along the Erewash to be taken onto the Soar Navigation for sale in the Leicester area. Interestingly, the dramatic increases in stone were partially the result of the increased demand for railway ballast. Charles Hadfield reported that large quantities of Cromford stone were used on the London & Birmingham Railway.

The canal was connected to the collieries and quarries by means of tramways varying in length from a $\frac{1}{4}$ to $1\frac{1}{2}$ miles. The Crich limestone quarries, for example, had a tramway outlet to the canal at Bullbridge. This line was later taken across the canal to meet with the Midland Railway. At Riddings, a donkey-drawn tramway was in operation, while at Birchwood and Swanwick, a steam-driven system was used. At Butterley, the ironworks, which were producing cannon and shot for the Woolwich Arsenal, were connected to the canal by means of two shafts down which cargo was lowered on to barges in the tunnel below.

In addition to the freight traffic, from September 1797, a packet boat for passengers, run by Nathaniel Wheatcroft of Cromford, operated twice weekly between Cromford Wharf and Nottingham, a distance of 38 miles. First class travellers paid 5s. for this trip while second class passengers paid just 3s.

By far the most important of the railway developments occurred between the canal at Lea Wood and the Peak Forest Canal at Whaley Bridge. The link was originally conceived as a canal navigation. Indeed, John Rennie had carried out a preliminary survey and presented an estimate of £650,000. The line was thought by many, including the highly supportive Grand Junction Canal Company (now known as the Grand Union), to be a way of shortening the route from London to Manchester. Naturally the Trent & Mersey, whose business was most under threat, was antagonistic and it appears to have won the argument. But the concept stayed alive and it was not long before thoughts had switched to a railway line, now known as the Cromford & High Peak. Following an Act of 2 May 1825, Josias Jessop, son of William, was appointed engineer and work began. The line, which opened on 6 July 1831, included stationary engines to haul wagons up as many as nine inclines to the more conventional railways that ran along the summit. One of the inclines can be seen clearly from the A6 just south of Cromford. The line was an important conduit for various cargoes going north–south, including agricultural produce and even water. The company faced severe competition by the late 1840s from other railway lines but the route remained open until 1967. It is now part of a long-distance path: the High Peak Trail.

In the 1830s, competition from the Staffordshire collieries led the Cromford company to halve its tolls for cargoes of Derbyshire coal destined for the Grand Junction Canal. However, as this action had no effect on the level of traffic, the concession was withdrawn. But, in the 1840s, the more direct competition of the new railways was beginning to be realised and toll reductions were made in order simply to compete. By 1843 the effects of doing this were already being felt: toll revenue had decreased by a quarter in just two years and the company was forced to halve its annual dividend.

There was also a growing problem with water supply during the 1840s. The original main source, the Cromford sough, which ran from local lead mines through Arkwright's Mill and into the canal at Cromford Wharf, was stopped when the mines were worked at a lower level. The company were forced to build a pumping station near Wigwell Aqueduct to lift water from the Derwent up to the canal summit level.

Although there appears to have been some thought to cooperation with its canal neighbours, the Cromford Canal Company saw the inevitability of the railway age fairly early – in fact some time before the main competitive line, the Erewash Valley line, was even finished. The company sought and reached agreement with the Manchester, Buxton, Matlock & Midlands Junction Railway, the company which planned to build the line from Cheadle to Ambergate. The MBM & MJR bought the Cromford for £103,500, a deal which has since been seen as excellent salesmanship by the Cromford board. The sale was completed, after the passage of an Act, on 30 August 1852. Shortly thereafter the MBM & MJR leased the line jointly to the Midland Railway and

The 200 yd long Bullbridge Aqueduct in 1900 carried the canal over the River Amber, the railway and the road near Ambergate. After the canal closed, the aqueduct was demolished to make way for a road-widening scheme

The Boat Museum archive

the London and North Western. By 1870 it was a wholly owned part of the Midland.

As soon as the canal entered into railway hands, the level of traffic along the line declined. With the Midland Railway running alongside the waterway for its entire length, the canal was never in a good position to compete. From traffic of 300,000 tons in 1849, only 145,814 tons were carried in 1870 and 45,799 in 1888. These figures also disguise the fact that traffic was becoming increasingly local, with long-distance transport being taken almost entirely by rail.

The end of the canal as a going concern began in 1889 when subsidence in Butterley Tunnel forced temporary closure and the expenditure of £7,364 on repairs. By May 1893, the line was reopened but the traffic through the tunnel was down to less than 4,000 tons p.a. On 5 July 1900, a second collapse in the tunnel led to its final closure and the route was cut in half permanently. In 1904, Rudolph de Salis reported that the headroom was very low and the brickwork lining the tunnel was in a very dangerous condition. The whole issue was sealed with a third collapse in February 1907. The stone traffic from Whatstandwell to the south was transferred to the railway at no extra cost to the shippers and, in 1909, a Royal Commission pronounced the tunnel beyond economic repair.

Below the tunnel, traffic remained reasonably vigorous with some 39,000 tons being moved in 1905. North of the tunnel, a small amount of traffic continued with coal being shipped into Cromford and lead being taken from Lea Wood along the private arm to High Peak Wharf where it was transhipped on to rail. With the closure of the Lea Wood arm in 1936, the London, Midland & Scottish Railway (which had become the owners after the great railway amalgamations of 1923) announced their intention to close the canal on 13 March 1937. Although there were some objections, with the canal at one stage being offered free to the Grand Union Canal Company, all but the southernmost $\frac{1}{2}$ mile was officially abandoned in 1944. The last $\frac{1}{2}$ mile was finally closed in 1962.

Although the southern section of the Cromford has mostly disappeared into the undergrowth, in 1974 the Derbyshire County Council bought the northernmost $5\frac{1}{4}$ miles from Ambergate to Cromford for use as a public amenity. With the assistance of various voluntary groups, the last $1\frac{1}{2}$ miles of the canal has been made into a popular linear country park and is used extensively by towpath walkers and holiday makers.

The Walk

Start:	Ambergate station (OS ref: SK 348516)
Finish:	Cromford (OS ref: SK 299571)
Distance:	$5\frac{1}{2}$ miles/$8\frac{1}{2}$ km
Map:	OS Outdoor Leisure 24 (The White Peak)
Return:	Train Cromford to Ambergate. Also runs on Sundays. Telephone: (0332) 32051
Car parks:	At picnic site at Cromford end of canal (signposted from A6), at Ambergate station or at Whatstandwell station
Public transport:	BR to stations on Derby–Matlock line

As some of the most interesting features of the canal are towards the northern (Cromford) end of the line, this walk can be conveniently shortened to 3 miles by walking to Cromford from Whatstandwell station (OS ref: SK 333541).

For the full walk, leave Ambergate station car park to return to the main road (the A610 to Nottingham) and turn left to go under the railway bridge. Turn right along the A6 to pass the Little Chef roadside restaurant. After about 250 yd, turn right along Chase Road, following a footpath sign that promises Bullbridge and Fritchley. Go under the railway to a small bridge over the canal. Turn left to go along the towpath which is on the left-hand bank.

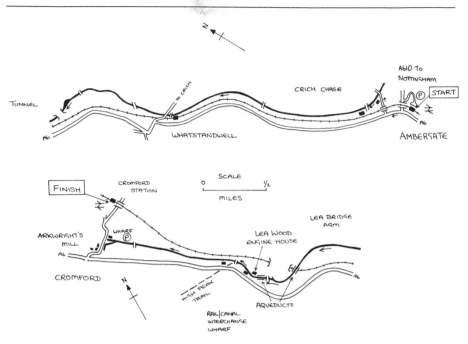

The Cromford Canal

The canal here is very overgrown with weeds suggesting that we have yet to reach the manicured, tourist section which is within easy reach of Cromford. The line soon goes under a bridge and we pass the first of several lengthman's cottages to be seen along the waterway. The job of lengthman must appear idyllic to many modern-day canal enthusiasts who have donned rose-tinted spectacles. The lengthman walked his allocated stretch of the line daily: opening sluice gates to release excess water, unblocking weirs, keeping a wary eye for potential breaches, maintaining the towpath, inspecting the culverts, bridges and tunnels. Keeping the water level up and ensuring that all the feeder channels were clear and operational were the particular concerns in summer. The lengthman would also have looked after the canal's boundaries and maintained the hedges and ditches. It was a poorly paid job and, without doubt, much tougher than our romantic back-glances might suggest. It must be said, however, that if you were going to be one then this was as good a place as any.

The canal is now on the side of a hill with both the railway and the road down the slope to the left. The right-hand side, meanwhile, is lined with a dense stand of alder trees. Just before the second cottage (used by the local St John's Ambulance group) is a small milepost announcing that we are still 4½ miles from Cromford Wharf. The woodland that now lines both sides of the canal is known as Crich Chase. In the early nineteenth century, the concern over poaching was such that a by-law was introduced which prohibited boatmen from mooring here overnight.

The way winds quietly and pleasantly onwards to Whatstandwell. Those joining the route here should leave the station and turn right to reach the Derwent Hotel. Turn right along a road signposted to Crich and the Tramway Museum. Within 100 yd, the road crosses the canal. Turn left to join the walk.

The canal runs along the base of a steep hill on top of which once stood Duke's quarries. At one point, the abutments of a tramway bridge, coming down from the Crich limestone quarries, still stand but now bear nothing but the weight of a small footbridge. The canal just here is stone-lined and is bordered by stop plank grooves, both suggestions of persistent breaches in the line. For a short distance the canal narrows so that the towpath appears wider than the waterway. It then passes a fine house before bending gradually to the left. The views to the left now show that the canal has retained the northern side of the valley whereas the river, railway and A6 have taken the southern side. Shortly, the canal goes under a bridge (with stop planks in position) and then broadens at Gregory Dam before entering a short tunnel with a convenient towpath that goes all the way through.

The line passes the Derwent Nature Reserve amid dense woodland before running around the contours of the hillside to reach an aqueduct which takes the canal over the railway. The Cromford now bends right around the hill to reach Lea Wood Junction. On the right is another derelict lengthman's cottage buried in the undergrowth and just beyond are the remains of the Lea Wood branch. This cut was built by Peter Nightingale, Florence's father, who shared the cost with the Cromford Canal Company. It was opened in early 1802 and ran just a $\frac{1}{2}$ mile to his own wharves at Lea Wood where there were two lead-works, cotton mills, a hat factory and a group of quarries. The Lea Wood cut was closed in March 1936 and is now dry. An iron aqueduct which carried the branch over the railway has sadly been removed.

Although it is possible to follow the Lea Wood branch, we stay on the main line. Cross the canal at the swing bridge and turn left to walk over Wigwell Aqueduct. Wigwell crosses the River Derwent on three arches, the centre one of which spans 80 ft. The aqueduct as a whole is 200 yd long and 30 ft high. Although it's solid enough now, it must have been a moving source of embarrassment to William Jessop. In September 1793, cracks appeared in the outer walls: a fault which Jessop blamed on the quality of the lime used in the mortar. Instead of setting, the cement stayed soft and allowed the masonry to bulge outwards. Jessop accepted responsibility for the fault and had the good grace to pay for the repairs out of his own pocket. The structure was made good by incorporating iron bars into the stonework in order to prevent the walls from crumbling. Given that Jessop was earning £350 p.a. as engineer and that he had already paid £650 for the repair work on Amber Aqueduct, his work on the Cromford can be seen to have been anything but a profitable exercise.

The Lea Wood pumping station and Wigwell Aqueduct

Continue along the path and up to the fine Lea Wood pumping house. The building was put up in 1849 to raise water from the River Derwent to the canal. This proved necessary when supplies of water from the Cromford sough feeder were restricted through changes in mining practice in Cromford. It is constructed of local gritstone. The chimney is some 95 ft high and has a rather curious splayed parapet which is made of cast iron. The building houses a Boulton & Watt type single-acting beam engine that was built by Graham & Co. The beam is nearly 33 ft long and the pump can raise between 20 and 30 tons of water per minute through a hidden tunnel system that passes underneath the building. The engine has been restored by the Cromford Canal Society and is periodically open to the public. At the time of writing, however, this situation is somewhat confused (see below).

The walk continues along the well-restored towpath to pass High Peak Wharf where a canal/railway interchange shed still stands with its awning overhanging the line. This was the terminus of the 33 mile long Cromford & High Peak Railway. From here the railway ran alongside the canal for a short distance before rising up an incline to a 12 mile long summit level at 1,271 ft. In the days of horse-drawn trains, the journey to the Peak Forest Canal at Whaley Bridge took over six hours. Although the prime traffic along the line was limestone, there was also a small number of passenger carriages. The line was opened fully on 6 July 1831 and closed as recently as 1967.

A little further on are various other buildings associated with the railway. The railway company had its workshops here including a servicing shed which has an inspection pit laid with original 4 ft fish-bellied rails lettered 'C & HP Railway'. There is also a charming (to look at rather than live in) row of railway workers' cottages.

If you cross the canal at the next swing bridge, you can visit High Peak Junction. The first inclined plane leads off from here and can be seen disappearing up the hill to the left. Originally this incline consisted of two separate planes with individual rises of 204 ft and 261 ft. They were combined in 1857 and were worked as one 1,320 yd long slope with a gradient of between 1:8 and 1:9. Until 1964, trucks were hauled up the hill by a steam engine that was housed in another fine engine house. This was replaced by an electric motor for the final three years of operation. The problem was that during the period when the new motor was installed, the prime user (Middleton Quarry) found the replacement road traffic simpler and cheaper to use. In working the system, trucks were attached to a wire rope to form trains in which ascending wagons were balanced by those descending. In the early days, runaway trucks hurtling uncontrollably downhill were a regular and highly dangerous problem. The situation was alleviated (if not cured) by the construction of special pits into which runaways were diverted so that they could smash themselves to pieces in relative safety.

At the junction there are a couple of restored trucks and, in the small workshop building, an information centre, a display of various artefacts, and a small shop. Around the corner are toilets and a picnic area.

Return across the swing bridge and turn left. It is now just a short walk, under a small accommodation bridge, to Cromford Wharf. This first announces itself on the canal line with a winding hole, where boats were able to turn, and to the right with a large, recently constructed, car park. On the top of the hill to the left is Sir Richard Arkwright's home which was appropriately called the Rock House. The archway near the winding hole is the boathouse for Rock House.

The entire area of the wharf was formerly Arkwright's garden and was bought from him by the canal company. It should be said that Arkwright himself had every intention of moving and was building himself a mansion just up the road. The wharf was opened in 1794 and, away from the canal to the right, there are various buildings which were originally (from right to left): a sawpit, a smithy, a

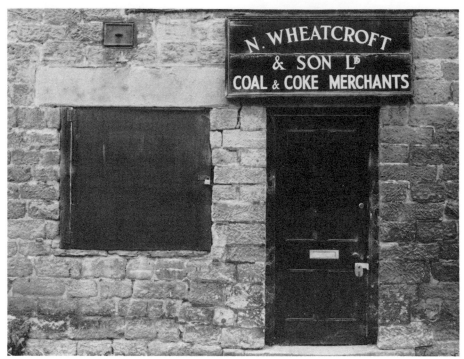

Nathaniel Wheatcroft's offices at Cromford Wharf

stable and some canal workers' cottages. The canal itself forks to two wharves which the canal company divided in function. The area opposite the winding hole was reserved for incoming coal shipments. The right-hand fork of the canal was used for incoming timber and other goods while the left-hand fork was used for outgoing cargo such as limestone and lead. The first canalside buildings were those on the far left-hand side: a warehouse and, beyond, a counting house or company office. These were both built in 1794. The warehouse with the covered area on the right was built in 1824 by Nathaniel Wheatcroft who at the time was a major canal carrier. The Wheatcroft family also acted as wharfingers and were at one time coal merchants. In the 1830s, German Wheatcroft & Sons operated passenger services to various parts of the country, including London and Bristol, three days a week. German also operated a passenger 'fly' up and down the Cromford & High Peak Railway. Nathaniel Wheatcroft operated services to Nottingham and the Potteries on Tuesdays and Fridays.

Continue past the right-hand warehouse and turn left to walk around the back of the wharf. Here it can be seen that the left-hand fork of the canal extends to the road from where it narrows considerably to bend left around the rocky outcrop that forms the foundations of Rock House. This is the feeder channel that carries water from Cromford sough via Arkwright's Mills which stand beyond.

Arkwright's Mills have now been taken over by the Arkwright Society whose intention is to restore the main part of the building to its original working state while letting the rest as office space for small companies. The mills are open to visitors and there is a small gift shop and café on site.

To return to Ambergate, turn right when leaving the wharf area to pass St Mary's parish church. This leads to a bridge which crosses the Derwent. Continue along the right-hand road to a point where a sign points left to Cromford station. This fine, sadly neglected station offers regular trains to Whatstandwell and Ambergate, even on Sundays!

Further Explorations

There's not a lot of Cromford Canal left to explore but it is worth having a quick look at Butterley Tunnel. The eastern portal can be reached from Golden Valley (OS Landranger map 120, ref: SK 423512). This point is on the Newlands Road between Codnor and Somercotes. At the Newlands Inn, take a path that goes down the left-hand side to reach the very overgrown canal. This area has been converted into a nature reserve and walk. After about 200 yd, the path rises slightly and those with penetrative sight should be able to make out the barred portal of the tunnel over to the right.

The tunnel is 3,063 yd long and was built narrow, in other words unsuitable for barges. The locks to the south were all wide enough to allow Erewash barges to serve the collieries up to the tunnel but not beyond. As there was no room for two boats to pass each other mid-tunnel, traffic through it was carefully regulated. Westward boats were allowed entry between 5 and 6 a.m. and 1 and 2 p.m., and those going east between 9 and 10 a.m. and 5 and 6 p.m. In 1802, when traffic had increased, the tunnel was worked at night and additional entry times were allowed at the east end between 9 and 10 p.m. and the west from 1 and 2 a.m. Three hours were allowed for passage which may seem a little excessive but was due to the fact that this was a 'legging' tunnel. Here men known as 'leggers' lay on their sides on the boat and literally walked their way through with their feet pressing against the walls.

The channel that comes in from the right of the tunnel is a feeder from Butterley reservoir. The stepped weir down to the canal is typical of situations where the supply reservoir was built at a level considerably above that of the waterway.

After returning to the Newlands Inn and crossing the road, a public footpath follows the course of the canal to Ironville and beyond. It is possible to follow paths to Langley Mill although the canal disappears long before reaching the Erewash.

Further Information

The Cromford Canal Society is not at present operational. Hopefully a restructured group will emerge to safeguard the interests of this wonderful waterway and, perhaps, even strive to rejoin it to the Erewash.

The history of the Cromford Canal, together with the Erewash and the Nottingham, is described in:
Hadfield, Charles, *The Canals of the East Midlands*. David & Charles, 1966.

A shortened history of the Cromford and other canals in Derbyshire can be found in:
Smith, M., *Derbyshire Canals*. J.H. Hall & Sons Ltd, 1987.

The Arkwright Society published a local history trail leaflet on the canal. This is available at the Arkwright Mill in Cromford.

Those tempted by the High Peak Trail along the Cromford & High Peak Railway should refer to a cousin of this volume:
Vinter, J., *Railway Walks: LMS*. Alan Sutton Publishing, 1990.

5
THE GRANTHAM CANAL

Bottesford to Grantham

Introduction

Grantham is now more popularly famous for its grocer's daughter than for its commercial activity but it was once an important market town for the rich farmlands of Lincolnshire. Its importance was heightened when the local entrepreneurs sponsored the construction of a canal for the import and export of coal and agricultural produce. Although the new waterway forged a tangible link with the industry and collierics around the city of Nottingham, the Lincolnshire countryside remained remote from industrial development, as it does to this day. The canal, now long-closed and virtually boat-free, retains a feeling of loneliness and isolation, which it probably always will, despite being subject to an enthusiastic restoration programme.

The 33 mile long Grantham Canal leaves the River Trent at West Bridgford within sight of Trent Bridge and the Nottingham Forest football ground. After going through the now closed Trent Lock, the canal soon becomes impassable as various new bridges and embankments block the original line effectively to seal this route forever. The situation remedies itself as the canal winds south and then east to pass to the north of Nottingham airport. Here the waterway is carried over Pulser Brook on a small aqueduct. If restoration plans are enacted, the canal will use the northerly course of this brook to take it to a new junction with the Trent at the Holme Pierrepoint water sports centre.

After passing to the north of the colliery village of Cotgrave, the line turns south to go past Cropwell Bishop to Hickling. Here the canal reverts to a north-easterly course as it meanders through the Vale of Belvoir to Redmile. Broad zigzags now take the canal near Bottesford and Woolsthorpe before passing Denton reservoir to come to an end at the A1 embankment. On the other side there remains just a short stretch to Earles Field Bridge beyond which the original line into Grantham Basin has been filled in and, perhaps appropriately given the location, privatized.

The Grantham has no dramatic scenery or startling architectural structures but it does offer a pleasant, peaceful walk through the arable countryside that typifies the Vale of Belvoir. It also has the additional interest of a canal in restoration, the seemingly impossible challenge that tests the ingenuity and tenacity of the most devout of enthusiasts. Walk it before it becomes popular.

History

The last decade of the eighteenth century, the period often referred to as the time of canal mania, saw considerable waterway construction and improvement both in and around the city of Nottingham. Most of this activity concerned the building of canals to take the products of collieries and quarries on to an ever more navigable River Trent and from there to markets in the south. Although it had no quarries the small town of Grantham, some 21 miles to the east of Nottingham, clearly wanted to be involved. The businessmen of Grantham, who were obliged to pay high rates for land transport of their goods to Newark-on-Trent or to the port of Boston, liked the idea of a canal to connect the town with the ever expanding waterway network. The appeal was in the potential for cheap carriage of their own goods and in the access to a plentiful and cheap supply of coal from the Derbyshire fields.

William Jessop, who had been busy in the area surveying the Nottingham Canal, was approached by the Grantham businessmen and asked to make a survey of a line from the Trent to Grantham. In a plan that was announced on 27 August 1791, Jessop proposed a contour canal leaving the River Trent at Radcliffe, a point to the east of Nottingham city centre. The enthusiasm generated for the line was such that all £40,000 of the estimated cost was raised at this single meeting. A bill based on Jessop's plan was presented to Parliament in 1792 but was rejected due to objections from local landowners and from those who feared loss of overland trade to Newark. Waterway-related opposition was also evident from those with interests in the prospective Grantham to Newark canal and from those who feared that the River Witham would lose water.

Jessop re-surveyed the junction with the Trent and devised a new line to West Bridgford, just 1 mile from the modern Nottingham city centre. There was also to be a $3\frac{1}{2}$ mile long branch to Bingham, a village about 8 miles due east of Nottingham. With less opposition, this Act was more successful and was passed on 30 April 1793. Capital of £75,000 was authorized with £30,000 in reserve if required. On top of the specified toll rates, an additional charge of $2\frac{1}{2}$d. per ton was permitted for all cargo entering or leaving the River Trent. The proprietors of the Trent Navigation were obliged by the Act to keep the Trent Lock entrance dredged to a minimum of 3 ft in order to guarantee access to the new line.

By 1910, hardly any cargo traffic used the canal and so pleasure boating and fishing became popular, as here at Earle's Field Bridge in Grantham

The Boat Museum archive

The canal was to be one of 33 miles, not including the branch to Bingham which seems to have been quietly dropped. Jessop's rather old fashioned contour canal meandered through the Vale of Belvoir passing small villages and gradually rising by about 140 ft to reach Grantham town centre. Although Grantham businessmen occupied most of the seats on the board, the principal shareholders were from Nottingham. Two engineers were appointed to build the line. James Green was asked to take charge of the line from the Trent to the Leicestershire boundary near Hickling. Green was employed by Lord Middleton at Wollaton Hall and came with the approval of Jessop with whom he had worked on the Nottingham Canal. For the rest of the line into Grantham, the board appointed William King, agent to the Duke of Rutland at Belvoir. The Duke owned that stretch and clearly must have had a say in his appointment. King was also made responsible for the construction of the two main reservoirs: the Denton and the Knipton, the latter of which was formed after damming the River Devon. Jessop was called in as consultant, presumably to keep a check on the otherwise comparatively inexperienced engineers.

By February 1797, King's eastern end was reported to be navigable and the first traffic probably passed along the route during the following April. There was some minor delay at the western end, caused by problems with the gypsum under rock near Cropwell Bishop. Despite this, the whole line was opened for traffic during the course of the summer. The main engineering feature was the long cutting through Harlaxton Drift, near Grantham. This was originally only able to take narrowboats and had two passing places so that traffic wasn't overly inconvenienced. The line here was widened later. The completed canal had cost £118,500, the extra funds having been raised by additional calls on shares and by obtaining loans. The line had eighteen broad locks able to take craft measuring 75 ft by 14 ft. The only tricky portion of the line was at the Trent entrance where to get into the Trent Lock (No. 1) a rope and capstan had to be used to haul boats round.

Almost immediately the canal company started its own trade in coal and coke. Others started shipping lime, building materials and groceries from Nottingham and various types of agricultural produce (corn, malt, beans and wool) back into town. Before too long Grantham was to become a port for the wide area of rich, Lincolnshire farmlands that surround it. By 1798, the canal was also carrying people, a passenger boat having been introduced between Nottingham and Cotgrave on Saturdays

In 1803, the company was able to pay its first dividend, albeit only of 2 per cent. But by 1815 a sizeable proportion of the debts had been repaid and the dividend had risen to 5 per cent with total toll revenue never exceeding £9,000 p.a. Traffic continued to grow and reached a peak in 1841 with toll receipts of £13,079. To maintain this relatively high revenue level the company kept its toll charges up despite the protestations of the canal users. At one stage the company was accused of maintaining too high a toll for coal traffic and was indicted for conspiracy to enhance prices. Such overcharging appears to have had the affect of making some traders seek alternative routes into and out of Grantham. One trader, for example, found it cheaper to import coal overland from Newark than along the canal. The Oakham and Melton Mowbray Navigations were also able, by reducing their own tolls, to capture a lot of the grain traffic from the Grantham line. Despite the accusations and the loss of traffic to competitors, the canal maintained its pricing policy and appears to have remained moderately successful.

This situation continued until 1845 when the company found itself confronted with railway competition and decided (along with the Nottingham Canal) that its best interest was to sell out to the potential rival, namely the Nottingham, Vale of Belvoir & Grantham Railway. This it duly did on the proviso that the railway company built and opened its Ambergate–Grantham line. Before the railway had got around to this, however, it had undergone a merger to become the Ambergate, Nottingham, Boston & Eastern Junction Railway.

The rail line from Nottingham to Grantham was opened on 15 July 1850, the rest being left for a more favourable economic climate. However, the planned take-over was held up by the railway company's unwillingness to part with any funds. The problem was partly related to a cash-flow problem and partly because the ANB & EJR had plans to link with the Great Northern Railway (which owned and operated the main line from London to Grantham). The GNR was known to be unwilling to waste funds on money-losing canals and the ANB & EJR must have been eager to show that it was of similar mind. The Grantham and Nottingham Canal Companies were thus forced to take legal action in order to compel the railway company to honour its agreement and pay up. This it eventually did on 20 December 1854. The company now became the Ambergate, Nottingham, Boston & Eastern Junction Railway and Canal Company. Three months later the ANB & EJR & C Co. obtained its longed for agreement with the Great Northern Railway Company who would now work the Grantham to Nottingham line. Having giving up all pretence of ever reaching Ambergate the company changed its name in 1860 to the Nottingham and Grantham Railway and Canal Company and a year later leased itself to the GNR for a period of 999 years.

The GNR now had canals that it didn't want and, indeed, canals that were in open competition with its own railway. As a result the GNR did little to promote or develop the canal and trade steadily declined. Despite this, some 18,802 tons

The Cameron sisters took advantage of the traffic-free canal by hiring out pleasure boats from Grantham Basin. This picture was taken in the 1920s. The warehouses in the background were demolished in October 1929

G. Knapp, The Boat Museum archive

of cargo were recorded in 1905, comprising mostly of manure and roadstone, yielding tolls of just £242. By 1924 (by which time the canal had become just a minute part of the massive London and North Eastern Railway) this had reduced to 1,583 tons and most of the regular traffic had long gone. By 1929, trade had stopped almost completely and the line was legally abandoned in 1936 under the London & North Eastern Railway (General Powers) Act. The LNER argued that many of the lock gates needed replacing and that the expense of undertaking the work simply wasn't warranted.

Fortunately for us, the canal has remained mostly in water as it is used as a kind of linear reservoir. The various canal structures have usually only been demolished when they have fallen down, become subject to vandalism, or where road-works have enforced bridge strengthening or road widening. Unfortunately this latter activity has led to two major obstacles: a series of blockages in West Bridgford and the A1 dual carriageway near Grantham. These two sets of road-works have effectively sealed both ends of the canal. In addition all of the locks, other than Trent Lock, have been de-gated and converted into weirs.

In 1969, the Grantham Canal Society, now called the Grantham Canal Restoration Society Ltd, was formed with the aim of restoring the waterway to cruising standards and reconnecting it to the rest of the canal network. This noble intention will need some major works at the two sealed ends. At the Nottingham end, current plans are to build a wholly new line to the Trent. This would take the canal north after Lock 4 to join Pulser Brook as it passes under the A52 to Holme Pierrepoint. In among the gravel diggings in the area, a 2000 m rowing course has been dug and the new canal junction and marina are being considered at SK 396630. The A1 may be even more of a problem as digging a fresh tunnel through the road embankment would be prohibitively expensive. However, where there's life . . .

The Walk

Start:	Bottesford station (OS ref: SK 810393)
Finish:	Grantham Old Wharf (OS ref: SK 905350)
Distance:	10½ miles/17 km
Map:	OS Landranger 130 (Grantham)
Outward:	Train Grantham to Bottesford. Also runs on Sundays Telephone: (0476) 64135
Car park:	Grantham railway station
Public transport:	BR Nottingham to Grantham line

Grantham has a busy railway station for such a small town, mostly as a result of

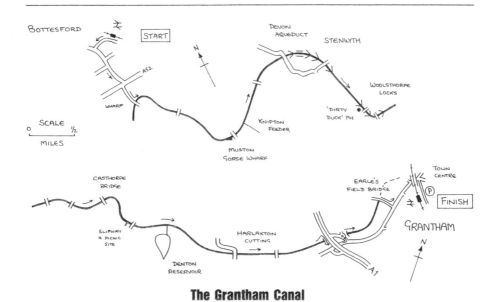

The Grantham Canal

the fact that it is on the main line from London to York and Scotland. The line from Grantham to Bottesford is a side issue in comparison but can be used for the outward journey having parked the car in the station car park.

From Bottesford station turn right to follow the road around left with the church to the left. Turn left at Rectory Lane (a T-junction) following signs for Belvoir. This leads to Grantham Road. Pass the Bull Inn and then bear left along Belvoir Road. This passes some houses to reach the A52 Grantham to Nottingham road (Bottesford bypass). Cross this busy road and follow signs to Belvoir Castle. As this minor road bends left and rises towpathers will sigh with relief as they reach Bottesford Bridge and the Grantham Canal.

Before starting on the walk, go up to the bridge and look down right. Here, as the towpath widens to bend under the bridge, is the site of the former Bottesford Wharf. To start the walk turn down left to take the towpath which is on the left-hand bank. Immediately to the left is a small notice board describing the route taken by the canal together with a milepost which indicates 24¼ miles from the River Trent at West Bridgford. Further notices and mileposts occur at regular intervals between here and Grantham.

The canal shortly enters a cutting through Toston Hill around which the sounds of the kennels to the right echo ominously. The cutting soon shallows and, a ¼ mile further on, the canal reaches the culverted Easthorpe or Middlestile Bridge. A number of the bridges along the route have become culverts since the canal was officially abandoned in 1936. To reopen the line each bridge will have to be raised with all the consequent roadworks involved in building the embankments required on both sides of the road.

Muston Gorse Wharf with Belvoir Castle in the distance

The canal now passes through open country with Belvoir Castle visible over to the right. After the canal has gone through a double bend, it widens on the right-hand bank to form Muston Gorse Wharf. From here a private tramway, owned by the Duke of Rutland, formerly took coal and other goods up to the castle which is approximately $1\frac{1}{2}$ miles away.

The canal continues past Muston Gorse Bridge. A little way further on, the feeder channel from the Knipton reservoir enters the canal. The reservoir is some 3 miles to the south near the village of Knipton. Chris Cove-Smith points out that for 1 mile of its journey it is carried through a tunnel under Belvoir Castle. The canal, meanwhile, continues on for a $\frac{1}{2}$ mile to Langore Bridge, an original accommodation type bridge built to permit the local farmer to pass from one side of the canal to the other. The sides of the bridge still bear the marks of the many towropes that have passed this way. A $\frac{1}{4}$ mile further on, the canal leaves Nottinghamshire to enter Lincolnshire near a winding hole or turning point. The line then reaches the relatively new Muston Bridge followed by the first of the Woolsthorpe flight of locks: Woolsthorpe Bottom or Muston Lock. All the locks along this stretch of the canal have been de-gated and converted into weirs. Their renovation is one of the first priorities of the Grantham Canal Restoration Society. The dilapidated state of some of the locks, however, means they will need complete rebuilding – a difficult job when virtually all the labour to be used will be volunteers who will only be able to work at weekends or during summer camps.

To the left from here until we reach the outskirts of Woolsthorpe the left-hand side of the canal is occupied by the line of the former ironstone railway which joins the course of the canal from Belvoir Junction on the main line near Bottesford. This line was built by the Great Northern Railway after they had taken control of the canal in 1861. The line (known as the Belvoir branch) was opened in 1883 much to the annoyance of the local boatmen and the canal carriers who had been confidently predicting an increase in canal traffic when the Stanton Ironworks Company started their excavations for iron ore at Brewers Grave (a ½ mile east of Woolsthorpe) in 1879. Traffic along the line appears to have stopped as recently as 1973. The course of the tramway can still be seen as a clear track between two rows of trees and bushes on the other side of the fence to the left. At one point, just before the next lock, the trestles are all that remain of an old railway bridge across the tiny River Devon. This also marks the position of a small aqueduct which carries the canal over the same river.

The next lock is Stenwith Lock and it has an associated keeper's cottage, now a private house. After the crumbling Kingston's Lock, towpathers go under Stenwith Bridge to reach Woolsthorpe Middle Lock. This heralds the start of a ½ mile straight section known as Half Mile Pond. This ends at Carpenter's Lock and Woolsthorpe Bridge where the Rutland Arms (locally known as the Dirty Duck) offers sustenance. Those who are tempted may wish to know that there are about 5 miles to go to Grantham. Just after the bridge is Willis' Lock and a keeper's cottage. Beyond this is the site of the Woolsthorpe Wharf where, for a brief period before the railway was opened, iron ore from Brewers Grave was loaded on to narrowboats *en route* to Stanton Ironworks at Ilkeston.

The old railway now crosses the canal by means of a low bridge and towpathers have to make their way up the steps to the right, on to the line and back down the other side. Here, virtually immediately, is Woolsthorpe Top Lock. In the spring of 1991, the Grantham Canal Restoration Society started restoration work on this lock by dewatering it and digging out some 5 ft of sludge, a prerequisite to assessing the amount of work that will be needed to reopen the structure.

The line continues on to go under Longmore Bridge which is closely followed by two more problematic bridges: Casthorpe Bridle Bridge and Casthorpe Bridge itself. The latter, which carries a minor road from Sedgebrook to Denton, is the centre of a campaign by the restoration society. The aim is to persuade Lincolnshire County Council to raise the height of the bridge sufficiently to make some 3 miles of this eastern end of the waterway navigable. It is the focus of attention because it is the last obstruction to navigation between here and the end of the canal near Grantham. The use of this section of the waterway has been stimulated by the provision of picnic sites and, just before the next bridge, a slipway to allow small craft to be launched on to the canal.

Within a ¼ mile a feeder from the Denton reservoir enters the canal on the right-hand bank. The reservoir feeds the top pound of the canal from a sixty-one million gallon supply just a couple of hundred yards to the right. Even though the canal has been unused since the 1930s both feeder reservoirs (this and Knipton) are in good condition and still supply water to the canal. Shortly after the feeder the line passes a winding hole, bends right and then left to enter the Harlaxton cutting. Here the canal is overhung by sultry willow trees and the whole character of the line alters. The waterway is notably narrower than hitherto and for many years after the opening of the eastern canal in February 1797 only one boat could pass through the cutting at a time. Luxury came in 1801 when two passing spaces were added.

The canal goes under Harlaxton Bridge mid-cutting to reach the site of Harlaxton Wharf (on the right-hand bank) and then on to Vincent's Bridge. After this last bridge, the cutting gradually fades away and the view opens out to reveal the dual carriageways of the A1 trunk road riding high on an embankment directly across the line of the canal. As we approach the road, it becomes clear that here is a restorer's nightmare. This major road, effectively equivalent to a motorway, acts as a very effective dam and the waterway channel is culverted for some 300 yd through both the bypass and the slip-road which runs beyond it. The cost of driving a new tunnel through the embankment would be prohibitive even if the Department of Transport would allow such a

The problems faced by restorers are no better exemplified than by the A1 embankment on the Grantham Canal

venture. Various schemes have been devised because of this. The prime option is that a new basin (apparently with the suggested name of the Jessop Basin) be built on this side of the A1. This would leave a short stretch of water on the Grantham side which would then be made into a kind of linear water park. However, all hope of continuing the line on to Grantham has not been abandoned. Plans to widen the A1 may necessitate some rebuilding of the embankment and may yet allow a tunnel to be built through the line. Watch that space!

For the moment towpathers have to take a slightly devious route to reach Grantham. Having reached the embankment, turn right to pass over the culvert and on to the A607 road. Here turn left to go under the A1 bridge and on past the squash courts to the A1 slip-road which goes left. Turn left along here to reach a roundabout. Cross the road towards a hotel and pass down the left-hand side of the canal which has reappeared from its culvert for the final stretch into Grantham. The towpath is now broad, dry and well-made as it passes between some recent housing and a series of industrial units. It finally ends at Earle's Field Bridge where the view from the far side of the road is of a car park. Formerly the line continued on from here for about 500–600 yd to the town wharf. To return to the station, turn right along Earle's Field Lane to the A607 where you should turn left to continue to the railway bridge. Here Wharf Road crosses the A607. The site of the old basin was to the left but is now privately owned and not accessible. Turn right after going under the railway bridge to reach the station or straight on for the centre of town.

Further Explorations

As the entire 33 miles of the Grantham Canal is open for walking, if not boating, it should in theory be swarming with towpathers. However, the dearth of public transport makes the organization of one-way walks tricky. Grantham enthusiasts should therefore consider the possibility of short two-way walks, circular routes that make the most of the highly convoluted line, or a two-car trick with some like-minded fellows.

Having visited the eastern (Grantham) end of the line, you may like to walk the western end. The line can be followed using a combination of OS Landranger 129 (Nottingham & Loughborough) and Chris Cove-Smith's guide to the canal (available from the restoration society). For two-car trickers one suggested route of 8 miles runs from Cropwell Bishop, a small village off the A46 about 7½ miles south-east of Nottingham city centre, to the River Trent.

The walk starts at Cropwell Town Bridge. Here the canal runs through an area of extensive gypsum quarrying, a rock that has caused a number of problems in maintaining the canal bed. There always were difficulties in keeping the line in water along this stretch (William Jessop came in for a fair amount of criticism because of it) and there is a 2 mile long dry section to this day. However, the course of the canal can still be followed without difficulty.

The 20 mile pound from Woolsthorpe comes to an end just after Cropwell Bishop with a spate of eight locks in under 3 miles (three 'Fosse' locks and a further five at the colliery village of Cotgrave). Polser Brook, the possible future route of the canal to the Trent at Holme Pierrepont, runs off to the right from the canal after the last lock in this series.

The current line continues westwards across Thurlbeck Aqueduct and past Nottingham airport to West Bridgford. Here towpathers have to scale the heights (and the hazardous crossing) of the A52 ring road embankment before descending to pass two locks. A little further on the canal disappears for a while under a road junction. If you maintain your course and aim for the floodlights of the Nottingham Forest football ground, you will shortly be able to rejoin the canal and to reach the River Trent at the aptly named Trent Lock. For those wanting to get into central Nottingham, turn left and cross the Trent at the bridge. This road (the A606) will take you into the city centre.

Further Information

The Grantham Canal Restoration Society has been pressing the case for the line for many years but they now appear to be gathering steam and what was once a dream could become reality. Undaunted by the not-inconsiderable task ahead of them, the society is enthusiastic and would welcome the membership of anyone willing to help further the restoration of this fine waterway. Those interested should contact:

Roger Cook,
28 Kendal Road,
Cropwell Bishop,
Nottinghamshire,
NG12 3DX.

The society sells (for a modest sum which goes to funds) a brief history and guide to the canal:
Cove-Smith, C., *The Grantham Canal Today*. 1974.

Although this covers the history of the line, further information about this and the other canals in the Nottingham area can be found in:
Hadfield, Charles, *The Canals of the East Midlands*. David & Charles, 1966.

6
THE LEICESTER LINE
Market Harborough to Foxton

Introduction

The hill at Foxton has become one of the most popular sights on the whole canal network and it's not hard to see why. The waterway interest is intense and accessible. There's a nice view. There's a place to eat and drink. There's plenty of activity. And there's 'fun for all the family'. Of course, this can have its negative side. Blackpool beach could be less crowded on some summer Sundays. Odd then that just a couple of hundred yards to the north, south or east, the Leicester line is one of the most peaceful and remote of any in the country.

The Leicester line is, strictly speaking, an integral part of the Grand Union Canal, a line that runs from London to Birmingham along one arm and to Leicester along the other. The Leicester branch started life as a series of wholly independent lines that began, in the south, at a junction with the former Grand Junction Canal at Norton, just 1 or 2 miles to the north-east of Daventry. From there the canal follows the M1 to pass the Watford services to reach the seven Watford Locks. After the 1,528 yd long Crick Tunnel, the canal continues north along a highly convoluted course past Yelvertoft and through the empty Northamptonshire farmland. It enters Leicestershire while crossing an aqueduct near Welford. Here a short arm leaves the main line for Welford Wharf just outside the centre of town. After passing through another tunnel (1,170 yd), the canal winds past Husbands Bosworth and heads north-east to reach Foxton, with its flight of ten locks and the remains of its famous canal lift. At the bottom of the flight a 5 mile long branch leaves for Market Harborough, while the main line bends north-west and goes through the 880 yd long Saddington Tunnel to pass down a long chain of locks to reach Blaby to the south of Leicester. It is now just a short distance to the centre of the city. The line curves around the south-western suburbs and then heads north to join the River Soar Navigation at West Bridge. After this, the river navigation takes the line to join the River Trent near Nottingham, a total distance from Norton Junction of 66 miles.

Don't be put off by the crowds. Foxton is worth visiting and, after all, you won't see a soul along most of the walk.

History

For at least a century the good people of Leicester had eyed the River Soar as a potential trading route to Nottingham, the River Trent and all ports north. But from 1634, when Charles I gave them permission, nearly a century and a half passed before the waterway was converted into a navigable route. Even then it was on an initiative from Loughborough, approximately 10 miles to the north of Leicester and 9 miles (by river) south of the Trent. An Act was passed in April 1776 and the canalized Soar was opened from the Trent to Loughborough just two years later.

The navigation was soon returning a good profit and by 1780 was providing its promoters with a 5 per cent dividend. Cargo, which included coal from the Erewash Canal, was taken up the Soar and transhipped for onward road carriage. The success of the line was noticed in Leicester and the idea of extending the navigation was again raised. This time action was taken. An Act of 1791 authorized the extension of the Soar navigation for a further 16 miles and the line was formally opened in October 1794.

The success of the Soar Navigation and the arrival of canal mania soon stimulated talk about taking the line even further south. At a meeting in Market Harborough in 1792, the possibility of a canal from Leicester to the town was discussed. Such was the enthusiasm of local tradesfolk and dignitaries, and the over-optimism that typified the period, that the meeting soon got around to extending the line to Northampton where the new canal, it was said, could link up with the River Nene as well as with the new Grand Junction Canal. These links, it was agreed, would open up trade to London and the east.

An initial survey was carried out by Christopher Staveley who, quite sensibly, divided the project into two halves: Leicester to Market Harborough and Market Harborough to Northampton. In the summer of 1792, William Jessop was appointed engineer and John Varley was engaged to carry out a more detailed survey. By this time the Grand Junction Canal had already decided to build its own branch to Northampton and the Nene. It was clearly prudent not to build another line into Northampton and Varley decided simply to form a junction with the GJC instead. This had the additional benefit of providing more direct access to London.

The final proposal for what was to be the Leicestershire & Northamptonshire

Union Canal (popularly known as the Old Union) took the Soar navigation from Leicester to Aylestone and then cut a canal via Saddington, Foxton, Theddingworth, East Farndon, Kelmarsh and Maidwell on to Hardingstone and the GJC. This was a line of about 44 miles plus a branch of 4 miles from Lubenham to Market Harborough. The main construction work was to be in the form of four tunnels: Saddington (880 yd), Foxton (1,056 yd), Kelmarsh (990 yd) and Great Oxendon (286 yd). This plan was accepted by the promoters and was duly passed by Parliament on 30 April 1793, the same day as the GJC Act. The Old Union was authorized to raise £200,000 plus a further £100,000 if needed.

Activity started at West Bridge, Leicester. In October 1794, the canal was opened to Blaby, just over 5 miles to the south. By March 1795, the line had moved 3 miles closer to Market Harborough but already there were financial and labour problems. The company had great difficulty in obtaining funds from its shareholders and this shortage was beginning to affect its work. There were also pressures from local landowners who demanded inflated prices for the land required by the canal company. Although Saddington Tunnel was started in July 1795, the committee decided in the autumn to curtail its ambitions by only taking the line as far as Gumley, 3 miles short of Market Harborough. With the problems they already had, it must have come as a serious blow to discover that Saddington Tunnel had not been built straight and needed some expensive rebuilding at several places to allow wide vessels to pass through. By 7 April 1797, 17 miles of broad canal, from Leicester to Debdale Wharf near Gumley, was open for traffic. For over twelve years this was the southernmost terminal of the canal. It was not until October 1809 that the line finally reached Market Harborough which was by now officially recognized as the terminus. The Old Union had cost £205,000 and, despite the failure to reach the GJC, the opening of the line was celebrated at the Angel in Market Harborough with a fine dinner for all the local VIPs involved.

The route to the south wasn't totally forgotten however. By June 1808 the GJC saw the potential for an alternative route east of Birmingham to Manchester and actively supported a meeting in London where a new line was discussed. Benjamin Bevan was asked to consider the many routes that had already been proposed and to pick the most appropriate. He chose a route which passed through Foxton, Welford and Crick to join the Oxford Canal at Braunston. However, as the GJC were involved they applied pressure for a direct link with them and the route was altered so that the two lines met at Norton. It was this route that received parliamentary approval in May 1810 under the title of the Grand Union Act together with authorization to raise £245,000 with £50,000 more if required. The proposed line included a branch to the village of Welford. The required sums were, perhaps surprisingly, quickly raised and Bevan was appointed engineer.

The work began on the Grand Union (now known as the Old Grand Union to avoid confusion with the present GU) at Foxton where ten locks in two staircases of five were built to take the line up to its summit pound. These were opened on 1 October 1812 by which time the line had reached Husbands Bosworth Tunnel. The 1,170 yd long tunnel was finished in May 1813 and the line opened as far as Stanford, 10 miles south of Foxton and just over 13 miles from Norton.

Problems arose in 1812 at Crick where the proposed tunnel was found to pass through an area of quicksand. Bevan was forced to revise the line through an even longer tunnel to the east. This was compensated for by the avoidance of a tunnel at Watford where the line was diverted through a private park following a suitable compensation agreement with the local landowner.

The Old Grand Union as opened on 9 August 1814 was just over 23 miles long with seventeen locks. The locks were built to a narrowboat gauge, although all the tunnels and bridges were built to a broad barge width. The line from Market Harborough had cost £292,000.

By 1812, the Old Union Canal was already paying its investors dividends and the line had a steady volume of local traffic. Dividends reached the peak of 6 per cent in 1837 although, as was the case with most canals, levels declined from then on. The Old Grand Union, however, was never a real success. The expected through traffic was not as great as forecast and it wasn't until 1827 that the first dividend was paid. Even by 1840, the dividend was still only $1\frac{3}{4}$ per cent.

It was inevitable that the two lines would work together and, by 1863, co-operation was such that the two companies were as good as one. At this time the canal carrying company Fellows, Morton & Clayton operated steamers between London and Leicester with Boots the Chemists in Nottingham as an important customer. Fly-boats were able to do the trip between London and Leicester in just three days. There were also fly-boats operating from Market Harborough.

By the 1890s the Leicester line as a whole (i.e. the Old Union and Old Grand Union Canals) was beginning to deteriorate and the companies were virtually bankrupt. Through coal traffic had declined from 125,000 tons p.a. in the 1850s to around 5,000 tons and comments were widely made that trade could only improve if the canals were dredged and the locks on the Old Grand Union widened. The chief protagonists in this were Fellows, Morton & Clayton who suggested that both canals should be bought out by the Grand Junction Canal Company and then modernized. This it duly did, following an Act of 1894, for a price of approximately £17,000.

The GJC immediately implemented a dredging programme and negotiated lower tolls with the navigation companies to the north. In fact, the GJC guaranteed toll levels to the Leicester and Loughborough Navigations and to the Erewash Canal. This agreement later converted to an option to buy the three in 1897. The GJC also decided, with prompting from Fellows, Morton & Clayton, to improve the Watford and Foxton Locks. The argument that was forwarded said that wider locks would enable the carriers to operate larger vessels, reduce carrying costs and increase traffic.

The GJC took a radical approach to the problem at Foxton. Here it decided to replace the ten locks with an inclined plane lift which would take a 70 ton barge or two narrow boats up and down the hill. The plane was developed by Gordon Thomas who built a test model at Bulbourne. The system consisted of two counterbalancing watertight tanks, resting on carriages that ran on rails. Power was supplied by means of a steam engine. Construction work started on the site early in 1898 and the incline was open for traffic in July 1900 at a cost of £37,500. Appropriately, the first working boat through was a Fellows, Morton & Clayton steamer. The lift had the benefit of taking around twelve minutes to pass up the hill compared with about an hour for the locks. It also saved a considerable volume of water, an important commodity along a line which suffered during periods of drought.

A similar plane was considered as a replacement for the seven Watford locks but the capital and running costs involved dissuaded the company from proceeding. Instead it was thought that the Watford locks should be widened. The decision to go ahead was never made and they were merely refurbished as narrow locks during the winter of 1901. During the course of 1900, the

The Foxton inclined-plane lift in about 1905. This view shows the upper docks and the engine house, now the site of the trust's museum. The caissons went down the hill to the right

British Waterways

The newly completed Foxton inclined-plane lift is here being inspected by a group of dignitaries as the northern caisson descends into the lower canal arm

British Waterways

company also dropped its options to take over the Loughborough and Leicester Navigations and the Erewash Canal. The increase in traffic promised by Fellows, Morton & Clayton simply hadn't materialized and the company must have been losing faith in the Leicester line. This situation culminated in the closure of the Foxton lift. The cost of maintaining the engine in steam together with the three members of staff needed to operate the works simply made the otherwise splendid structure uneconomic. Although the incline remained in place until 1926, it was unused after March 1911. These attempts at reinvigorating the Leicester line had cost the GJC nearly £80,000 and there had been little return for its money.

From 1910, when 40,767 tons of cargo passed through Foxton, trade went into a gradual and then dramatic decline with the greater competition from the railways. By 1924, less than 10,000 tons went along the Leicester line although the mainline GJC traffic from London to Birmingham had remained good. In 1925 the company was again considering ways of expanding its traffic and a plan was launched to take over the three Warwick Canals (the Warwick & Birmingham, the Warwick & Napton and the Birmingham & Warwick Junction Canal) as well as the Regent's Canal. Following the passage of an approval Act, the

Grand Union Canal Co. Ltd, which incorporated all these lines (and later the Leicester and Loughborough Navigations and the Erewash Canal), came into being on 1 January 1928.

The new Grand Union started with vigour, as befits a youngster, and plans were made to widen the Foxton and Watford Locks. As a grant was refused by the government, the scheme was dropped and all thoughts of restoring trade to pre-First World War levels were forgotten. Some coal and timber traffic still took the line (some seventy pairs of narrowboats a week passed through Foxton during the Second World War) but by the time of nationalization in 1948 this had virtually stopped. By 1968, the Leicester line was scheduled for abandonment and was only rescued by the Transport Act which designated it as a cruiseway suitable for pleasure boat use. It remains this today.

The Walk

Start and Finish:	Market Harborough Basin (OS ref: SP 727879)
Distance:	9½ miles/15 km
Map:	OS Landranger 141 (Kettering & Corby)
Car park:	At the basin (near The Six Packs pub) or in a lay-by just a little further along the A6 towards Leicester
Public transport:	BR Market Harborough on Leicester to London line

The first part of this walk goes along the towpath with a shorter return cross-country. The footpaths on the return journey are accessible but appear to be little used and can become obscured and overgrown. This is particularly the case in late summer. If you want to return along the towpath, this will extend the walk to 12 miles. A second alternative is to catch the Midland Red Fox bus which runs sporadically between Foxton and Market Harborough. Do enquire first on (0533) 313391. A third alternative is to skip the walk entirely and simply to visit the Foxton Locks from the Leicester County Council car park. This is signposted from the A427 at Lubenham.

The full walk begins at The Six Packs pub on the main road out of Market Harborough towards Leicester. An unmetalled track leads to the Market Harborough Basin where there are a number of boatyards in various states of disrepair. The towpath begins by passing to the left and around the back of the old Harborough Marine buildings before setting off between the back gardens of Market Harborough suburbia.

The canal to Market Harborough is in truth no more than a branch although at one time it was planned to be part of the main line. If that plan had been instigated the canal would have continued beyond the Harborough Basin along

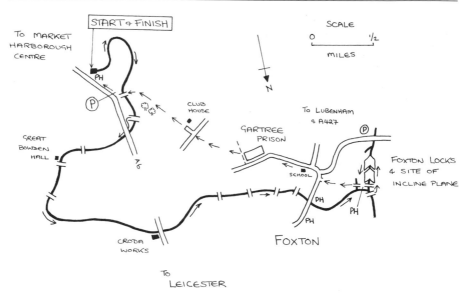

The Leicester line

the Welland Valley past Lubenham and on to Northampton. In the event the line to Market Harborough was built as a terminus when the old Leicestershire & Northamptonshire Union Company ran out of money in 1809. The line then became a branch when the proposal from the company's engineer, John Barnes, was enacted and the main line was taken up a flight of locks at Foxton to join the Grand Junction Canal at Norton.

Although it looks a little drab these days, the basin at Market Harborough was the venue of the 1950 Inland Waterways Association national rally. The event was not only the first of its kind and a major focus for the developing canal restoration movement but was also the scene of one of the more notorious incidents in the row between Tom Rolt and Robert Aickman. Those who enjoy a good gossip should read David Bolton's and/or Tom Rolt's books on the event (see below).

The canal makes a huge loop around the houses to the right. At intervals the line narrows to stop-plank grooves or sites of missing swing bridges. After going around the loop, the canal bends sharply left to reach a small footbridge. Those returning cross-country will see this again. At the next bridge, go underneath and then up left to cross the canal to the right-hand bank. The line now enters a shallow cutting to go under the A6. This cutting then deepens to about 40 ft before reaching the fine arched Saunt's Bridge. After a sharp bend the canal leads to Bowden Hall hidden amid the trees on the right. Go under Bowden Hall Bridge, with its associated pipe, and continue along an increasingly overgrown towpath to another bridge and a bend left.

The canal now stands above the valley to the right where a tributary of the Welland winds its way east. From here on we can enjoy a steady, quiet walk through butterfly and bird rich arable land to reach the slightly incongruous Croda factory. At one time the factory rendered animal bones into meal and tallow. From the bridge (Gallows Hill Bridge) the canal winds around the contours to a brick-arch bridge, followed by a fenced off spill weir that takes surplus water down to the river. After passing under a footbridge the village of Foxton can be seen through the hedge to the right. Just before the following bridge a post indicates that there is just 1 mile to the Leicester (Grand Union) main line.

The canal now swings to the right to reach the Foxton swing bridge and Foxton Bridge. Just after this latter bridge a gap in the hedge to the right gives access to Foxton village where there are two pubs and a shop. The Black Horse, which is visible from the canal, can be reached by walking up to the road and turning right. The Shoulder of Mutton, meanwhile, can be reached by turning

View up the Foxton flight
of ten locks on the
Leicester line

left down the slight hill. The small supermarket is further down the hill after The Shoulder of Mutton.

Back on the towpath, there is now just a short stroll before the entire character of the walk changes. If here on a weekend or during the summer months, prepare yourselves! The first sign is that the canal becomes crowded with moored boats as we reach the bottom of the Foxton flight. This is first seen on the left with a small bridge that passes over a canal arm. This arm, which is now a private mooring spot, was once the bottom entrance of the Foxton inclined plane, a canal lift mechanism which replaced the locks for a period early in the twentieth century. More of this later. The towpath, meanwhile, continues onwards to Foxton Junction where the Harborough arm joins the main line.

On the left is the complex of buildings which surround the old wharf at the bottom of the Foxton flight. Originally the buildings included a carpenter's shop, a blacksmith's forge and the offices of the canal company as well as the old lock house. Nowadays there is a pub (Bridge 61), a shop and a boat yard. In addition, on a warm day, there will be, quite literally, hundreds of people swarming all over the locks and the remains of the inclined plane. On the canal, meanwhile, there will be some highly confused boaters who, having gotten used to the solitude of the rest of the Leicester line, have suddenly arrived at the canal equivalent of Trafalgar Square. Perhaps more alarmingly from their point of view they also find themselves to be the centre of attention as they attempt to take their craft up the ten locks.

Continue round the towpath to reach the roving bridge and cross the main line. From the bridge there is a fine view to the left up the flight of locks that climb the Foxton hill up to the lock-keeper's cottage at the top. At the bottom a small brick bridge goes over the first lock. To reach it continue over the roving bridge and walk onward towards the locks.

Although paths go virtually everywhere, take the towpath that runs up the right-hand side of the locks. The Foxton flight consists of two staircases of five chambers each. A staircase is simply a set of locks in which the top gate of one acts as the bottom gate of the next. Clearly it is not possible for boats to pass in this kind of set up, so boats are well organized by the resident lock-keeper and there is a convenient central pound where boats can pass each other. Each lock is provided with a large side pound into which water is passed when a lock is evacuated. These can be seen over to the left. These provide extra water to each lock thereby avoiding complete drainage of the system at any one point.

After a stroll up to the lock cottage at the top of the flight, cross the canal at the lock bridge and pass down slightly left to go along a clear path in between two of the side pounds. This path is part of the inclined plane trail and leads around to the top of the incline. To the left is what remains of the old boilerhouse (now converted into a museum) and to the right are the remains of the top lock and the slope itself with its rail grooves still largely in position.

The inclined plane at Foxton was built to reduce the amount of time taken to pass up and down the slope and to reduce the substantial loss of water down the hill. A boat using the incline could be moved up or down the 75 ft in twelve minutes, compared with forty-five to sixty minutes for passage along the locks, and with no loss of water from the upper pound.

The system, which was engineered by Gordon Cale Thomas, consisted of two 307 ft long parallel slopes, with a gradient of 1 in 4, on which ran huge tanks or caissons, each large enough to hold two narrowboats. The tanks were made of steel plates and were 80 ft long, 15 ft wide and 5 ft deep. They weighed about 250 tons each and were supported on eight wheels which ran on four rails that went up and down the slope. The position of the rails is still visible on the incline and a small section of the track (which turns out to be identical to that used on Brunel's Great Western) has been fitted into place, together with an explanatory notice.

The lift was operated by having the tanks on the two slopes linked by a steel wire rope so that they counterbalanced each other and thereby reduced the amount of effort needed to move them. The power needed to overcome the inertia was provided by a steam engine: a double-cylinder, high pressure jet condensing type steamed using two 'Lancashire' boilers, one of which was kept in reserve. The engine was coupled to the tanks by a 7 in steel wire haulage rope which passed over a winding drum in the engine house. The ropes were carried on rollers which were set into the face of the planes – an example of which can be seen further down the hill.

At the bottom of the incline, the tanks simply sank into the lower canal arm and the boats floated in or out. At the top, the situation was a little more complex. Here the tanks fitted flush into a lock and were then forced against it by means of steam-powered hydraulic rams which applied their pressure through buffers at the other end. This, hopefully, produced a water-tight seal. To seal the system during use, the tanks and the docks were fitted with guillotine gates at each end.

After two years work by the contractors, Messrs. J. & H. Gwynne of Hammersmith, the lift was opened on 10 July 1900 at a cost of £37,500 including the price of the land. Three men were employed to work the lift: one on the steam engine and one each to operate the top and bottom gates. For the first six months, the lift cost about £1 4s. 6d. a day to run, the main expenses being manpower and coal. With full use this worked out at about one twentieth of a penny per ton of cargo.

When working, the system was a good one but in the long run the plane was not a success. The key mechanical problem was that the rails repeatedly gave way under the weight of the tanks. In retrospect it was said that they were never seated properly. The ¾ in bolts regularly snapped and the rails subsided. This, surely, was a problem that could have been solved but the main reason for the failure of the lift was that the traffic over the Leicester line was too irregular to justify the cost of maintaining the engine in steam and of paying the staff.

The northern hill of the Foxton inclined plane on the Leicester line

In 1909, the locks were reinstated, initially to allow traffic to pass at night when the plane wasn't working. But on 26 October 1910 it was announced that the plane was to be abandoned and that all traffic was to use the locks. Although the plane was used sporadically thereafter, it was last operated in March 1911. Maintenance continued until 1914 during which time various alternative methods of powering and working the incline were considered. But in 1914 the dismantling work began. The boilers were removed and the chimney demolished between 1920 and 1924. The demolition was finally completed in 1927–8 when what was left was sold for scrap. One boat load is said to have sunk in the lower arm and apparently could be seen for many years sticking out of the water.

Part of the old engine house has been rebuilt and now contains the Foxton Canal Museum. Here the Foxton Inclined Plane Trust have their various artefacts including a range of photographs and a working model. Entrance to the museum is by donation.

From the top of the incline, take the path that goes down to the left of the main slope. This passes one of the railway carts that was used during the construction of the plane and an original rope pulley. The brick bridge at the bottom of the hill goes over the canal arm which runs to the bottom of the incline. The arm has been privately owned since 1969 and is now used for residential mooring. Cross the bridge and follow the unmetalled lane out to the

Tyrley Locks on the Birmingham & Liverpool Junction Canal

Canal buildings at the bottom of the Foxton Locks on the Leicester line

The northern portal of Telford's Harecastle Tunnel on the Trent & Mersey Canal

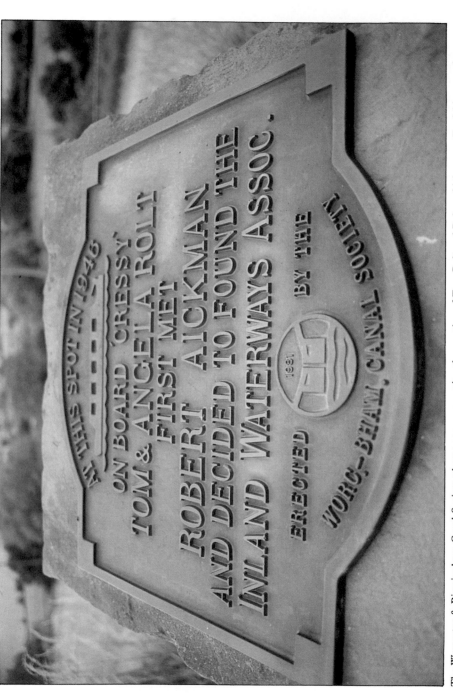

AT THIS SPOT IN 1946
ON BOARD 'CRESSY'
TOM & ANGELA ROLT
FIRST MET
ROBERT AICKMAN
AND DECIDED TO FOUND THE
INLAND WATERWAYS ASSOC.

ERECTED 1981
BY THE
WORC.-B'HAM. CANAL SOCIETY

The Worcester & Birmingham Canal Society plaque commemorating the meeting of Tom Rolt and Robert Aickman at Tardebigge on the Worcester & Birmingham Canal

Kidderminster Locks and St Mary and All Saints Church, Kidderminster on the Staffordshire & Worcestershire Canal

Narrowboats crossing Pont Cysyllte Aqueduct on the Llangollen Canal

road. If you wish to revisit Foxton (for the pubs, shop, bus stop or towpath route back to Market Harborough), turn left. To take the shorter cross-country route, turn right.

Follow this minor road around left (ignore the turning to the right) and keep straight on past a junction where a road sign points right to Market Harborough. This quiet country lane passes a school and then, most dramatically, Gartree prison. As the prison buildings come to an end, a footpath sign on the left-hand verge of the road indicates right down a metalled track. Turn right to follow the lane to a point just before the main prison walls end. Across the grass verge, a stile takes the footpath into a field on the other side of which is a yellow-painted post which marks the line of the public right of way. The course between the two posts should now be continued straight on to pass to the left of a barbed wire fence and along the edge of another field. This leads to a metalled lane.

Continue on, maintaining the same straight course, to pass some huts (the home of a model aircraft club) and a clay pigeon shoot. With a hedge to the right, the path soon reaches a signpost which points in four directions (A6–A6–Lubenham–Foxton). Continue straight on (i.e. the same straight course that was originally set between the first stile and the yellow-painted post) to pass to the right of a short stand of copper beech trees. From here the line of the canal, as it winds around the suburbs of Market Harborough, can be seen ahead and to the right. If the original straight course is maintained, you will soon reach the canal embankment and, a short distance further on, a gap in the hedge up to the footbridge passed earlier. Cross the canal and follow the lane out to a lay-by. Turn right to reach the main Market Harborough–Leicester road. Turn right to walk back to The Six Packs. The short stroll back to the pub isn't wholly dull. Just outside the Knoll is a Grand Junction Canal Company boundary post marking the extent of the company's property surrounding the canal basin.

Further Explorations

The towpath on the Leicester line is often impassable and unwelcoming to walkers. It is passable, however, in the area of the Watford Locks and these are well worth a return stroll of about 1¼ miles from the B4036 just off the A5 near the M1 at Watford (Northamptonshire).

The canal at Watford is nervously wedged between the M1 on one side and the A5 and the main Euston–Rugby railway on the other. Park near the M1 bridge which is next to the appropriately named Watford Gap motorway service station. The towpath starts near the Stag's Head restaurant and takes the right-hand bank. The good, grassy path leads around to the locks which are announced by a brick pumping station, a set of stop planks and a bridge which proclaims to have been rebuilt in 1976. The first lock and the lock-keeper's cottage follow. The line then bends slightly right amid a rich thicket of trees to reach a staircase of four locks before reaching a wide pound, the top lock, a British Waterways hut and the M1 bridge.

The Watford Locks raise the line by just over 52 ft. As part of the scheme of improvements that brought about the Foxton Incline, there were plans to replace the Watford flight with a similar lift system. However, even before the Foxton Incline was open, the plan was dropped in favour of upgrading the locks. They were consequently partially rebuilt during the winter of 1901–2 but a scheme to widen them was abandoned and an inclined plane was not considered again.

After a brief picnic by the staircase, return along the towpath to the B4036 and the car.

Further Information

The Foxton Inclined Plane Trust not only runs the museum but fully intends to restore the lift to its former glory. The trust was formed in 1980 and membership is open to anyone interested in the plane and who wants to help in any way. They can be contacted at:

Foxton Inclined Plane Trust Ltd,
Bottom Lock,
Foxton,
Market Harborough,
Leicestershire,
LE16 7RA.

The Department of Planning & Transportation Leicestershire County Council in association with the trust publishes two books that are crammed with information about the lift and the canal:
Foxton Locks and Inclined Plane. 1985.
Foxton Locks and the Grand Junction Canal Co. 1988.
Both are available in the museum or in the bottom lock shop.

The history of the Leicester lines are included in:
Faulkner, Alan H., *The Grand Junction Canal.* David & Charles, 1973.

For details of the success of and conflicts surrounding the 1950 IWA Market Harborough rally:
Bolton, David, *Race Against Time.* Mandarin Paperbacks, 1990.
Rolt, L.T.C., *Landscape with Canals.* Alan Sutton Publishing, 1977.

7
THE LLANGOLLEN CANAL

Chirk to Llangollen

Introduction

In many ways the Llangollen Canal was a bit of a failure. The original canal of which it was a part, the Ellesmere Canal, was intended to be a north–south line that joined the Rivers Mersey, Dee and Severn. Instead, it ended up being an east–west line that never really fulfilled any particular strategic function. But there again if failures are made like this then long may they be built!

The Llangollen Canal wasn't known as such until a British Waterways booklet was published in 1956. The Shropshire Union, of which it is really a part, consists of a line that runs from Autherley Junction (near Wolverhampton) to Ellesmere Port via Chester. What is more correctly known as the Llangollen branch leaves that main line at Hurleston, about 2½ miles north-west of Nantwich. From there it passes through rural Cheshire before entering the remoter parts of Shropshire just north of Whitchurch at Grindley Brook. The land now seems even emptier as the canal turns south and west to go past the peatlands of Whixall Moss. After skirting around the southern edge of Ellesmere, the line reaches Frankton Junction. From here the Montgomery Canal once ran 35 miles to Newtown and, who knows, may do so once again within a few years. The Llangollen meanwhile turns north-west to reach Chirk where, half-way across a splendid aqueduct, the line enters Wales. After going through two tunnels, the canal passes over one of the most spectacular sights on the inland waterways network: the magnificent Pont Cysyllte Aqueduct. At the northern end, the canal passes Trevor Basin and turns abruptly west, reaching Llangollen 5 miles further on. It is now just 2 narrow miles to Llantisilio and the Horseshoe Falls, the feeder and beginning of the canal.

If anything can tempt the armchair towpather out of his seat then this walk must be it. The splendid scenery around Llangollen coupled with the incomparable Pont Cysyllte should be enough to get anyone up and going. There's none better. Do it!

History

A proposal for what became the Ellesmere Canal was originally advanced in 1789 by a group of Ruabon industrialists who sought an improved freight route to the Mersey. The plan, as launched to a public meeting at the Royal Oak Hotel in Ellesmere on 31 August 1791, was a grand one: a north–south route to link the Rivers Mersey, Dee and Severn. The line that was agreed ran from the Mersey at Netherpool (now known as Ellesmere Port) to the Dee at Chester and thence via Overton to the Severn at Shrewsbury. Branches were proposed to Ruabon and Llangollen, to Llanymynech and possibly to Whitchurch and Wem. Progress was agreed and the line was surveyed by two local men, William Turner of Whitchurch and John Duncombe of Oswestry. They estimated the cost of the new canal to be £171,098 of which about two-thirds would be for the branches.

There clearly must have been some debate as to whether the two local lads were up to the job for not long after, following a suggestion from John Smeaton, the promoters called in William Jessop, whom they described as 'an Engineer of approved Character and Experience', to proffer more learned advice. He duly reported back in August 1792. The Jessop line followed the original route to Chester but then it ran to Wrexham and Ruabon, through a 4,607 yd long tunnel, over an aqueduct at Pont Cysyllte through a tunnel to Chirk and then on to Shrewsbury. Jessop was man enough to draw attention to the difficulties with the terrain but he proposed the line at £176,898 or £196,898 if branches to Llanymynech and Holt were added.

Reassured by Jessop's considerations, the promoters tested the financial waters. In September 1792, subscriptions were invited and, as the country was amid canal mania, the response was overwhelming. Some 1,234 subscribers from as far afield as Derby and Leicester offered a total of £967,700. The company had the welcome problem of having to scale down the bids to accept just £246,500. The Ellesmere Canal was under way.

On 30 April 1793, the Act was passed to authorize a narrow canal (originally it was to be a broad line but this was changed during the passage of the bill in order to save money) with powers to raise funds of £400,000 plus £100,000 more if required. A number of amendments were added during the course of the bill including one which took the route along a slightly higher level near Ruabon to avoid the necessity of the tunnel. William Jessop was appointed engineer with Duncombe, Turner and Thomas Denson as assistants. This team was later, and most notably, extended when a young Thomas Telford was appointed as 'General agent, surveyor, engineer, architect and overlooker of the works'.

Work on the Ellesmere Canal began on the Wirral line in November 1788. By 1 July 1795, packet-boats were passing up and down its $8\frac{3}{4}$ miles and the line soon became both a busy and popular route. For a shilling passengers could ride between Ellesmere Port and Chester and be served with tea and cakes while doing so. Perhaps more importantly from the canal's point of view, the first coal boats from the Mersey arrived at Tower Wharf, Chester, in early 1796. Toll revenue was coming in!

Meanwhile, on what is now the Llangollen Canal, cutting started somewhat later and at various points along the line. In 1794, work started at Hordley, near Frankton, moving westwards towards Llanymynech (on what is now known as the Montgomery Canal). The aim was to gain access to the limestone quarries at Llanymynech. The 11 miles from Frankton to Carreghofa (24 miles short of the Montgomery terminus at Newtown) were completed in the autumn of 1796.

At Fron Cysyllte, near Llangollen, the challenge to span the River Dee was both daunting and exciting. The original proposal was to lock the canal down the valley to an aqueduct 50 ft above the Dee. This design was in keeping with the technology and experience available to engineers at the time. The relatively junior Thomas Telford, who had been asked merely to produce the working drawings, did not approve of this rather unambitious project. He had recently had experience of working with Thomas Eyton on the iron Longdon-on-Tern Aqueduct on the Shrewsbury Canal and he suggested that a similar structure could be used at Fron Cysyllte. He argued that it would be just as cheap for the company to use embankments to carry the line to an iron trough which would then take the canal across the valley on 125 ft high stone pillars. It was a bold and risky suggestion but the flair and imagination which it encompassed was there to be recognized. Clearly Jessop and the committee agreed and approval was gained almost immediately. On 25 July 1795, the foundation stone for the aqueduct was laid by Richard Myddleton, the local MP. Ten years later the great aqueduct was opened with the due pomp and ceremony that befits a monument to the genius that built it. Even today, Pont Cysyllte is impressive and it is no wonder that Sir Walter Scott described it as the greatest work of art he'd ever seen.

During the course of the Pont Cysyllte construction work, efforts were made to take the line from Chirk to Weston. Work started on the Hordley to Weston section separately from the Hordley to Chirk stretch and was finished earlier, in 1797. The whole stretch was navigable in 1801 when Chirk Aqueduct, a fine structure that would receive considerably more attention if it wasn't for its neighbour, was opened.

With all this activity in the centre of the line, it was now time to begin on the problematical course north from Ruabon to Chester. Telford was given the job of re-surveying the route and, following some minor brushes with local landowners, had his proposals approved by Jessop and authorized by a new Act of 1796. From Pont Cysyllte, the Telford line rose 76 ft by locks to Plas Kynaston and then ran level past Ruabon and Bersham towards Chester.

A narrowboat waits at Black Park Basin near Chirk where cargo was transhipped to and from the Glyn Valley Tramway. The railway was an important wharf for local collieries and for the granite and slate quarries along the valley. It was removed in 1935

D. Llewellyn Davies, British Waterways

Construction work started in June of 1796 and about $2\frac{1}{2}$ miles of the line was built between Moss and Gwersyllt.

By 1800, the formidable terrain, coupled with the difficult economic situation in the midst of the Napoleonic Wars, had rendered the northern venture financially impossible. Jessop was forced to report to his committee that it was now wholly inadvisable to build a canal between Pont Cysyllte and Chester, and further progress was stopped. This, of course, left the company in a position where it had a series of odd bits of canal totally detached from the rest of the system and even from its key objective of the Rivers Dee and Mersey.

A cheaper route into Chester was therefore sought and this was calculated as being one which struck eastwards from Frankton to the Chester Canal at Hurleston near Nantwich. The branch to Whitchurch had already been started in February 1797, so all that was needed was a connecting stretch between Whitchurch and Hurleston. This was agreed and work began in November 1802. The line to Whitchurch was finished in 1804 and the line to Hurleston was finally ready on 25 March 1805. The continuing line to Shrewsbury was never completed.

The decision to continue building Pont Cysyllte Aqueduct was a difficult one; what was once the main north–south line now merely took a small branch into the otherwise insignificant town of Trevor. The promoters must have taken a gamble on the carriage of coal from the Ruabon collieries and Telford clearly saw the importance of the Dee at Llantisilio as a water supply. We can only be thankful that they made the decision they did.

The Llangollen Canal from Hurleston to Trevor and to Weston appears to have been opened in late 1805. The first passage was recorded in the local press when five vessels from the Montgomery branch, all laden with oak timber, arrived at Tower Wharf in Chester. Once the connection had been made, trade grew rapidly with a flourishing business in coal, limestone, lime and building materials. The line as opened was 29 miles long from Hurleston to Frankton, with a further 11 miles to Pont Cysyllte and another six to the Llantisilio canal terminus. There are twenty-one locks altogether including six at Grindley Brook and four at Hurleston. The necessary water was obtained under an Act of 1804 which authorized a navigable feeder to be built from Trevor along the Vale of Llangollen to the Dee at Llantisilio. Here Telford built an impressive semi-circular weir known as the Horseshoe Falls. This feeder line, now a popular holiday route to Llangollen, was opened in early 1808.

On completion of the line, the company had raised £410,875 and received some £48,586 in tolls and other receipts. No dividends had yet been paid. Jessop had left in 1801. Telford, however, remained as part-time general agent with Thomas Denson as resident engineer.

There was no doubt that the Ellesmere Canal was going to be a moderately successful line but its dependency on the Chester Canal and the fact that it wasn't linked with the rest of the canal network always meant that its potential was limited. The company was clearly aware of this and, in order to ensure continuity, the Ellesmere attempted to take over the Chester in 1804. This attempt collapsed when the two parties could not agree on a suitable evaluation. By 1813, however, the Chester was heading towards a more stable financial state and further overtures from the Ellesmere were not so rebuffed. On 1 July 1813, the two companies merged to form the United Company of Proprietors of the Ellesmere and Chester Canals.

The future of the new company was always going to be limited by its lack of outlets to the south. Initially, the ECC paid court to the Trent & Mersey Canal Company suggesting a more tangible link at Middlewich. But the level of co-operation between the canal companies was inevitably low and the T & M refused as it was convinced that the EEC would take some of its own trade. Eventually, following the passage of the Birmingham & Liverpool Junction Canal Act, agreement was reached and the line from Barbridge to Wardle (near Middlewich) was opened in September 1833. This was soon followed by the B & LJ which forged a new line from Autherley, on the Staffordshire & Worcestershire near Wolverhampton, to the ECC at Nantwich on 2 March 1835 (for more information on the B & LJ see Chapter 2). The ECC now had two links with the canal network and prospects must have looked good.

Although there was an increase in trade, the opening of these routes to the south did not have the expected impact on activity along the Llangollen line. Limestone was carried from Llanymynech, coal from Chirk and iron from Ruabon but these all hit heavy competition from the quarries and collieries in the Midlands. As an example, of the 60,000 tons of iron carried by the ECC to Liverpool, only 11,000 came from North Wales, the rest coming from Staffordshire and Shropshire. And, as elsewhere, railway competition was beginning to bite.

It was this baring of teeth that prompted the next move. In May 1845, the ECC and B & LJ merged to form what was to become the Shropshire Union Railway and Canal Company. The plan was a bold one. In deference to the new age, the new company would convert their canals into railways. Their engineer, W.A. Provis, assured them that the cost of converting a canal to a railway was just half that of building a railway from scratch. This would provide them with a significant advantage over the many lines that were already springing up throughout the region and competing heavily for traffic. With this objective, the new Shropshire Union incorporated the Montgomery Canal and the Shrewsbury Canal, and took a lease on the Shropshire Canal. Work started not, interestingly enough, on a canal bed but on a line between Shrewsbury and Stafford. This was a joint venture with the Shrewsbury & Birmingham Railway that was seen as a quick solution which wouldn't involve any loss of trade on the operating canal lines.

This period of far-sightedness by the new SU seems to have been short-lived. By the autumn of 1846, the newly formed London and North Western Railway, seeing that the SU could become a serious competitor to its own activities, offered to lease the SU. Perhaps rather unimaginatively, the SU committee agreed. The Act authorizing the move was passed in June 1847. By July 1849 (by which time the Shrewsbury to Stafford line had opened), the LNWR had persuaded the SU to give up all thoughts of further railway building in return for having their debt serviced. With this, the SU's railway ambitions effectively evaporated.

Despite this, the SU's canal activities were thriving. In 1850, the revenue on the line was over £180,000 with a useful profit and a very vigorous carrying business. But the threat of increasing railway competition was being felt. From an operating surplus of £45,000 p.a. in the late 1840s, profits dropped to around £11,000 by the late 1860s even though the level of receipts were similar. The SU were forced to accept lower rates in the face of railway competition. An agreement with the Great Western Railway, for example, forced canal rates to below those of the railway for traffic to Llangollen, Ruabon and Chirk. At this time, most of the company's income was, in fact, from their canal carrying business rather than from tolls: in 1870 the company operated some 213 narrowboats of which fifty-six regularly plied the Llangollen line.

By now the profitability of the Llangollen Canal was becoming a cause for concern. In 1873, the new engineer, G.R. Jebb, even went to the lengths of suggesting that the line from Llangollen to Weston should be converted into a narrow gauge railway. Although reprieved, things did not improve. In the 1880s, a number of works near Pont Cysyllte began to close and by 1905 there was little coal or limestone being shipped from Chirk. By the time of the First World War, the canal as a whole was running at a loss and the LNWR saw no reason to keep it solvent. It forced the SU to make economies. On 1 June 1920, the SU announced that it was to give up the carrying business that still ran some 202 boats up and down the line. At the end of 1922, the SU company was absorbed by the LNWR which in turn was itself absorbed into the London Midland & Scottish Railway the following year.

In 1929, the canal as a whole carried 433,000 tons of cargo but by 1940 this was down to 151,000. In the now notorious LMS Act of 1944, the entire lengths of what today are known as the Montgomery and Llangollen Canals were

The Llangollen line has always been prone to breaches and has recently been subject to extensive (and expensive) renovations by British Waterways. This breach at Bryn Howell occurred in September 1960

British Waterways

abandoned. No commercial traffic had passed along the lines since 1939 and prospects of it returning in the face of rail and road competition were slight. Although the Montgomery line from Frankton was soon closed, in the event, the line from Hurleston to Llantisilio was kept, albeit only as a water channel to feed the Hurleston reservoir which supplies water to the taps of Crewe.

While this situation continued, the line deteriorated still further. In the summer of 1947, the canal was virtually impassable: Tom Rolt gave up trying to fight his way through the weed and dereliction and failed even to get as far as Chirk. By 1955, the official position (as defined by the Board of Survey to the British Transport Commission) was that there were no prospects of the canal generating enough commercial activity to justify its continued maintenance for navigation. Luckily, local enthusiasts, led by Trevor William, clerk of Wrexham Rural District Council, disagreed. They rallied support for the canal to be restored as a holiday route and their actions led to the reclassification of the line in 1968, when the Llangollen was recognized as a cruiseway under the Transport Act. Since then, and at great cost, the line has been gradually improved into the wonderful holiday route that it is today.

The Walk

Start:	Chirk (OS ref: SJ 291376) or Fron Cysyllte (OS ref: SJ 272413)
Finish:	Llangollen (OS ref: SJ 215422) or Horseshoe Falls (OS ref: SJ 195433)
Distance:	Minimum 5 miles/8 km Maximum 11 miles/18 km
Maps:	OS Landranger 117 (Chester), 125 (Bala & Lake Vrynwy) and 126 (Shrewsbury)
Outward:	Bryn Melyn buses from Market Street (Heol Y Farchnad), Llangollen, opposite central car park and public conveniences. Telephone: (0978) 860701. There are extra buses on Wednesdays and Saturdays but, sadly, none on Sundays
Return:	To Llangollen from Horeshoe Falls: Llangollen (steam) Railway from Berwyn station. Telephone: (0978) 860951
Car park:	Central car park in Llangollen
Public transport:	Chirk has a BR station

This excellent walk can be sub-divided to suit all needs. The full walk goes from Chirk to the end of the canal at Horseshoe Falls from where, if you arrive on

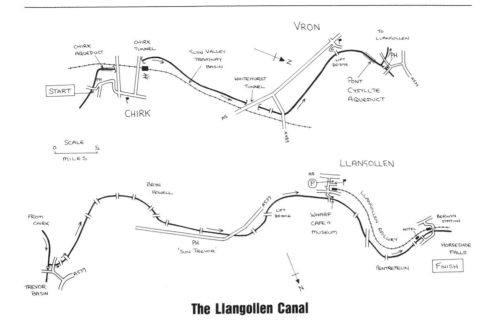

The Llangollen Canal

time, you can return to Llangollen by steam railway. If this proves impossible or undesirable, the walk can be halted 2 miles short at Llangollen Wharf. The start of the walk at Chirk is reached by bus from Llangollen. However, the stretch between Chirk and Fron Cysyllte (locally known as Vron) involves passing through two very dark and claustrophobic tunnels which some walkers may wish to avoid. If this is the case, I suggest that you take the bus only as far as Vron and walk from there, a distance of 5 miles to Llangollen.

Chirk to Fron Cysyllte

The bus from Llangollen stops in the centre of Chirk from where walkers should continue along the road signposted to Shrewsbury. This bends right down a hill and then swings around left. Cross the road to reach a footpath sign pointing over a stile to the right. This isn't the way of the walk but it is worth taking the path down into a field where there is a fine view of the Llangollen Canal's Chirk Aqueduct and the BR viaduct which is just beyond and above it.

Return to the road and continue on to 'the last pub in England', The Bridge. Take the small road (signposted to Weston Rhyn) that rises up by the pub to a post office and canal bridge. Turn right to take the towpath which is on the right-hand side of the canal.

After passing a house festooned with canal memorabilia, the line turns abruptly right to cross the 696 ft long Chirk Aqueduct. The aqueduct, although eclipsed in fame by its near neighbour at Pont Cysyllte, is nevertheless a splendid structure. Jessop, in proposing that the aqueduct be built, said 'instead of an obstruction it will be a romantic feature in the view'. And he was right. The contract for the work was let in January 1796 to William Hazeldine, John Simpson and William Davies but the design is Telford's. The aqueduct consists of a masonry structure within which is a canal bed made of iron plates bolted together. The plates are secured to the masonry sides in such a way that they tie the walls together to resist the lateral pressure of the water which could have led to blow-outs. The use of the cast-iron plates significantly reduced the weight of the structure (compared with masonry and clay puddling) and enabled the aqueduct to be built some 70 ft above the River Ceiriog below. It was finished in 1801 at a cost of £20,898. The railway viaduct, which carries the line from Shrewsbury to Chester (stopping at Chirk station), was built in the 1840s. What could have been a scenic disaster was happily avoided and the two structures complement each other splendidly to form a memorable spectacle.

Chirk Aqueduct and Viaduct

Half way across the aqueduct we return to Wales (is there any finer border crossing?) and from there we reach Chirk Tunnel. Opened in June 1802, the tunnel is unusual in having a towpath along its right-hand wall, a contribution to the development of canal tunnels made by Telford who plainly thought the practice of 'legging' to be rather old fashioned. Those who suffer even the remotest hint of claustrophobia should not attempt to go through the tunnel as the centre is pitch black and progress is only possible by going blindly onwards while clinging desperately on to the railings. Those with young children should certainly not attempt it as the railings do not require limbo dancing skills to pass under. Instead, take the path that goes up to the right of the tunnel entrance to a road which runs parallel with the tunnel and on to the railway station from where the towpath can be rejoined. For those who do walk through, it is a curious experience and one can only wonder what the horses made of the sudden plunge into darkness.

The tunnel opens into the fresh air and a cutting. For those who have followed Tom Rolt's progress along the Llangollen in *Landscape with Canals*, it was the collapse of this cutting which finally ended Rolt's attempt to reach Pont Cysyllte in 1947. The cutting leads shortly to a winding hole on the far bank. At one time here on the right was Black Park Basin, a canal-railway interchange where cargo was transhipped between the canal and the Glyn Valley tramway. This narrow gauge line was built to link with the slate and granite quarries in the Ceiriog Valley. Coal was also transhipped here from the local collieries. Sadly, the line was closed in 1935.

After the canal narrows (note the stop planks and slot designed to separate the section in the event of repair or breach), the line passes the newly built (June 1991) Chirk Marina with mooring spaces and a small shop (not accessible to towpathers). The line now passes through a second tunnel, the 191 yd long Whithurst Tunnel, which goes under Telford's later engineering venture, the Holyhead Road or A5. This tunnel, although much shorter and lighter than that at Chirk, should also be treated with caution by those with small children.

The canal now passes under two bridges to a point where it bends left. Here the Offa's Dyke long distance path joins the Llangollen Canal and the quality of the towpath improves somewhat. The canal enters a concrete-lined section as it follows the hillside on the south-western side of the Dee Valley. In this short stretch the line passes through a narrowed section with stop plank slots and then crosses a small aqueduct before reaching the site of an old quay. This huge site was formerly connected with the Pen-y-Craig limestone quarries. The large edifice which fronts the canal here contains a series of lime kilns which open out at canal level so that their product could be loaded directly on to boats for subsequent shipment. From here it is just a couple of hundred yards to the lift bridge (No. 28) which marks our arrival at Fron Cysyllte.

Fron Cysyllte to Llangollen

For those starting at Vron bus-stop, follow the advice of a signpost to the aqueduct (or Pont Cario Dwr) which points down the hill to the left. This soon reaches the aforementioned lift bridge. Cross the bridge and turn left.

From the lift bridge, the canal turns sharply right through a heavily wooded section which marks the start of the 97 ft high embankment built as the preliminary for the crossing of the Dee Valley. At the time of its construction, this was the highest man-made embankment in the country and was itself a significant achievement. But, as the wood thins, the staggering engineering

The pillars of Pont Cysyllte Aqueduct

prowess that was needed to span the Dee is evident as the ground suddenly falls away to reveal Pont Cysyllte (pronounced pont-kur-sulth-ter) Aqueduct, surely one of the greatest sights of the entire British canal network. Designed by Telford and Jessop and constructed by the ironmaster William Hazeldine, it took ten years to build and was finally opened on 26 November 1805. The aqueduct consists of 418 cast-iron plates bolted together to form a 1,007 ft long metal trough carried on nineteen stone piers across the valley. The design therefore differs from Chirk Aqueduct where only the bed was made of cast iron. Tom Rolt reports in *Landscape with Canals* that the plates used to make the trough were cast at Plas Kynaston on the site now occupied by the massive chemical works whose chimneys are to be seen on the other side of the Dee Valley at Trevor. The plates were then moved to the site by means of a temporary inclined tramway and assembled, rather precariously, *in situ*. Seemingly the opening of the aqueduct was greeted with great celebrations which included a salute from a cannon and various musical tributes from boatloads of floating bandsmen. The aqueduct cost a cool £47,018 (about £8,500 for the embankment, £17,300 for the iron work and £21,150 for the masonry). But wasn't it worth it!

The towpath across the aqueduct is cantilevered over the trough to provide a somewhat airy experience that may alarm those with no head for heights. There is, however, a strong handrail which can be clung to as the holiday-makers, coach parties, cyclists and joggers struggle to pass one another some 126 ft above the River Dee. But the structure is firm enough. The piers are standing on solid sandstone and were themselves made hollow towards the top in order to reduce their weight and save on masonry.

Pont Cysyllte is drained every eight years or so for maintenance by blocking off both ends with stop planks and then drawing a plug in the trough. The resulting cascade of water, which appears like a high free-falling waterfall, must be a somewhat alarming sight for those below who don't realize what precisely is going on.

At the far end a notice, in both English and Welsh, details the aqueduct's statistics. Almost immediately opposite, on the left-hand bank, a narrow channel heads off towards Llangollen, whereas a wide channel, in what is now Trevor Wharf, lies directly ahead. This is the only remaining remnant of the line that was originally intended to continue on to Chester and it runs for just a couple of hundred yards before coming to a stop at a former railway interchange basin. A number of boats moor at Trevor Wharf where there is a small canal shop, some public conveniences and the highly convenient Telford Inn.

Our route continues through the horse tunnel of the small road bridge and up right to the road. Here turn right to cross the bridge. From the bridge there is a good view of all that remains of Trevor Basin: two wharf arms that go either side of what was formerly the site of a railway terminus. Having abandoned the idea of a line to Chester, a short canal (of about a $\frac{1}{3}$ mile) called the Plas Kynaston

Canal was built to the Plas Kynaston iron foundry and factory, now the site of the chemical works over to the right. This ran from the right hand arm as seen from the bridge. The Plas Kynaston Canal closed in about 1914.

To continue the walk, go along the road to a T-junction. Turn left to pass the Telford Inn to another small bridge (which was being rebuilt when I passed) which goes over the canal. If you wish to get an excellent view of the aqueduct (best on a sunny late afternoon or early evening) continue down the road for a short distance when the view opens out to the left. If not, turn right at the bridge to take the right-hand bank. At the first bridge, the towpath changes sides.

This channel on from Trevor Wharf is notably narrower than the Llangollen Canal before Pont Cysyllte and was never intended to be a prime navigable route. Indeed the only reason it was built was to act as a feeder from the Dee at the Horseshoe Falls, some 2 miles beyond Llangollen at Llantisilio. An indication of its use can be seen by the strong current which flows from the direction of Llangollen, an unusual sight in an artificial waterway. This line along the Vale of Llangollen could have developed, however, as in 1801 there was a proposal for a line called the Merionethshire Canal which would have taken the canal on to Barmouth on the west Wales coast. If built, it would have been an important conduit for the many quarries and mines in that part of Wales.

Whether wide or narrow, the route between Trevor and Llangollen is a contour-hugging line high above the River Dee on the northern hillside. It must be one of the most delightful stretches of canal in the country: quiet, wooded and with magnificent views down to the Dee on the left. When looking at the terrain, it is remarkable that the entire length between Chirk and Llangollen is free of any locks. And yet this stretch has been a constant trial to British Waterways. The proof of this is seen after passing under three bridges (Nos. 33–5) and following the canal right to a point where the view to the left opens out. This stretch, near a BW noticeboard, has been breached three times since the 1950s: in 1960, 1982 and 1985. As a result, the entire length has been completely rebuilt with a reinforced-concrete trough overlying a specially developed high-drainage bed which hopefully should prevent any further movement of the underlying hillside. In fact, the whole line between Chirk and Llangollen is gradually being upgraded by BW at a cost of over five million pounds. Such, albeit expensive, rebuilding should hopefully keep this superb waterway open into the forseeable future but one can guess what the response of the authorities would have been had these slips occurred before the Second World War rather than after it. Surely a simple pipe would have been easier and cheaper, and we can only be grateful that one was never installed.

The line continues under bridges 36 and 37 to reach a rather angular winding hole, a sizeable hotel and bridge 38 (Bryn Howell Bridge). Bridge 39 is an old concrete affair which formerly carried the Llangollen Railway to Ruabon and, who knows, may do so again some day. Just before bridge 40 stands a 2 ft high concrete roofed shed which holds the stop planks which are used to slide down the slot near the bridge itself. Shortly thereafter, the canal is joined by the A539 Trevor to Llangollen road which runs close to the right bank for nearly 1 mile and past the Sun Trevor pub.

Sun Trevor Bridge (No. 41) marks the site of one of the first breaches in the post-war series. In 1945, the bank here slipped and washed down to the railway below, taking a hapless goods train with it. The engine driver was killed. Another collapse occurred a little further on in 1947, luckily with less serious results. However, the subsequent loss of water reduced levels throughout the length of the canal and nearly prevented Tom Rolt from escaping to the Shropshire Union main line after his aborted journey in 1947. Such unhappy memories should not, however, mar the superb views to the left across the Dee Valley to Llangollen which can now be seen in the distance approximately $1\frac{1}{2}$ miles away.

The stretch between bridges 41 and 42 is incredibly narrow and portends things to come. There are just three possible passing places in a $\frac{1}{3}$ mile, as the canal struggles to find a way between the precipitous drop to the left and the hard-faced rocks to the right. If you're lucky enough to be here at a time when it isn't pouring with rain (the author wasn't), the next stretch offers a delightful mixture of pastoral, mountainous and watery scenes. Just after the lift bridge (No. 44), the canal is forced against some sheer cliff faces overhung by trees before finally winding its way through the outskirts of Llangollen. Again the line narrows so that there is only room for one boat at a time. Navigation guides suggest 'tact and restraint' to boat captains although from the look of it not all have read the right books. To the left, below the towpath, the bank drops sharply to the road below. Soon, through the trees to the left, the outskirts of Llangollen appear. After a while the canal widens and is lined with moored boats as we come out from under the trees to enter the town.

We soon reach Llangollen Wharf which is positioned high above the town, with a fine view down to the river and the main street. The wharf building has been converted into a café and a small canal museum at one end, and into stabling at the other. On appropriate days, the occupants of the stables can be seen plying their trade with boatloads of, not always cheerful-looking, holiday-makers being towed upstream towards the Horseshoe Falls in horse-drawn barges.

There are now some 2 miles to the origin of the Llangollen Canal. If you wish to check on the availability of a steam train return, take the paths either to the left or right of the wharf building that go down to Abbey Road from where the Llangollen Railway station is easily spotted. If you've decided to stop here, Market Street can be found by crossing the Dee Bridge and following the parking and public convenience signs.

Llangollen to Llantisilio

This final section of the walk could also make a fine afternoon stroll in its own right. From the wharf (well signposted from Abbey Road), turn left to pass along the tree-lined canal. Touring boats visiting Llangollen are only able to go about 50 yd beyond the wharf, where there is the last winding hole that could allow them to turn. However, the towpath continues through the outskirts of the town and past the famous International Eisteddfod ground, which is on the left-hand side before bridge 46. Train line and canal then follow each other around the contours to reach Pentrefelin, now the site of a motor museum. From 1852, this was a wharf from where a tramway went $4\frac{1}{2}$ miles up the Horseshoe Pass (to the right) to the slate quarries at Oernant and Moel y Faen.

Eventually the path runs round to reach the Chain Bridge Hotel. From here the intrepid towpather ignores a rather begrudging Shropshire Union notice and goes on under a pleasant stone road bridge (Kings Bridge) to a small hut built by the water authorities in 1947 to control the flow of water into the canal. Six million gallons per day flow into the line which starts here. Beyond the hut a path leads into a field and to the Horseshoe Falls, an arc-shaped weir built by Telford to supply water to the canal.

To reach the railway station return back under Kings Bridge to the hotel where there is a metal footbridge which goes over the canal and up the cliff to a road. Turn left to go round and over the road bridge. This eventually reaches Berwyn station from where trains can be caught into Llangollen.

Further Explorations

Comparatively close to Chirk and Llangollen is the Montgomery Canal, currently the scene of much excitement and activity as enthusiasts strive to restore it to its former glory.

The Montgomeryshire Canal Act was passed in 1794 and took a line from the Ellesmere Canal at Llanymynech to Welshpool and Newtown. Later the stretch from the Llangollen at Frankton also became known as the Montgomery Canal and this means that the waterway today follows a total of 35 miles. Sadly, following a serious breach in 1936, the London, Midland & Scottish Railway, who then owned the line, successfully had the canal abandoned by an Act of 1944. It quickly decayed and would undoubtedly have disappeared from the landscape had not a group of enthusiasts started to restore the line in 1969. By

1987, an Act of Parliament was passed which should enable full restoration of what Charles Hadfield considers to be Britain's loveliest canal.

Although it is possible to walk most of the 35 miles to Newtown, a flavour of the Montgomery can be had with a brief walk and a couple of visits. The line starts at the Frankton Junction with the Llangollen. This can be found by taking the A495 which runs between Oswestry and Ellesmere/Whitchurch. At the village of Welsh Frankton, a signpost points south to Lower Frankton. This small road twists and turns around a series of farms eventually to reach a canal bridge (over the Llangollen Canal) and some verge side parking. For a walk of about 1 mile, go back to the bridge and turn right through a gate. After about 20 to 30 yd, you will reach Frankton Junction. If you turn right here you will shortly come to the first locks on the Montgomery Canal, a staircase of two – where the bottom gate of the first is the top gate of the second. Some of the original buildings still line the canal, although none perform their original task of toll-house, keeper's cottage or inn. Two further locks, restored in 1987, are passed before this short stretch comes to an end and the weeds take over. Continue along the clear footpath which runs to the right of the line to a point where a reed-ridden basin is reached near a small, but high, road bridge (Lockgate Bridge). To the left, a straight line of plant growth can be traced running off to the east. This is the former Weston branch which went on for 5 miles to Weston Lullingfields. It seems impossible to believe that this would have been part of the main-line route from Chester to Shrewsbury had the original Ellesmere Canal plan been carried out. To return to the car, you can either retrace your steps or go up by the side of the bridge to the road. By turning right, you can then follow the lane back to the parking space.

From Llanymynech, the line can be followed south to the newly (1986) restored Carreghofa Locks and then on to Vrynwy Aqueduct. Some 6 miles of waterway are also open for use around Welshpool.

Further Information

The Llangollen and Montgomery Canals are both part of the old Shropshire Union and are ably served by:
   The Shropshire Union Canal Society Ltd,
   Mrs Mary Awcock,
   'Oak Haven',
   Longden-on-Tern,
   Telford,
   TF6 6LJ.

The society's immediate aim is to reopen the Montgomery Canal and members are actively involved in this work. Should you wish to become involved in any capacity then new members are very welcome.

The history of the canal can be traced through:
Hadfield, Charles, *The Canals of the West Midlands*. David & Charles, 1969.

If you are interested in following the entire line:
Pellow,T. and Bowen, P., *Canal to Llangollen*. Landscape Press, 1988.

The trials and tribulations of trying to reach Llangollen in the late 1940s can be followed in:
Rolt, L.T.C., *Landscape with Canals*. Alan Sutton Publishing, 1977.

8
THE NORTH WALSHAM & DILHAM CANAL

Honing to Tonnage Bridge

Introduction

For those who find the industrial nature of the British canal system a bit grating, the North Walsham & Dilham is heaven sent. A neglected line, way out on a limb on the northern fringe of the Norfolk Broads, the NW & D is more of a nature reserve than a commercial waterway and my guess is that it will probably stay that way. There are no boats on the NW & D nor, in some places, any water. But perhaps most strangely, in these days when virtually anything that's derelict has somebody trying to breathe life back into it, it doesn't have a restoration society either. Or have I spoke too soon?

The North Walsham & Dilham Canal is on the northernmost fringes of the Norfolk Broads. From Barton Broad (a couple of miles north-east of Wroxham), the River Ant can be traced north and west, away from the town of Stalham to Wayford Bridge, a point about 1 mile east of the village of Dilham. Here the $8\frac{3}{4}$ mile long North Walsham & Dilham Canal starts life by taking a course that runs north and then west to reach the small village of Honing. At Briggate, the line returns to a more northerly course to skirt around the eastern fringes of North Walsham towards Ebridge, Bacton Wood and Swafield. From there, the canal formerly ran a short distance to its terminus in a basin adjacent to Antingham Mills. Antingham Ponds, to the north-west of the basin, was the ultimate limit of navigation but primarily acted as a reservoir and feeder to the canal line.

There's only sporadic industrial archaeology interest along the NW & D but it makes for a fine place to spend a warm, sunny afternoon among the butterflies, dragonflies, herons and sparrow-hawks. If you're more used to the glories of the Walsall Canal, you just might come here to see how the other half live.

History

In 1810, the River Ant was already navigable from the northern Norfolk Broads to the small hamlet of Dilham where a thriving local trade took goods and produce to and from the villages further north. Corn and flour from the mills around North Walsham and Antingham were shipped from Dilham to Yarmouth and commodities such as coal, marl and oilcake were imported via the Dilham staithes or wharves. With the success of the many canals around the country, it must have only been a matter of time before suggestions were made to extend the Ant artificially towards the town of North Walsham and, during the course of 1811, no fewer than three plans for a new navigation were widely circulated. The first was a scheme by William Youard and there were two by John Millington of Hammersmith. Interest was such that a meeting was held on 14 September at the King's Arms in North Walsham where it was immediately decided to petition Parliament for an Act to make 'a cut or canal for boats from the River Ant . . . at or near a place called Wayford Bridge near Dilham to the towns of North Walsham and Antingham'.

The bill was read on 18 February 1812 and despite opposition from the inhabitants of Dilham and Worstead, who feared loss of their wharf and carrying businesses, the Act received royal assent on 5 May 1812. The Act authorized the raising of £33,000 plus a further £10,000 by mortgage of the rates and dues if the original sum wasn't enough. The Act also authorized tolls of 1d. per mile for passengers, $\frac{1}{2}$d. per mile for cattle, horses and asses, and 6d. a score for sheep and pigs.

Despite the initial enthusiasm and the fact that the share issue was virtually fully subscribed, the final authority to start the construction work was not given until 15 December 1824. During the period of inactivity, the proprietors had spent some time fighting off claims for damages from those who feared loss of trade due to the new line. Digging, in fact, finally began on 5 April 1825 with a crew of some one hundred navvies from Bedfordshire under the control of the engineer John Millington and the clerk to the committee, William Youard. The first day of digging was one of great celebration with the navvies parading in the market place of North Walsham. With the committee and a band before them, they then marched off to Austin Bridge, where the first sod was cut by William Youard and the band played 'God Save the King'. The work for the day stopped shortly thereafter and everybody went back to the market 'to partake of some barrels of strong beer'.

Curiously, the NW & D was John Millington's only venture into canal engineering and he later emigrated to America, where he apparently used his enormous experience to write books on the subject. It must also be said that actual day-to-day supervision of the works wasn't carried out by Millington at

By the turn of the twentieth century, the canal had little or no traffic and maintenance levels were declining. Honing Lock, here photographed in the 1920s, remained operational until about 1935 but has since been converted into a weir and, as towpathers will have noticed, has become considerably overgrown

Norfolk County Council

all but by a contractor, Thomas Hughes, who officially succeeded Millington as engineer in 1827. Hughes was actually a much more experienced civil engineer having worked on the Caledonian, Dingwall and Edinburgh & Glasgow Union Canals, as well as on various harbour and river works.

Although there were few significant construction works along the line, the land through which the canal passes is very boggy in places and Hughes had some problems in establishing a firm line through the peat and in building embankments. Despite this, he had completed enough to allow the first laden wherries to reach Cubitt's Mill on 14 June 1826. In what seems to have been typical reaction throughout the construction of the line thousands of spectators were reported to have assembled to witness this interesting scene with the day finishing with a treat for the workmen of Mr Sharpe's strong ale and Barclay's brown stout. The official opening of the completed line took place not long after on 29 August 1826. This time there was a flotilla of boats that spent the day going up and down the 8¾ miles of the canal, no doubt with the consumption of yet more strong ale. The final cost of the new line was approximately £29,300, one of the few canals that was actually built to estimate. Hughes needed six locks, built at Honing, Briggate, Ebridge, Bacton Wood and two at Swafield to take the canal up 58 ft from Wayford Bridge to Antingham.

Conventional narrowboats were never used on the NW & D. Instead, a type of small Norfolk wherry of about 18 to 20 tons and measuring 50 ft long by 12 ft 4 in beam with 3 ft draught was the norm. There was also, for example, one small 12 ton wherry, the *Cabbage Wherry*, which regularly ran vegetables from Antingham to Yarmouth. In addition because of the shallowness of the canal, there were some vessels of a type known as slip-keel wherries in which the keel could be unbolted and removed when the craft was about to enter shallow water.

The NW & D was primarily used to ship corn, flour, mill offals, feed cake, manure and wood. Interestingly although some coal traffic went along the line, it still proved to be cheaper to drop coal from the collieries of north-eastern England on to the beaches of the Norfolk coast at Mundesley or Bacton, and then to transport it overland in carts. The canal's share of the coal traffic was further limited later by the introduction of the railway which ran from Norwich to Cromer.

In common with many other canals whose prosperity relied primarily on shipping agricultural produce, the NW & D was never a financial success. Even as early as 1830, the price of a £50 share had dropped to £10 and there seemed to be little prospect of increasing trade. In 1866 the proprietors gained powers (from an Act) to let them sell the canal. In fact, nothing happened until 17 December 1885 when the company clerk, James Turner, informed the Board of Trade that the Bacton Wood miller and wherry owner, Edward Press, was buying the line. The deal, worth just £600, was completed on 16 March 1886.

It was required by the 1866 Act that the sum received from the sale be divided among the known shareholders in proportion to their holdings in the company. James Turner was entrusted with the task but after paying out just a few of the shareholders he absconded with the balance which was never seen again. The canal committee thought themselves to be morally if not legally bound to reimburse the money and, in 1896, each of the five proprietors, including Edward Press, donated nearly £111 to make up the sum necessary for the payments.

During the course of 1887, further problems within the operation of the company were spotted by Walter Rye, a solicitor who had become principal clerk. There had been 'great irregularites in the management and direction of the Company . . . amounting to an almost total disregard to the provisions of the Act'. The Act, for example, did not allow anybody holding a place of profit to serve on the committee but Edward Press had done so and, although a treasurer should have been appointed and security given by him, no such appointment had been made. Mr Rye also pointed out that there was no way of formally winding up the company without both huge expense and revealing the many irregularities. Interestingly, although these were fairly serious concerns, they appear to have been quietly forgotten about.

Meanwhile, traffic on the NW & D continued although the roughly 1½ miles between Swafield and Antingham were abandoned in 1893. In 1898, 400 tons of local traffic, 5,000 tons of exports and 6,386 tons of imports were moved along the line. But by 1906 receipts were under £400 p.a. and, on 2 July of that year, Edward Press died. Although trade appeared to be in terminal decline, the canal was sold by auction on 11 September 1907 to a Mr Percy, a director of the General Estates Company, for £2,550. The company already owned the rights of the Gorleston ferry and of the tolls of Selby Bridge and were associated with the Yarmouth & Gorleston Steamboat Company. They didn't keep the canal for long. In 1921, the canal was taken over by E.G. Cubitt and G. Walker (for £1,500) who then promptly sold it to a newly formed company, the North Walsham Canal Company Ltd.

The canal itself was, by now, in a poor state of repair. It had been badly

Honing Lock, 1991

damaged in floods in August 1912 when a bank was breached above Bacton Wood Lock. Attempts were made during the 1920s to improve the canal below Swafield but there was little or no improvement in traffic and the section from Antingham to Swafield Bridge was closed in 1927 by warrant of the Ministry of Transport. The last commercial boat on the canal was the wherry *Ella*, owned by the descendants of Edward Press, which loaded cargo at Bacton Wood Staithe in December 1934. Soon after this the canal became heavily silted and eventually derelict. The bottom three locks were in use until 1935 but there appears to be no record of them being used subsequently.

The only section that is even vaguely navigable today is the stretch from Wayford Bridge to the tail of Honing Lock and large stretches above that are dry. Interestingly, the canal wasn't nationalized and it remains, at least nominally, in the ownership of the North Walsham Canal Company.

The Walk

Start and finish:	Weaver's Way car park, Station Road, Honing (the old Honing railway station (OS ref: TG 317276)
Distance:	5¼ miles/8¼ km
Map:	OS Landranger 133 (North East Norfolk)
Car Park:	As above
Public transport:	None. The closest public transport gets is North Walsham which is on the BR line from Norwich to Cromer. As the Weaver's Way passes through North Walsham, it is possible to start the walk from there by following the course of this reasonably well signposted long-distance path. A return walk will, however, add an extra 8½ miles to the distance

To reach the Weaver's Way car park leave the main A149 (North Walsham to Stalham road) near Lyngate where there is a signpost towards Honing. After about a ½ mile, the road goes over the canal at Briggate. It's worth a quick stop here to see the surprisingly well-preserved remains of Briggate Lock which are right next to the bridge. Continue along the road towards Honing for another 250 yd to a point where a signpost indicates left for the Weaver's Way car park. This was formerly the car park of Honing railway station that has been sequestered by Norfolk County Council for users of the long-distance path. The Weaver's Way, opened in 1980 and so-called to commemorate the former local industry, originally ran for 15 miles between Blickling Hall and Stalham. Since then, various bits have been added and you can now walk a total of 56 miles from Cromer to Great Yarmouth. Should you wish to do so that is.

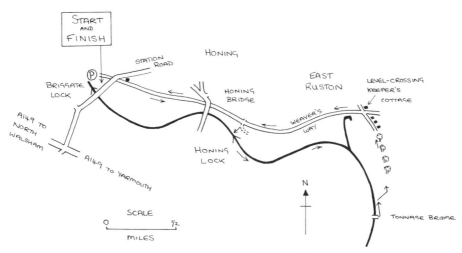

The North Walsham & Dilham Canal

The walk starts by circumnavigating the station platform to arrive down at railway track level and then turning right. The Weaver's Way uses the track of the former Midland & Great Northern Joint Railway (the M & GNJR), a line that went from North Walsham to Great Yarmouth which was closed in February 1959. Most of the line has been converted to road and, if one looks at a map, it seems amazing that this bit survived untarmacked. The clear, if somewhat overgrown, track leads immediately to a gate to cross Station Road, through another gate and back on to the line. Just by the second gate is a level-crossing keeper's cottage, now a private house.

The railway path leads through open country to reach a crossroad where the promisingly named Canal Farm Lane leads right to Canal Farm. We, meanwhile, continue along the railway with views to Honing to the left. Within a couple of hundred yards, the route enters woodland. The path remains clear and is bordered on either side by an almost hidden railway boundary fence. This soon reaches the rather fine iron Honing Road Bridge. Just before the bridge on the right is the site of the former Honing Wharf or Staithe. A short 110 yd long cut was made from the main line to the wharf so that cargo could be unloaded as near to the village as possible. After the bridge, the main course of the North Walsham & Dilham Canal joins the track to the right. The canal which is heavily overgrown with reeds then turns right while the railway path continues past a house to reach another crossroad track. Turn right here and, following a footpath sign, immediately right again to take a narrow path over a small ditch to a bigger bridge which crosses the canal at the site of an old lock. Honing Lock is now nothing more than a weir but still bears the long redundant winding gear

Briggate Lock on the North Walsham & Dilham Canal

and high lock walls. Honing was the first lock (of a total of six) on the canal from the River Ant at Wayford Bridge.

Cross the bridge and turn left to follow a clear path that gives the impression of moving away from the canal. It doesn't. Within 10 to 20 yd, the path arrives on the right-hand bank of the line that is now passing through dense woodland. Shortly, the path goes through a gate to reach an open field. This stretch of the canal sports the finest growth of yellow water-lilies outside of a stately home.

The canal continues around a large arc for the next 2 miles. At one point a short branch (of just under a $\frac{1}{2}$ mile) runs off to the north from the left-hand bank in the direction of East Ruston but other than that the canal and the country are quiet, relaxing and in perfect harmony. It is an excellent linear wildlife park and towpathers can only be grateful that the Anglian Water Authority (as it then was) stopped pumping effluent into the canal in 1980. It's not until the towpath is pushed away from the canal by a dense hedge near Oaks Farm that the mind is forced back into action and some thought processes are required once again.

A gap in the hedge leads to a lane. Turn left to reach Tonnage Bridge. When the canal was in use, there was a small wharf and a cottage here and the bridge was used as a convenient spot to monitor the flow of traffic along the line and to levy tolls on goods entering and leaving the canal. The cottage and wharf have now disappeared and so did the bridge, nearly. It was restored in 1982 by the local landowner in the same style as the original, if marginally wider. A noticeboard, put up by the Broads Authority, explains the history of the bridge and, usefully, provides a map for the next section of the walk.

Cross the bridge and turn left to go over a stile. Keep near the canal until the next stile where a brook going off to the right is marked by a row of small trees. Take a diagonal course across this field (and away from the canal) to a gate. Cross the stile here and walk on for nearly a $\frac{1}{2}$ mile, crossing another stile, to reach a gate. This goes into an unmetalled lane past a small number of houses to reach a five-bar gate and the Weaver's Way. This is another old level-crossing and the keeper's cottage is to the right. Turn left to walk along the railway path. Within a short distance, an expanse of water to the left marks the terminus of the East Garston branch of the canal.

Peace and solitude near Tonnage Bridge

Keep on the Weaver's Way path which will take you back under Honing Bridge to the car park at Honing railway station.

Further Explorations

Unlike most of the canals in this book, the NW & D is not under the stewardship of British Waterways and thus there is no broad-brush permission to walk along the towpath of the canal. Walkers must stay on public rights of way where available and it has to be said that few are. Regrettably, therefore, it is not possible to walk the entire length of the NW & D nor even to visit some of the more interesting features. We can only peer across fields or over bridges in the hope of being able to follow the line or identify the remains of the occasional lock.

The ultimate end of the canal at Antingham Ponds can be explored by a short stroll of about 2 miles around the virtually traffic-free lanes north of North Walsham. The walk starts at Lyngate (OS ref: TG 275316 on map 133).

From North Walsham, take the Mundesley Road (B1145) which runs north-east from the A149. After about a ½ mile, a signpost indicates Lyngate (not to be confused with the Lyngate near Honing). Follow this lane until you meet a road which comes in from the left. Park on a verge near here. Continue along the road and turn right. After about a ¼ mile, the lane goes over a former canal bridge. Antingham Basin is to the left and the final locks at Swafield are away to the right. The lane shortly turns sharp left and continues past a turning to Bradfield to reach a now closed-off turning to the left. This was formerly the entrance to the Antingham bone mills and the canal terminus basin which were both in the field over to the left. The lane continues round to a point where a road comes in from the left. Continue straight on for a short distance and look left to see Antingham Ponds, the feeder reservoir for the canal. Return to the junction and turn right. This lane crosses two bridges over streams. Both were formerly navigable although only by small lighters. If you turn left at the next turning, you will shortly arrive back at Lyngate.

Further Information

There is little or no information on the NW&D and it would seem to be a good area for some fresh research. Most of the information included here came from: Boyes, J. and Russel, R., *The Canals of Eastern England*. David & Charles, 1977.

9
THE STAFFORDSHIRE & WORCESTERSHIRE CANAL
Stourport to Kidderminster

Introduction

The Staffordshire & Worcestershire Canal, or the Stour Cut as it became known to many of the boatman who worked it, was a true model for canal builders everywhere. As part of Brindley's grand plan to join the four corners of the country, it was one of the earliest canals built and yet it must rank among the very best. Undoubtedly the fact that it ran along such a vital route was important but the positive approach to its planning and execution must have helped. The proprietors knew precisely what they wanted and why. The S & W was efficiently built at a cost that was barely different from the budget. It immediately attracted a large amount of cargo and it yielded high and continuous dividends for over a century. All this and good looks too! It's the sort of thing that would make a mere mortal canal silt up with envy.

The S & W starts at a junction with the Trent & Mersey Canal at Great Haywood, a village about 5 miles east of Stafford. From here the canal widens to form Tixall Wide, a mini-Norfolk Broad on the outskirts of Cannock Chase. After Milford, the line goes under the M6 to reach Penkridge from where it continues south, twisting and turning along a typical Brindley contour-following line to Hatherton Junction. Here is the former Hatherton branch to the Birmingham Canal Navigations. After squeezing through a narrow sandstone cutting, known as the Pendeford Rockin', the S & W reaches the outskirts of Wolverhampton. Here are two important junctions just a ½ mile apart. First is the Autherley Junction with the Shropshire Union and then Aldersley Junction with the Birmingham Canal Navigations.

The course now passes Tettenhall and Compton to Wombourn where the locks and bridges around Bratch are a constant source of interest. After more tortuous bends, the S & W meets the Stourbridge Canal, another important route into central Birmingham. South of Stourton, the sandstone rocks force the canal into some amazing contortions: at Austcliff, near Cookley, a cliff overhangs the waterway; at Debdale, the lock seems to be built out of solid rock; a little further on the canal almost doubles back on itself. After all this excitement, it comes as a relief to reach the town of Kidderminster with its interesting array of canal-side buildings. After more fun with sandstone at Caldwall Lock, the line passes under the arches of the renascent Severn Valley Railway to continue gently down to Stourport, a town whose existence is solely due to Brindley's decision to join the River Severn there. Here are a marvellous collection of basins and buildings that would make a fine end for any canal.

There is no doubt that people seem to develop a deep and lasting affection for the S & W. It is a much loved canal and, consequently, one of the busiest.

History

Although Josiah Wedgwood and his colleagues saw the advantage of joining their proposed Trent & Mersey Canal with the River Severn, the decisive meeting at Wolseley Bridge in December 1765 concluded that the venture would be best left to others. So it was that a group led by James Perry met in Wolverhampton on 20 January 1766. By 19 March, they had decided to go ahead with the plan and authorized Hugh Henshall (James Brindley's brother-in-law) and Samuel Simcock to carry out a preliminary survey.

On 14 May 1766, the Act for a canal from the Trent & Mersey at Great Haywood to the River Severn at Little Mitton (now Stourport) was passed. It was to cost £70,000 (with £30,000 more if needed) and many local dignitaries chipped in. Earl Gower of Trentham, who had been a major player in the construction of the Trent & Mersey Canal, headed a list of the rich and famous which included Thomas Anson of Shugborough (whose land the northern end of the canal passed through), members of the Molineux family, the Earl of Stamford, Sir Richard Wrottesley and Sir Edward Littleton. The star canal builder of the age, James Brindley, was appointed surveyor and, with Perry as treasurer, building began. In fact, although Brindley laid out the line, the presumably grossly overworked maestro left the supervision of the engineering to Samuel Simcock and Thomas Dadford.

The construction of the S & W appears to have been both swift and efficient despite the fact that Brindley had not actually built locks before (the first is said to have been at Compton). By November 1770, following a supplementary Act to raise a mere £10,000, the canal was open from Stourport to Compton, near Wolverhampton (about $2\frac{1}{4}$ miles south of Aldersley Junction), and cargo from the Severn was already finding its way into the Midlands. The basin at Stourport

An early engraving by James Sheriff of the Stourport Basin as seen from the opposite bank of the River Severn. The Tontine Hotel can clearly be identified on the right, while the entrance to the canal can be seen in the top left-hand corner of the basin

British Waterways

was finished in 1771 and the whole line open for traffic on 28 May 1772 at a cost of about £100,000. The 46 miles of canal contained twelve locks from Great Haywood to the summit at Gailey and then thirty-one down from Compton to Stourport. Apart from two barge locks into the Severn, the canal was built to the Midlands narrowboat width of 7 ft. To add a further string to its bow, shortly after completion, on 21 September 1772, a line to central Birmingham became available when the Birmingham Canal was opened from the junction at Aldersley. It was thus now possible to navigate all the way from Bristol to central Birmingham for the first time.

The Staffordshire & Worcestershire Canal was a success from the beginning. Only eighteen months after opening, it was fully able to pay off its interest and provide its shareholders with a dividend of £4 a share. By 1775 the dividend had increased to £12. It is no wonder that, in 1783, £100 shares were trading for £400.

Further increases in trade followed the opening of the Dudley and Stourbridge Canals in 1779. Business at Stourport was such that it was rapidly becoming a major port for Staffordshire coal and all manner of manufactured goods such as ironware, glass, pottery and textiles from Birmingham, the Potteries and Manchester – much going to Bristol for export. In return, imports from Bristol and agricultural goods as well as cargo from the Coalbrookdale area were landed at Stourport on its way north. The affect that all this had on the previously insignificant spot now called Stourport was dramatic. The canal company built the basins and warehouses but there was a concomitant growth in houses, inns and various forms of industry, such as vinegar works, tan-yards, iron foundries and spinning mills.

The canal company was keen to build on this success and took efforts to promote connecting navigations. It actively supported the Stroudwater Navigation (from Framilode below Gloucester to Stroud) for example, and many of the personalities from the S & W subscribed to the Thames & Severn Canal which opened in 1789. This latter venture offered the potential of a somewhat lengthy line of 269½ miles from London to Birmingham. This was a very short-lived hope as the Oxford Canal, opened in the same year as the Thames & Severn, reduced the distance to 227 miles and by 1805 the Grand Junction shortened it even more.

The main object of the S & W company's concern, however, was the River Severn. In 1784, it commissioned William Jessop to try to sort out problems with low water in summer months as well as to improve the towpaths. Although proposals on how to improve the river were forthcoming, the necessary powers to do so were not. Strong opposition came from those who now saw the opportunity for a more direct line between Worcester and Stourbridge. A second attempt to make improvements in 1790–1 similarly came to nought even though the necessary powers and funds were available. The situation came to a head with the proposal for the Worcester & Birmingham Canal which threatened to avoid the tricky bits of the Severn altogether and to take a large amount of the central Birmingham traffic. The S & W fought the proposal hard and initially had some success in defeating the bill of 1790. But with the passage of the 1791 Act, the W & B emerged as a potential threat to S & W business.

Despite this challenge, the following years further demonstrated the basic soundness of the S & W. A branch to Stafford along the Rivers Penk and Sow was opened to Radford Wharf in February 1816 and trade on the S & W continued to be good. In the years around 1815 the average annual dividend was £43 16s. With the completion of the Worcester & Birmingham in December 1815, much of the coal and other cargoes were permanently lost but the underlying strength of the company's trade meant that toll receipts were comparatively unaffected. Dividends fell initially to £33 a share but rose again to £40 p.a. within three years.

Although the current storm had been weathered, the forecast remained unpromising. More and more competition was coming from an ever improving transport network. In 1825, for example, Pickfords were advertising a road 'conveyance' between Kidderminster and Birmingham in direct competition with the S & W. In response, canal hours were extended with some boats being allowed to work overnight for the first time. By March 1830, the S & W had to follow the Birmingham canals in allowing all boats to work locks through the

Canal boats unloading coal on to a (rather posed) series of Baggeridge's carts at Kidderminster Wharf in the 1920s with St Mary and All Saints Church in the background. On the far left is a steam dredger

British Waterways, The Boat Museum archive

night. Nightmen were employed and lock-keepers paid extra for their night-time labours. Another consequence of increased competition was the introduction of a complex toll policy designed to encourage particular cargoes on particular routes to particular places. The discounts became ever more complicated and unpredictable and led to a series of complaints from short-distance carriers who often had to pay more than those going further.

The building of the Birmingham & Liverpool Junction Canal (B & LJ) was also to be a threat to the fortunes of the S & W. The B & LJ, now part of the Shropshire Union, ran from Autherley Junction to the Chester Canal at Nantwich. Shippers from Birmingham went along the Birmingham Canal to Aldersley, then along a ½ mile of the S & W to Autherley and into the B & LJ. This line threatened the S & W's traffic from Autherley to the Trent & Mersey at Great Haywood and there was also concern that precious water would be lost down the new canal. Following their vigorous complaints, the S & W agreed to accept compensation for every lockful of water that flowed into the B & LJ. However, when the S & W ran short, permission was withdrawn and the B & LJ had to buy extra supplies from the Wyrley & Essington Canal (the northernmost of the Birmingham Canal Navigations). By the time the B & LJ opened on 2 March 1835, their promoters and users were growing rather tired of the S & W's high tolls and the pedantic fuss over water supplies. They decided to go over the S & W's heads. The plan was to build a mile long aqueduct 'fly-over' from the Birmingham Canal to the B & LJ; it would bypass the S & W altogether. The S & W were so shaken that they promptly agreed to reduce their tolls. So successful a trump card was this, that it was used twice more on different issues with similar effect in 1842 and 1867.

Throughout all these skirmishes, the S & W remained a solid, profitable company, further demonstrating that it was the right canal in the right place at the right time. In 1838, it carried 680,479 tons of goods and paid a £38 dividend on a £140 share. With such profits, the S & W was able to fund the Hatherton branch. This ran from Hatherton Junction about $5\frac{1}{2}$ miles north of Autherley to Churchbridge and was an important route for the collieries near Wyrley. The branch was further extended in 1863 to join the Wyrley & Essington Canal (part of the Birmingham Canal Navigations).

Following the first rumblings of railway competition in 1830, the S & W maintained a fiercely anti-railway stance and almost succeeded in forming an anti-train alliance with the other canal companies of the region. It was vigorous in its condemnation of schemes to convert canals to railways and insisted that the closure of any one canal was a blow to all canals. But such forthright resolution cracked in 1847. Perhaps the damage to the waterways network had already been done by then but there was a dramatic U-turn and the company declared that it would welcome an amalgamation with a railway company in order to guarantee continuing dividends. In the event, the company didn't succeed in this and it didn't really need to. Toll receipts remained high and 14 per cent dividends continued into the 1870s. Various schemes were devised to modernize the line and to expand its influence. It formed its own carrying company, became part of the trust to work the Thames & Severn Canal, and bought the Severn Towpath Company. But all these ventures were fighting the inevitable. Although the canal was still carrying some 722,000 tons of cargo in 1905, the best days were clearly over.

The story of the first half of the twentieth century is one of gradually declining trade and influence. Remarkably, however, the company was still independent and still able to pay $2\frac{1}{2}$ per cent dividends right up to the time when it was nationalized in 1948, a situation that was mostly thanks to a continuing coal trade. But the gradual incursion of the railways and the improvement of the roads eventually had their effect. The last regular commercial traffic was the carriage of Cannock Chase coal to Stourport power station. This business, which apparently involved some seventy pairs of narrow boats, came to an end in the 1950s.

The Report of the Board of Survey to the British Transport Commission had mixed feelings about the S & W. It concluded that there was no need to retain both it and the Worcester & Birmingham Canal, which in its view formed alternative routes from the Severn to the Midlands. In 1959, the Bowes Committee of Inquiry recommended that the S & W be closed. This bureaucratic decision, which mirrored the prevailing view about the country's waterways at that time, prompted action from local canal supporters. The S & W Canal Society was formed in the same year with the aim of developing and maintaining the line. Luckily for us, the wolves were held off long enough by both the society and the Inland Waterways Association to take the canal network into a more sympathetic age. In the Transport Act of 1968, both the Worcester & Birmingham and the entire length of the S & W were declared a cruiseway and British Waterways were given the authority to maintain them to a navigable standard. The future of the S & W was thus assured. Today, the line is not only open but full of vigour and a much-loved cruising route through the west Midlands.

The Walk

Start:	At Stourport Basin (OS ref: SO 810710)
Finish:	Kidderminster Lock (OS ref: SO 829768)
Distance:	4¾ miles/7 km
Map:	OS Landranger 138 (Kidderminster & Wyre Forest Area)
Return:	Midland Red West Buses 11–16. Very regular, even on Sundays. Leave from Town Hall, Kidderminster to Stourport Basin (York Street). Telephone: (0562) 823631
Car park:	At Severn Bridge, Stourport or signposted multi-storey in central Kidderminster
Public transport:	Kidderminster has a BR main-line station which runs to Birmingham New Street

The walk starts at Stourport and goes to Kidderminster but could equally easily go the other way. I chose to park in Kidderminster, take the bus to Stourport and walk back but, with a comparative wealth of buses, alternative combinations can be devised.

From the York Street bus-stop in Stourport, turn right and then left along Bridge Street. Just before the road passes over the River Severn, follow the slope down left to a car park and (Crown) basin. This wide area of tarmac eventually leads to the river. Turn left along the river towpath to go past a funfair. Within a short distance this arrives at the splendid Stourport Basin.

Stourport was effectively created at the behest of James Brindley and the S & W Canal Company. Before they arrived, the tiny hamlet of Little Mitton

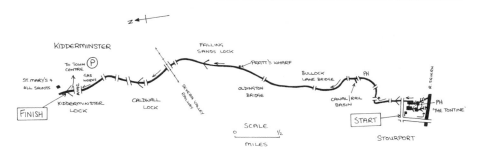

The Staffordshire & Worcestershire Canal

had just a handful of inhabitants. Less than thirty years later (in 1795), the population was more than 1,300 and the town consisted of a series of basins, warehouses, workers' cottages and a developing subsidiary industry. Lower Mitton wasn't Brindley's first choice. He had originally wanted to make his junction with the Severn at Bewdley, an established river port some 4 miles upstream. But partly because of local objections and partly because of the very convenient sandstone levels between Kidderminster and Lower Mitton, modern-day Stourport became what it is.

The two entrances to the canal are now ahead of you. The first can be seen from a humpback bridge over the entrance to a narrow lock. Above it, in a staircase, a second lock lifts the canal boats to the Bottom or Lower Basin. On the left of the second lock is a covered dry lock. Two further staircase narrow locks lift craft to another basin, the Clock Basin. This lies just out of view and to the west of the Clock Warehouse which sits proudly at the centre of the complex. The site of the funfair, by the way, was formerly that of a steam engine which pumped water up into Clock Basin in order to maintain the levels.

Continue along the Severn towpath to reach the 15 ft 4 in wide Barge or Trow Lock. This was the original entrance to the canal from the Severn; the narrow locks were not added until 1781. Ahead of you is the Tontine, a large hotel and inn built in 1788 for the many traders who came here to do business. It was also the place where the S & W management committee periodically met its shareholders. The building is now divided into apartments and yet also still functions as an inn. The gardens afford an excellent spot for a summer's lunch. Ian Langford, in his fine, scholarly towpath guide (see below), explains that a tontine is a type of investment in which the shares become null and void when their owners die. Eventually the sole survivor takes ownership of the property. Whether this father-time version of Russian roulette was used with the Tontine Hotel is unknown.

Cross the barge lock at the top gates to go left of the Tontine to the second barge lock. Ahead is the Upper (sometimes called the Middle) Basin. This is the biggest in the complex and also the oldest, built between 1768 and 1771. On the

Stourport Basin

left is the Clock Warehouse, now home to the Stourport Yacht Club. The clock was apparently a gift from the locals (although the company did chip in) in 1812.

Pass to the right of the Upper Barge Lock and in front of the curved walls of the British Waterways yard (formerly the offices of the Severn & Canal Carrying Company). This shortly leads to a road (Mart Lane). Turn left to a viewing point (formerly the site of a warehouse known as the long room) where on a warm day the seats can be used to good advantage. On the other side of Mart Lane is a timber yard that is on the site of what was formerly the 'Furthermost' Basin. This was used for the unloading of coal destined for Stourport power station (situated on the Severn just a little way downstream). This traffic was finally transferred to the railway in 1949 and the basin filled in.

Our route continues along Mart Lane. The row of terraced workman's cottages on the right was built by the canal company and, although they look rather dilapidated, are listed ancient monuments. The entrance to the S & W is now over to the left. The canal passes York Street by Wallfield Bridge. One of the delights of the S & W is that since the 1830s, all the bridges bear fine oval, cast-iron name and number-plates. They are, without doubt, a useful aid to navigation for both boaters and walkers.

Immediately past the bridge is York Street Lock with its 1853 toll cottage (now converted into a shop). The canal follows a secluded line away from the streets of Stourport, wedged between a high brick wall and the sad remains of the canal company's maintenance yards. We go under Lower Mitton Bridge almost in secrecy, catching just the odd glimpse of the world (and Tesco's) outside. The fine warehouse on the left-hand bank formerly belonged to an iron foundry, the works itself having been demolished in preference for the police and fire stations.

After the busy Gilgal Bridge, the canal seems to be seeking out some more rural resort. The line taken is high above the Stour Valley on a course chosen to be along an easily worked sandstone ledge. As a consequence, the line bends and twists around the contours and there is no more splendid example than the next sudden sharp turn which gives the left-hand bank a cliff-like profile as it rises up to St Michael's churchyard above. An unusually inclined, small pedestrian bridge (Mitton Chapel Bridge, No. 7) crosses the canal on the bend.

The path continues on to pass some canal workers' cottages that date from 1800. After them is the Bird in Hand pub and opposite, after a second look, an allotment full of gravestones, part of St Michael's churchyard. The canal now passes under a wide brick arch bridge that once carried the Severn Valley line. Passenger services between Stourport and Hartlebury were withdrawn on 5 January 1970 although coal traffic continued to use this section until March 1979 *en route* for Stourport power station. On the left just after the bridge is the basin where goods were moved from the railway on to the canal. Steel was imported from South Wales and coal from Highley Colliery near Bewdley and shipped to the Wilden (just across the Stour Valley) and Stourvale (just north of Kidderminster) ironworks. The old pulley wheel, on the towpath side, was used to haul the boats in and out of the basin.

After a while we reach Upper Mitton Bridge (No. 8). Here the canal widens slightly at the site of Upper Mitton Wharf. To the right, through the hedgerow is the River Stour which meanders across the valley almost to run into the towpath; a spill weir provides an overflow from the canal. After Bullock Lane Bridge (No. 9), the canal runs quietly through a pleasantly wooded stretch high above the Stour Valley with the red sandstone occasionally outcropping along the left bank. Oldington Bridge (No. 10) is a fine and typical Brindley accommodation bridge made of brick but with a sandstone block stringer course and coping.

After a short distance, the towpath passes over a small bridge opposite a winding hole or boat-turning point. This is Pratt's Wharf. The bridge is now blocked but it was the entrance to a branch down to the River Stour. In among the trees, shrubs, dead fish and old Coke tins on the right is a derelict lock, built in the 1840s to link the canal to the Stour. At that time, the river was navigable for about a mile downstream to Wilden ironworks. Before the junction was built, the coal and iron were transhipped here into smaller river boats which then completed the journey. The arm was last used in 1949. The Wilden ironworks closed down shortly thereafter and the river is no longer navigable although a towpath can still be followed south to the site of the works which has now become a trading estate.

After a pipe crosses the canal to the sewage works on the left bank (hidden from the eye but not from the nose), the woods thicken and the canal bends around to Falling Sands Lock (No. 4). Before admiring the lock itself, take note of the wrought-iron, cantilever bridge that crosses in front of the lock gates. It has a gap between the two arms to allow the tow line to pass through it, a neat idea which must have saved the bargemen quite a lot of time. Falling Sands Lock is well named as it has been built on an area of unstable wind-blown sands.

After the lock, the canal follows the contours around to reach Falling Sands Bridge which is now, sadly, living up to its name and has been superceded by an ugly, temporary structure. Both are completely upstaged, however, by the magnificent Severn Valley Railway Viaduct which carries the steam railway into Kidderminster. The line was opened on 1 June 1878 to provide a link between the Oxford, Worcester & Wolverhampton Railway (the OW & WR) and the

The Severn Valley Railway Viaduct and Falling Sands Bridge, near Kidderminster on the Staffordshire & Worcestershire Canal

Severn Valley line at Bewdley. On 5 January 1970 it went the same way as the Stourport line. However, this line has been splendidly restored by enthusiasts from the Severn Valley Railway and now runs up to Bridgnorth. The viaduct spanning the Stour can be seen on the right.

The town of Kidderminster is now becoming increasingly obvious: a car scrap-yard occupies the right-hand view while the path consists of a series of concrete slabs each labelled 'Danger 132,000 volts'. Fortunately the scene is saved by the marvellous Caldwell Lock which appears to have been hewn out of solid rock. A split, cantilever, bridge again precedes the lock but there was also, at one time, a house built into the cliff. A small remnant of this looking like a fireplace can be seen just above the downstream lock gate. The house collapsed in the 1960s following years of vandalism and neglect. One can only wonder how the lock was built into the sandstone cliff which rises so strikingly behind it.

The stretch between here and central Kidderminster was the site of an interesting experiment in the 1920s. The company set up an electric barge, drawing power from overhead lines, to pull boats into town at a steady $3\frac{1}{2}$ miles per hour. Although successful at hauling the barges, the idea was never adopted, presumably because of the capital cost involved in putting up the overhead lines.

After a modern road bridge carries the A451 over the canal, the left bank contains the remains of Old Foundry Wharf and the last remnants of a brass and iron foundry that was closed in 1972. How long those remnants will remain is uncertain but a former coal wharf about 100 yd further along on the left has certainly reverted to nature. The canal passes under another bridge (Caldwall Mill Bridge, No. 14) to an old wharf now used as a car park and then to a metal girder bridge which carries Castle Street.

High walls line the canal and towpath. Here, as a path comes down from Castle Street, is the Kidderminster Public Wharf. The buildings to the right are one part of Kidderminster's famous carpet industry (in this case Brinton's). This stretch offers those interested in Victorian factory architecture a wonderful time. As the towpath swings gently left, the view is dominated by the terrific sight of the old gas works and its wonderful chimney. The area was a busy one for the canal as the towpath passes over a series of blocked-off factory arms. The timber yard across the waterway also used the S & W extensively.

We are now close to the centre of Kidderminster. After passing under a modern road bridge that carries the A442 to Bridgnorth, the canal reaches Town (or Kidderminster) Lock, which at 12 ft is one of the deepest on the line. The scene, overlooked by St Mary and All Saints Church, was once a busy one. Here was Mill Wharf, said to have been the centre of the canal's activities in the area, with a series of warehouses, a weighbridge and a stables. There was also a coal wharf for the carpet factory on the left-hand bank. From here the horse-drawn fly-boats left at 6 p.m. in the evening, loaded with Kidderminster carpet for the railway station at Wolverhampton, from where they were shipped out for distribution. It's remarkable that this activity continued until October 1950. Unfortunately, the most remarkable thing now is the traffic noise.

Before returning under the bridge and going left into the town centre, it's worth going a bit further along the canal to see the River Stour passing quietly, and mostly unloved, underneath a small aqueduct. Kidderminster itself is full of shops, cafés, take-aways and pubs. The Town Hall bus-stops are at the opposite end of the pedestrianized shopping centre.

Further Explorations

Virtually all 46 miles of the S & W are open to walkers (the only difficult stretch is that between Aldersley and Hatherton Junctions) and nearly all of it is worth a stroll or a visit.

The area between Kidderminster and Stourton (OS Landranger 139, ref: SO 861848) has some magnificent sandstone scenery and is full of interest. An excellent walk of 11½ miles goes from Kidderminster Lock to Stourbridge with return by train. The route initially passes through wooded stretches around Wolverley and heads on to Cookley. Just before this small town the canal does an almost 180° turn before reaching Debdale Lock with its intriguing cave storehouse and circular weir. After passing an ironworks, the canal goes under the town through what is thought to be the oldest navigable tunnel on the waterway system. Shortly after the tunnel, the canal bends around Austcliff, a sheer rock overhang, and on to the Whittington Locks and Kinver. After passing the Hyde, the canal goes through some delightful country to the diminutive Dunsley Tunnel. From Stewponey Lock, with its octagonal toll-house, the canal reaches the Stourbridge Canal. This line takes the waterway via the Dudley Canal into Birmingham, a through route first opened in 1792. Turn right to cross the bridge to go along the Stourbridge. This passes two locks, changes banks at a road bridge, passes two more locks and then continues to Wordsley Junction (about 2 miles). Here cross a bridge and turn right to follow the Stourbridge branch into town.

The area around Bratch (near Wombourne) is interesting and can provide a gentle stroll up and down the canal of about 2 miles. Park at the well signposted Bratch picnic-site car park (OS Landranger 139, ref: SJ 867938) and walk north (i.e. turn right at the canal). Virtually immediately, you reach Bratch Locks. Here is an attractive octagonal toll-cottage and three locks, each with a large side pound (to the left), built to ensure that there is sufficient water to operate the locks which are very close together. Walk north along the canal to the highly unusual bridge at Aw. The lock here also has one of Brindley's famous circular weirs. A notice back at the picnic site has information on the splendid Victorian waterworks pumping station (built here for easy access to coal from the canal) and the nearby Kingswinford Railway Path (ably described in Jeff Vinter, *Railway Walks: LMS*. Alan Sutton Publishing).

At the northern end, the S & W joins the Trent & Mersey Canal at Great Haywood Junction (OS Landranger 127, ref: SJ 995229). There is a good walk between the junction and Milford Bridge (968241), a distance of $2\frac{1}{4}$ miles each way. There is ample parking at Milford village green and some on street at Great Haywood. At the junction, the S & W is crossed by a proud and elegant roving bridge which carries the T & M's towpath. This is immediately followed by a toll-house with some fine windows and then two aqueducts, the second of which passes over the Trent. After a bridge, the canal widens to form Tixall Wide, a broad made to satisfy the demands of the local landowner, Thomas Clifford of Tixall Hall, who was also one of the original principal investors in the waterway. It's a very pleasant spot and some authors have described it as the most beautiful part of the waterway network. After Tixall Lock and a road bridge, the route crosses the River Sow by a typical Brindley aqueduct, the low arches producing a squat, heavy appearance. The walk ends at Milford Bridge, a turnover bridge where towing horses could change banks without needing to uncouple their tow-lines. Those with OS maps could devise a return across the neighboroughing Cannock Chase to meet the Trent & Mersey, and then back to Great Haywood.

Further Information

The S & W Canal Society can be found at:
   c/o Mrs A. Pollard,
   8 Frimstones Street,
   Wollaston,
   Stourbridge,
   West Midlands.

For more information about the canal, the following books can be recommended:
Langford, J. Ian, *Staffordshire & Worcestershire Canal, Towpath Guide No. 1*. Goose & Son Publishers, 1974.
Hadfield, Charles, *The Canals of the West Midlands*. David & Charles, 1969.

10
THE STRATFORD-UPON-AVON CANAL

Kingswood to Stratford

Introduction

Pick any tourist at random from the thronging hordes that pack around the Memorial Theatre in Stratford and ask them where the canal is and they probably couldn't tell you. Okay, so there's a basin with boats in the middle of Bancroft Gardens but a canal? And where do the boats go when they disappear under the bridge at the north-east corner? Don't they know that within just a few yards of the crowded streets lies a peaceful watery retreat? No, they don't – and long may it be so.

The Stratford-upon-Avon Canal is $25\frac{1}{2}$ miles of narrow canal from a junction with the Worcester & Birmingham at King's Norton to the River Avon at Stratford. The line is broadly divided into two halves geographically, historically and, for many years, legally. Built first, the northern end runs from the Worcester & Birmingham by a guillotine stop lock (designed to prevent excessive loss of water into the W & B) and through Brandwood Tunnel to Warstock and Shirley. From there it leaves suburbia to Earlswood where feeders from the canal's reservoirs top up the level. The last part of the $12\frac{1}{2}$ mile long northern canal runs past Hockley Heath and down the Lapworth Locks to Kingswood Junction where a short arm joins the Grand Union Canal (the former Warwick & Birmingham Canal).

The southern canal is 13 miles long and was built some ten years or so after the northern. By the early 1950s, it was largely derelict and was scheduled for abandonment. Responsibility for it was then taken over by the National Trust who ran it from 1960 until 1988, during which time it was gradually restored to life and water. The first part of the southern canal continues down the Lapworth Locks before running under the M40 and into some peaceful, rural countryside. There now follow three fine iron-trough aqueducts: Yarningale, Wootton

Wawen and Edstone (Bearley). After further lockage at Wilmcote, the line runs gently and secretively into Stratford to meet the Avon at Bancroft.

The southern Stratford will always hold a special place in the heart of canal enthusiasts as it was the first to be restored from dereliction in the early 1960s. But even for those who don't have the romantic associations, a walk along the Stratford makes a fine, if fairly long, day out and savours the charm of this fine waterway.

History

The town of Stratford in Warwickshire has always had its own waterway, the River Avon, which flows 42 miles into the River Severn at Tewkesbury. This waterway had been made navigable in 1639 and had become an important route for cargo coming to the Midlands from Bristol. Stratford was also on a major land route for grain passing north and for coal going south. You can imagine, therefore, the shock to the good peoples of the town when, in the late eighteenth century, the opening of the canal systems which joined Birmingham and Bristol, and Coventry and Oxford, caused the traffic that used the Stratford route to drop by 75 per cent. The townsfolk feared they were being bypassed and sought to be returned to life in the mainstream.

In the early 1790s, several different proposals to link Stratford with the burgeoning canal network were the talk of the town. One would have taken a line to the Coventry Canal at Warwick, a second to the Stourbridge Canal. In the event, the committee formed to consider the issue had a straight choice between making a cut to the Digbeth branch of the Birmingham Canal or a line to the, then still to be built, Worcester & Birmingham Canal at King's Norton. The issue finally resolved itself when the proprietors of the Dudley Canal decided to link with the W & B at Selly Oak. This new line, the Dudley No. 2 Canal, promised a cheap source of, and ready traffic in, coal heading from Netherton to the south.

John Snape of Birmingham surveyed a line from the proposed W & B at King's Norton to Stratford that ran close to the new Warwick & Birmingham Canal (now the Grand Union) at Lapworth. This survey included two quarry branches: one from Hockley to Tanworth and the other from Wilmcote to Temple Grafton. In these original plans there was to be no junction with the Avon. Instead, the proposals took the line to a terminus near the present site of Stratford railway station. Charles Hadfield explains that this, apparently strange, decision was taken to offset any possible antagonism from the Worcester & Birmingham who would have feared competition from an alternative River Severn to Birmingham route.

The Stratford-upon-Avon Canal Act was passed on 28 March 1793. It authorized capital of £120,000 with power to raise a further £60,000 if needed. In 1795, a further Act gave powers to raise another £10,000 to build the link between the Stratford and the Warwick & Birmingham at Lapworth. Construction work, under the engineer Josiah Clowes, began in November 1793 from the northern end and proceeded, at least initially, with some vigour. Although Clowes died in 1794, by 25 May 1796 the line was open from King's Norton to Hockley Heath – a distance of 9¾ miles. This stretch had no locks but did have the 350 yd long Brandwood Tunnel. Despite the comparative simplicity of the route, the works used the entire authorized capital.

With no finance available, building work slowed to a stop. It wasn't until 1799 that further progress was possible following the passage of another Act which authorized an extra £50,000. This same Act altered the course of the unbuilt section of canal near Lapworth to take it closer to the Warwick & Birmingham. With Samual Potter, formerly Clowes' assistant, appointed as engineer and Benjamin Outram called in to help, building was restarted. By 24 May 1802, the line reached Kingswood (near Lapworth) and the Warwick & Birmingham Canal. The cost to date had totalled £153,000 but, at last, there was a route from the W & B and Dudley Canals in Birmingham through to the south. There was also some prospect of toll income to bolster flagging fortune and enthusiasm.

Meanwhile in 1798, George Perrott, owner of the Lower Avon, made moves to join the canal with the Avon at Bancroft, Stratford. Such plans must have appeared somewhat fanciful at the time with no apparent possibility of raising the funds to get the line anywhere near the town let alone to the Avon. Attempts to raise the extra cash needed to do so failed miserably and it wasn't until 1810 that any further significant progress was possible. The stimulus was the would-be visionary plan to link the Wiltshire & Berkshire Canal with the Stratford to form a new west of central north–south canal line. The line was planned to run via Shipston on Stour, Bourton-on-the-Hill, Lower Swell, Bourton-on-the-Water, Great Rissington, Bampton and Fyfield. Although definitely lodged in the realms of speculation, the scheme took the fancy of one William James, a local land agent and businessman. James, who was already a shareholder in the Stratford Canal, saw the potential for making Stratford into the hub of a nationwide transport system. Whether this was a false vision or not, James' zeal was responsible for a burst of renewed enthusiasm over the ailing Stratford line. In 1810, money was again being raised successfully and, by 1812, construction work was under way once more. On 22 June 1813, the line was open to Wootton Wawen and supplying coal to Alcester.

In September 1813, the route south was adjusted and the plan to join the Avon officially lodged. By now William James was the owner of the Upper Avon Navigation and it was almost certainly with his encouragement that the link was forged in order to compete with the Worcester & Birmingham. This scheme was approved by Act of Parliament in May 1815. The plan, however, had already stimulated competition between the two companies. Initially this was seen simply as a race as to which line could open first, a contest won by the W & B which opened on 4 December 1815. The Stratford wasn't finally complete until the following year and opened on 24 June 1816 with full ceremony. Ironically, on the big day the canal committee had to borrow a boat from the rival W & B to take it from Wootton Wawen to Stratford, where the assembled population duly rejoiced.

The line as opened was 25½ miles long: the northern section from King's Norton to Kingswood was 12½ miles and the southern section into Stratford was 13 miles. The canal is level from its junction with the W & B to Lapworth, it then falls through fifty-four locks to the Avon. As the line from Kingswood to Stratford had cost £143,000, the total cost of the canal was about £297,000.

Although in heavy competition with its neighbours, the Stratford was successful in becoming the main route for the Netherton coal trade and for other goods from the Dudley No. 2 Canal. Iron, firebricks and other industrial products came mainly from the Stourbridge area and even as far away as Coalbrookdale. Salt from Droitwich arrived via the W & B to pass along the Stratford to Lapworth. The bulk of the trade to Stratford and the surrounding areas, however, was coal and return cargo was primarily agricultural produce (such as malt and corn). Although the branches to the quarries were never built, land transport brought limestone, paving stones and 'marble' to the canal. The Temple Grafton limestone was mostly shipped to Halesowen, where it was used in the ironworks, and to limekilns at Selly Oak and Haywood. There were also limeworks at Stratford.

The expected competition with the W & B for trade between Birmingham and the Severn was soon lost. The Stratford was simply too indirect and the Avon far too circuitous to attract any through custom and none was found. The whole issue was sealed in 1830 when the W & B Company leased the lower Avon, effectively closing off the through route for good, although competition remained for the carriage of coal to the Evesham area. Despite this, the Avon link was important to the canal. Throughout the 1830s, some 10,000 tons of coal p.a. were transhipped from the cut to the river at Bancroft. The canal also benefitted from the Stratford & Moreton horse tramway which carried coal (16–19,000 tons p.a. in the 1830s) on to Moreton-in-Marsh (14 miles to the south) and, later, to Shipston on Stour.

By 1838 the canal was carrying over 180,000 tons of goods p.a. and averaging annual gross receipts of £13,500. Dividends had been paid since 1824 and, in the 1830s, fluctuated between 30s. and £2 per share. But the 1830s were the peak years for the Stratford Canal and profitability was soon to decline in the face of railway competition. There had also been some significant positioning and counter positioning by the various canal companies, most notably the W & B and the Warwick & Birmingham, in order to compete for traffic and to promote or dissuade the various new schemes that were appearing – the most

Edstone Aqueduct on the Stratford Canal in the mid-1920s. The aqueduct consists of a 475 ft long cast-iron trough, carried on fourteen arches across the River Alne, two railway lines and a road. For enthusiasts, the car is a 21 hp Lanchester of about 1924
    Birmingham Museum of Science & Industry, The Boat Museum archive

notable of which was a new London to Birmingham line. But it was the direct competition with the railways which bit deepest into the company's profits. To compete with the London & Birmingham Railway, which opened in 1838, both the Stratford and the Warwick & Birmingham cut their tolls and thereby reduced their profitability. Despite this, Stratford in 1845 was a trading centre of note. There were twenty-two merchants operating from the town in coal, corn and timber: 50,000 tons of coal were still being shipped into the town, 15,000 tons travelling on via the tramway and 8,000 on via the Avon, 16,000 tons of limestone were still being shipped out from Wilmcote.

The company took an honest stance in the face of the increasingly overpowering railway competition. On 22 March 1845, it agreed to be taken over by the Oxford, Worcester & Wolverhampton Railway which planned to build a line from Birmingham, through Stratford and Moreton to Cheltenham. However, as part of this arrangement it was agreed that if the OW & WR could not get its Act to authorize the purchase then the Great Western Railway would be allowed to step into the breach to do a deal on its own behalf or as an agent.

In 1846, the OW & WR and the Birmingham and Oxford Junction Railway agreed that the latter would build a railway to the north of the Stratford whereas the former would build a line to the south. As a result, the OW & WR's Act of July 1846 included powers for the OW & WR to buy the canal for the B & OJ. The Act laid down clauses to ensure that the canal was maintained and that tolls should not be higher than they were at the time of the Act. Typically tolls for coal and goods taken between King's Norton and Stratford were 1½d. per ton per mile with goods taken as back carriage tolled at ¾d. The OW & WR, however, found that due to a problem with the Birmingham & Oxford Junction Bill passed in the same session, it was unable to sell the canal immediately to the B & OJ and hence delayed its purchase until the B & OJ had legal powers to own and run a canal. Unfortunately, repeated bills were thrown out of Parliament and the situation ran on unsatisfactorily.

In 1848, the B & OJ was bought by the Great Western Railway, changed its mind about the usefulness of owning the canal and became obstructive to the OW & WR, which suddenly found that it was committed to buying a canal it didn't want. The purchase was finally cemented on 1 January 1856 although the line was not worked by the OW & WR until 1 May 1857. In 1859, the OW & WR raised £160,000 to pay off the canal's annuities, mortgages and other debts still outstanding.

There was no doubt that the purchase was a mistake for the OW & WR. The canal was soon to begin its inevitable decline. Most of the traffic gradually seeped away on to the nearby railway line and little or no dredging was done to maintain the course. While not actively discouraging traffic, there were no positive efforts to attract through cargo, primarily as this would have competed with the railway traffic. The Stratford-upon-Avon Railway was opened on 10 October 1860 with stations at Bearley and Wilmcote and on to Birmingham. By 1861, the OW & WR had been renamed the West Midland Railway and railway traffic could pass straight through Stratford from Evesham. Canal receipts dropped by 20 per cent.

On 1 August 1863, the old antagonist, the GWR took over the West Midland and hence control of the Stratford Canal. As was inevitable with the GWR, canal traffic declined dramatically. Receipts fell from an average of £6,760 p.a. in the early 1860s to under £2,000 in the late 1880s. By the 1890s, traffic on the southern section was all but gone while that on the northern part was very much reduced. Traffic in the 1890s was mostly in manure with a small amount of coal, lime, grain and salt. When Temple Thurston passed this way in 1911, the locks were seldom used and rapidly weeding up.

Somehow the line survived until nationalization in 1948 although the southern half had been unnavigable since the Second World War. The last boat to reach Stratford did so in the early 1930s. In the Board of Survey report of 1955, it was recommended that the northern section be retained while the southern was put into the 'insufficient justification to keep' category. This official classification must have stimulated the Warwickshire County Council to consider the canal dead and, in 1958, to announce its closure so that repairs could be undertaken on Wilmcote Bridge. This act was the spur needed for the recovery of the waterway. Meetings of the recently formed Stratford-upon-Avon Canal Society were held to protest against the abandonment of the waterway and a use permit, issued for a canoe trip from Stratford to Earlswood, proved that the line was still a navigable canal.

Stopping the abandonment was one thing, restoring it was another. The canal society, in league with the Inland Waterways Association, proposed that the National Trust should take over the canal with the purpose of undertaking the restoration. The trust, who expressed interest in a number of canals at that time, agreed to take on the responsibility and, after the official withdrawal of the closure plan in 1959, took a lease on the southern section of the Stratford on 29 September 1960. With funds from the British Transport Commission, the Ministry of Transport, gifts from other supporters as well as from the National Trust itself, a programme of restoration was implemented under the control of David Hutchings. The sterling work of the volunteer effort (which included prisoners from Winson Green prison) was rewarded when the Queen Mother reopened the line on 11 July 1964. The line is a model restoration for all that have followed since.

The Queen Mother reopens the restored southern Stratford-upon-Avon Canal amid great celebration on 11 July 1964

Birmingham Post & Mail, British Waterways

Once fully restored, the role of the National Trust was always slightly anachronistic, particularly as its control was limited to the southern half. Responsibility was therefore transferred to British Waterways in May 1988. Since then the line has formed part of the highly popular Avon Ring circuit that allows boaters to motor from Stratford along the River Avon to Tewkesbury, then along the River Severn to Worcester, then up the Worcester & Birmingham to King's Norton and then back to the beginning along the Stratford.

The Walk

Start:	Lapworth railway station (OS ref: SP 188716)
Finish:	Bancroft Gardens, Stratford (OS ref: SP 204548)
Distance:	14½ miles/23 km (or shorter, see text)
Maps:	OS Landranger 151 (Stratford-upon-Avon) and 139 (Birmingham)
Return:	BR Stratford to Lapworth via Hatton. Some go direct, otherwise change at Hatton. Stay on same platform for train to Lapworth which goes in opposite direction. Telephone: (021) 2002601

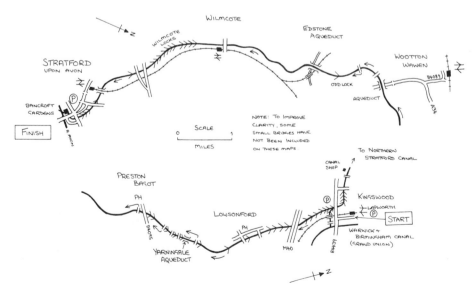

The Stratford-upon-Avon Canal

Car parks: Various and well-signposted at Stratford, including Stratford station or Lapworth station

Public transport: Lapworth is on the main Oxford to Birmingham New Street BR line

This long walk can be made shorter in a number of ways using stations at Wootton Wawen (a ½ mile from the canal) and Wilmcote. The route can be divided into two (7½ and 8 miles) using Wootton Wawen. As the most interesting canal structure along the route is between Wootton Wawen and Wilmcote, a 4 mile walk between the two stations will include this sight.

Lapworth to Wootton Wawen

From Lapworth station cross the line and turn right to pass along this residential road to a T-junction. Turn left for about a 100 yd to reach a bridge. Just before The Navigation pub, cross the road to turn right along the towpath which is on the right-hand bank.

This is the Grand Union Canal (formerly the Warwick & Birmingham) which, to the south, continues on to London and, to the north, goes to Birmingham where it splits at Bordesley Junction. One line continues to Digbeth. The other passes along the Birmingham & Warwick Junction Canal to join the Birmingham & Fazeley Canal at Salford Junction almost directly under the spaghetti junction of the M6.

If you're getting confused with all these junctions, life will shortly get worse because the walk soon reaches Kingswood Junction which comes in from the right. The main line of the Warwick & Birmingham now runs straight on south via the Hatton Locks to Warwick (see Further Explorations). Here we cross the junction roving bridge and turn right.

This short length of junction canal was opened on 24 May 1802 and, for ten years, was the southern exit of the Stratford. After passing under the railway, the canal bends right just before a lock. Originally the line ran straight on here through a lock, which was to the right of the building on the other side of the fence, to a basin. The lock had a fall of 6 ft thereby ensuring that the Warwick & Birmingham gained water at every downstream passage. When the Stratford company restarted construction work in 1812, they built a lock (now called lock 21) on the main line above the old basin, lowered the level of the basin itself to that of the branch (and hence of the W & B) and removed the top gates of the old lock. These were replaced by a stop lock with guillotine gates and a channel through which water could flow from the Stratford to the W & B. This revision angered the W & B who probably felt that they would gain less water and may even lose some if the level of the W & B were higher. The row rumbled on for several years before it was decided to build the present lock which runs to the level above No. 21. The original channel was then closed, filled in and converted into a private garden. The new line and lock were opened on 22 December 1818 and effectively restored the *status quo*.

After passing the lock, the vista opens out to the small lagoon which is Lapworth lower reservoir. The route to Stratford (13 miles) can be gained by turning left. However, it's well worth taking a short stroll up to see the Lapworth Locks and the northern Stratford. So cross the southern canal entrance roving bridge in front of you and bear right to pass a car park and picnic site on the left (toilets can be found here). The towpath passes a lock and the upper Lapworth reservoir to go under a road bridge that carries the B4439. The canal now bends around to the left to pass four locks. As the locks in the flight are very close together, each is provided with a sizeable side pound to ensure that sufficient water is available for continuous operation. Following a pipe bridge, there is a superb view up the Lapworth flight (Locks 8–14). After a photo-call and a visit to the small canal shop on the left, return to the entrance of the southern canal.

From the roving bridge, the canal runs down to pass lock 21. Immediately, the right bank of the canal widens to a reed-filled mire. This is the site of the original Harborough Banks Basin. The original route of the junction to the W & B can be seen on the towpath side of the canal where there is now a brick wall.

One of the most interesting characteristic features of the southern Stratford Canal can be seen from this point. On the towpath side of the canal just beyond the brick wall is a typical southern Stratford lock house – a unique design with its curved or barrel-vaulted roof built like the brick arch of a bridge. The barrel is held together with steel cross-ties that link the walls. It's said that these delightful little cottages were built by navvies who knew how to make tunnels and bridges but weren't too hot on the normal methods of building a house. Whatever the story, there are a number of them along the southern canal. Many have been extended in various ways although some remain, as the builders intended, a delightful and unique piece of architecture.

The towpath passes six locks altogether before becoming embroiled in the ugliest bridge on the line, on the new M40. The noise from the motorway engulfs what must have been a very remote part of the canal at one time. It's still very lovely now but we have to walk another $\frac{1}{4}$ mile before the rumble subsides and the birds become the loudest thing around. The second lock after the motorway is number 28. Here the residents of the barrel-roofed lock-cottage have set up a small canalware shop. These traditional designs are not to everybody's taste but there is a fine collection on display and for sale.

We pass the abutments of an ex-GWR railway bridge (which joined Lapworth with Henley-in-Arden), a lock, a bridge and another lock before entering

Lowsonford. Here, conveniently is the Fleur de Lys pub on the opposite bank. This can be reached by crossing at the next road bridge.

At the next lock (No. 31), there is another barrel-roofed cottage still owned by the National Trust. At the next lock, the towpath changes sides. Shortly thereafter is Yarningale Aqueduct, the first of three iron aqueducts between here and Stratford. This is the second aqueduct to be built on the site. The original structure was washed away on 28 July 1934 when the Grand Union burst its banks and caused the small stream that runs underneath the Stratford to flood. The new structure was made, in what must have been record time, by the Horseley ironworks in Birmingham and the canal was reopened on 23 August. It consists of a series of cast-iron sections which have been bolted together to form a kind of trough through which the waterway flows across the stream. In the restoration of the southern canal in the early 1960s, the brick abutments were restored by inmates from Winson Green prison. The sunken towpath along the side of the aqueduct gives walkers an unusual view as the

Yarningale Aqueduct

waterway comes up to eye level (depending on your height of course) before returning to normal on the other side.

Four locks further on (at lock 37), the canal runs past arguably the most extraordinary extension to a barrel-roofed cottage on the line. This one has been privately owned since before the National Trust took control of the waterway. Here also are the Haven Tea Rooms.

Just before lock 38 take the path through the hedge to the left and turn right along the road and then back down to the canal at the bridge. This is Preston Bagot. If thirsty turn right across the bridge to the Crab Mill pub. The towpath, meanwhile, passes under a road bridge to cross the canal at the footbridge a little further on. The line now enters some remote countryside and wanders around the contours for about 2 miles before reaching Wootton Wawen.

Wootton Wawen to Stratford

For those joining from Wootton Wawen station, make your way to the A34 main road and turn right. Walk on for about a ½ mile to a point where the road dips to go under an iron bridge which is in fact an aqueduct carrying the Stratford Canal. There is a small track which runs up the right-hand side of the aqueduct to the canal. Turn right for Stratford.

For those who have walked from Lapworth, Wootton Wawen is first announced by the large Anglo-Welsh boatyard with its tall office building which dominates the left-hand bank. The canal reached Wootton Wawen on 22 June 1813 and the village soon became an important inland port. The boatyard is situated on the site of the wharf built to tranship cargo before the line was fully opened. From September 1814 until June 1816, a horse-drawn tramway was employed to carry coal from here into Stratford. Just after the boatyard, the canal narrows to cross the busy A34 via another iron-trough aqueduct. Carried on two brick-built piers this structure has been here since October 1813. This longevity has been achieved despite the attention of various lorries that have ploughed into it over the years. One such accident in 1968 took two cast-iron girders out of the structure leading to closure of the road. Remarkably they were replaced by modern steel versions within two days thanks to the efforts of a variety of volunteers. Since then the bridge has been hit twice more, although with less significant effect. In all cases, the canal trough remained intact. One good thing about the M40 is that it has removed a large amount of the traffic so that such incidents may be less likely to happen in the future.

Although Wootton Wawen Aqueduct has charm, the best is yet to come. After bending left, the canal heads off south through a slight cutting (after which the towpath changes sides) to the isolated lock known as 'Odd Lock' (No.

Wootton Wawen Aqueduct and the Anglo Welsh boatyard on the Stratford-upon-Avon Canal

39). Shortly thereafter is Edstone or Bearley Aqueduct. Built in 1815, this is another iron-trough structure but on a much grander scale compared with the two already crossed. The Stratford-upon-Avon Canal Society inform me that Edstone Aqueduct is 475 ft or 145 m long (other sources, interestingly, quote anything from 521 ft to 754 ft). It was designed by William Whitmore who was the engineer responsible for the southern half of the canal. It's a splendid sight as its fourteen arches span over the River Alne, the two lines of the Stratford to Birmingham railway and the minor Bearley to Little Alne road. At one time another railway, a branch to Alcester, also ran underneath. Notes from the canal society say that the engine that worked the line was housed under the aqueduct and used water from the canal. As on the other two aqueducts, the towpath is lower than the canal to give walkers an interesting view of boats passing alongside.

After passing a cottage (the site of the former Bearley Wharf), the canal twists round the contours towards Wilmcote. Just before Wilmcote Bridge was the former stone wharf. Limestone was carried 4 miles from the quarries at Temple Grafton to the canal by means of a horse tramway. In 1845, some 16,000 tons of limestone were shipped in this way. The original plan had been for a canal branch but presumably the tramway was less of a financial risk for what was possibly a transient business.

The bridge at Wilmcote is Featherbed Lane Bridge or, more prosaically, No. 59. The poor condition of a former occupant of this site was given as the reason for the warrant for abandonment of the canal presented by Warwickshire County Council in 1958. From this came the restoration movement and the gradual reopening of the southern line. Such restoration includes this section of the towpath which was completely rebuilt by volunteers in the early 1980s. The bridge itself was rebuilt somewhat earlier but also looks to be in fine fettle. Those who wish to utilize Wilmcote station should go up to the road and turn left (turn right for Mary Arden's house).

Over the next 3 miles, the canal passes down the eleven locks that constitute the Wilmcote flight. Thereafter the canal enters Stratford by the back door, squeezed between various industrial estates and even Stratford Town football ground. After two bridges carry the train lines into Stratford railway station, the canal secretly eases its way into the centre of town.

The stretch around the next lock (No. 52) was formerly the site of Kendalls, a supplier of chemicals to the brewing industry. Their first works were on the towpath side of lock 52. However, a new factory was built on the opposite bank just after the next two bridges and before lock 53. This spot is now occupied by private dwellings. Both sites received supplies of chemicals via the canal, and Nick Billingham has reported, in *Waterways World*, that the trade continued at least up to the First World War.

The canal now descends rapidly via three locks to reach a bridge carrying the busy A46. Go under this and immediately over a small turnover bridge to the opposite bank where there are some new retirement flats. Follow the path to reach another bridge. Here the towpath ends, so go up some steps and cross the two lanes of the road to reach Bancroft Gardens. The canal basin here can be followed around to the final lock (No. 56) and a small bridge that was rebuilt by prisoners during the great restoration. Just beyond is the River Avon.

Bancroft was originally the site of two basins and a major port. The second basin was on the spot now occupied by the theatre and was filled in when the original building was put up in the 1870s (the present theatre dates from the early 1930s). The waterway link between the two basins ran straight on from the point where the canal enters the modern basin from under the bridge (i.e. parallel with Waterside). Cargo was unloaded here both for Stratford and for transhipment south via the Stratford & Moreton Horse Railway. A wagon that once ran on the line is permanently sited by the path on the side furthest away from the theatre. The tramway itself ran along this path and across the bridge, now used as a pedestrian route to the sports grounds and car parks on the other side of the Avon. The canal company buildings that lined the basin have been cleared and the area is more often than not awash with tourists and theatre-goers all wondering where the boats go after they turn right out of the basin.

Further Explorations

Sadly, the only portion of the northern Stratford that is walkable throughout is the $2\frac{1}{2}$ miles from Lapworth to Hockley Heath. It is not possible therefore to explore much of the length between Lapworth and King's Norton. Luckily the proximity of the Warwick & Birmingham Canal means that if you are in the area, there's plenty of good towpathing to be had.

The Warwick & Birmingham Canal Act received its Royal Assent on 6 March 1793. It authorized a line from Saltisford in Warwick to the Digbeth Branch of the Birmingham Canal, a distance of just over 22 miles. The canal was fully opened on 19 March 1800, the same day as the Warwick & Napton Canal which took the line on from Saltisford to the Napton Junction to Birmingham. On Canal. This then became the shortest route from London to Birmingham. On 1 January 1929, the W & B merged with the W & N and the Grand Junction and others to form the Grand Union Canal Company.

The recommended walk is of 4 miles from Warwick to Hatton railway station. There is free car parking at both stations and very regular trains between the two. Starting at Warwick, return from the car park to the main road. Turn left and continue for about a $\frac{1}{3}$ mile until the road goes over the canal. On the far side of the bridge, take the towpath left. From here the line passes through the outskirts of Warwick to the Saltisford arm and then bends around right to the start of the splendid Hatton Locks. There are twenty-one locks in the Hatton flight, traditionally known as the 'stairway to heaven', between Saltisford and Hatton station. One of the interesting things about them is that they were originally narrow locks. In 1934, they were widened as part of a scheme to attract Birmingham to London traffic back on to the canals. The narrow locks are clearly visible in the massive lock structures although they have all either been filled in or converted into weirs.

There is a brief respite at the top of the main flight when the canal reaches the British Waterways' yard. Those in need of refreshment after the up-hill stretch should go over the stile to the left of the BW yard and through the field into the garden of The Waterman. For those not so tempted, the towpath changes sides at the bridge next to the yard and then goes past the final four locks in the flight.

The line then passes through gentle countryside for about 1 mile before Hatton railway station can be seen on the left. Go up on the far side of the bridge and the entrance to the station is in front of you.

Further Information

The Stratford-upon-Avon Canal Society was formed in 1956 with the main aim of ensuring that the southern section of the canal was saved. This objective culminated in the reopening of the line in 1964. Even though its objective has long since been achieved, the society still thrives, meets regularly and publishes a newsletter. Among the activities has been the restoration of certain sections of the towpath. Walkers should therefore be pleased to support them. Their address is:

Stratford-upon-Avon Canal Society,
'Santa Monica',
Earlswood,
Solihull,
West Midlands,
B94 6AQ.

The society publishes its own forty-page guide to the canal which contains a lot of interesting snippets about the line, its history and its restoration.

For detailed history buffs, the Stratford, the Avon and the Stratford & Moreton Railway are described in:
Hadfield, C. and Norris, J., *Waterways to Stratford*. David & Charles, 1968.

The history of the Warwick & Birmingham Canals is ably described in:
Faulkner, A., *The Warwick Canals*. The Railway and Canal Historical Society, 1985.

11
THE TRENT & MERSEY CANAL
Stoke-on-Trent to Kidsgrove

Introduction

A stroll through the centre of some grubby eighteenth-century potteries, a steelworks and a range of old and derelict buildings doesn't sound very promising but, in this walk, we head directly for the heart of the Trent & Mersey Canal. Here is the industry for which the canal was built and the industry which grew on what the canal could feed it. Although large areas have been landscaped, evidence of just how busy the line must have been in its heyday abounds. The whole stretch between Stoke and Longport is littered with old wharves and redundant canal arms. Here are the earthly remains of a busy and profitable waterway, the very origins of the nation's canal network.

At its eastern end, the Trent & Mersey Canal leaves the River Trent at Derwent Mouth where the Derwent joins the Trent just a few miles south of Long Eaton. Within 1 mile, it reaches the canal port of Shardlow, a village effectively created by the T & M, and then continues west along the Trent Valley, past the former entrance to the Derby Canal, and on to Burton-on-Trent. At this point the T & M becomes a narrow canal built to Brindley's 7 ft 'gauge'. Heading south-west, the route passes through Alrewas to reach the first of the important junctions with routes to the south. At Fradley, the junction is with the Coventry Canal – the way, via the northern Oxford and the Grand Union, to London. The T & M now leaves the Trent Valley to turn west and then north-west. After going through Rugeley, the line reaches Great Haywood, the junction with the Staffordshire & Worcestershire Canal and the way, via the River Severn, to Bristol.

After Stone, the canal meets its *raison d'être* at Stoke. In the midst of the Potteries at Etruria, the site of Wedgwood's factory, the Caldon Canal leaves the T & M to run north-east to Leek and Froghall. Meanwhile, boats on the main line continue, past the many redundant pottery works and through Telford's historic Harecastle Tunnel, to reach Kidsgrove. Here the Macclesfield Canal leaves at Hardings Wood Junction only then to go over the T & M at Red Bull Aqueduct. The main line, however, continues through the Cheshire salt fields where, in many places, mining subsidence has severely affected the course of the canal. At Middlewich, a branch goes off to join with the Shropshire Union Canal. After passing through Northwich the line comes close to the Weaver Navigation at Anderton where, to link the two waterways, a massive boat lift was built. The canal then runs through three tunnels to join the Bridgewater Canal. One arm now goes on to Manchester and the other to Runcorn, where nowadays it sadly comes to a dead end, a metaphorical stone's throw away from the Mersey.

Finally, before you venture forth, you should perhaps be aware that, according to Jean Lindsay, the Trent & Mersey Canal Company's motto, as inscribed on its seal, has the natural and idiomatic translation of 'May it flood the countryside and drown the population'. Don't say I didn't warn you.

History

It was the Bridgewater Canal which inspired Josiah Wedgwood, potter and entrepreneur extraordinaire, to see the value of building an artificial waterway to

The pottery industry thrived along the Trent & Mersey corridor in and around Stoke. It used the canal both for the import of china and clay and flint, and for the gentle export of its goods to the world. This picture from Longport in the early 1900s shows the huge numbers of bottle kilns that spread around the canal

British Waterways

connect his new factory at Burslem with the outside world. The Bridgewater, built to service the coal mines around Worsley, was the first man-made canal to be both of utilitarian value and highly profitable. The shrewd Wedgwood foresaw the immense benefit of having a line of his own to bring in china clay from the West Country and to export his delicate products to the rest of Britain and the world.

A canal to link the Rivers Mersey and Weaver with the Trent had first been raised in 1755 (before the Bridgewater) by two northern businessmen who sponsored a survey. Little is known of their plan or why it foundered but the idea was revived in 1758 when James Brindley was asked to survey a line from Stoke-on-Trent to Wilden ferry (on the Trent near Nottingham) by Thomas Broade, Lord Anson and Earl Gower. Brindley's survey, of what he called the Grand Trunk, suggested the possibility of linking the 'Cheshire rivers with the River Severn and finally to connect the Counties through which the R. Trent flows with all parts of the Kingdom'. The first map of the proposed Grand Trunk, which was drawn by Hugh Henshall and approved by John Smeaton, was produced in 1760. It described a route from Longbridge, near Burslem, to Wilden ferry. Smeaton later produced his own survey in which two important features were added: a proposal to extend the line to the River Weaver and a plan to tunnel through the summit at Harecastle.

Whether any of these plans would have come to fruition without the skills and enthusiasm of Josiah Wedgwood is open to question, for nothing appears to have happened until the idea came to his attention in 1765. Wedgwood had discussed the potential for a canal network with two influential friends, Erasmus Darwin (in London) and Thomas Bentley (in the north-west), and had been in contact with Earl Gower concerning Brindley's survey. He now became the guiding light in the venture and it was his abilities and, perhaps a new skill in those times, public relations efforts that drove the project forward. In April 1765, a plan was issued and touted by those interested in the scheme. The canal was to be 76 miles long from Wilden to the River Weaver at Frodsham. It was to be built on the Bridgewater model for boats 6 ft wide and 70 ft long.

There now followed a period during which the precise line taken in Cheshire was under much debate. The original proposal to enter the Weaver was challenged by a new idea that the canal should join with an extension of the Bridgewater Canal, a link which would facilitate a convenient line to Manchester and possibly to Liverpool. It was this latter feature which swung the matter. At a meeting at Wolseley Bridge on 30 December 1765, under the patronage of Earl Gower, a line through to the Mersey (at or near the Runcorn gap) was described, with the cost estimated as being £101,000. Although branches, including one to the River Severn, were discussed, it was concluded that they were not to be part of the matter in hand and were excluded from any proposals.

On 15 January 1766, a petition for the Trent & Mersey Canal was presented to the Commons and the Act received royal assent on 14 May. The company was given permission to raise £130,000 by 650 shares of £200 each. The Act also allowed an extra £20,000 to be raised if needed. On 26 July, Josiah Wedgwood ceremonially cut the first sod at Burslem and the work began both at the Derwent Mouth Junction with the navigable Trent (near Long Eaton) and at the Harecastle Tunnel.

By 1767, concerns were already being expressed about the health of James Brindley, whom Wedgwood, for one, thought was over-doing things. Despite this, Brindley was confident enough about the works to bet his committee that the entire canal would be open by Christmas 1772. Things were certainly progressing well for, by June 1770, the canal was open from Derwent Mouth to the Great Haywood Junction with the Staffordshire & Worcestershire Canal. On 12 November 1771, it was open to Stone (about 8 miles south of Stoke). Charles Hadfield reports that celebrations at the opening of the canal at Stone were rather over-exuberant. The repeated firing of a cannon caused a newly built bridge and lock to collapse. The canal had been open for traffic as early as September 1770 with vessels setting out from Great Haywood every Monday and Thursday morning for Weston-upon-Trent (about 5 miles up from Derwent Mouth). The T & M was thus a going concern and successfully raised a further £70,000 following the passage of a new Act.

On 27 September 1772, James Brindley, the father of the modern canal system, died and Hugh Henshall, Brindley's brother-in-law who had been involved with the T & M from the outset, took over as engineer. By now the canal ran from the Trent right through to the Potteries, a distance of 48 miles. But Brindley's bet would have been lost. The canal from Harecastle to Sandbach wasn't open until April 1775 and Henshall was experiencing severe geological problems at the Cheshire end between Middlewich and Acton. This matter was resolved by building two extra tunnels and the whole line was opened in May 1777. The final cost was £296,600. The finished canal was just over 93 miles long with a rise of 326 ft by thirty-five locks (later increased to thirty-six) from Preston Brook to Harecastle and a fall of 316 ft to Derwent Mouth. Most of the canal was built to Brindley's original 7 ft width but the stretches from Wilden ferry to Burton and from Middlewich to Preston Bridge were 14 ft wide in order to carry river barges of 40 tons burden.

The effect of the canal on the Potteries was both immediate and substantial. Prosperity came to what was once a relatively backward and remote area. Enormous quantities of china clay (from Devon and Cornwall) and flint (from Gravesend and Newhaven) were brought in via the Mersey for the pottery industry. Even the local farms benefitted from the now readily available supply of horse manure that was shipped out from the cities. In 1836, it was recorded that the outward traffic was 61,000 tons to Liverpool (of which 51,000 were earthenware and china), 59,500 tons to Manchester (of which 30,000 were bricks and tiles and 25,000 coal) and 42,000 tons to London (30,000 tons coal and 12,000 tons earthenware and china).

As time went on, various branches and connections were added to the canal. The Caldon arm took the line to Leek, Froghall and, for a while, Uttoxeter. There were also short lengths at Hall Green (leaving at Hardings Wood to join the Macclesfield), Dale Hall (from Langport in Burslem), Bond End (to Burton at Shobnall), Wardle Green (which joined the T & M with the Ellesmere & Chester's Middlewich branch), the Derby Canal (from Swarkestone) and the Newcastle-under-Lyme Junction Canal which ran from the T & M at Stoke.

Although the company's funds had run perilously short towards the end of the construction stage, by June 1781 they were already in a position to pay a 5 per cent dividend. This was partly helped by the profits coming from the company's own carrying company, Hugh Henshall & Co. Tolls, and the profit on carrying rose encouragingly, especially after the through connection with the Coventry Canal (at Fradley Junction) was opened in 1790. By 1812 the annual revenues totalled £114,928, enabling the company to distribute £71,500 in dividends. By 1820, the company's debt had been completely cleared and business continued to increase. By this time, shares were changing hands at eight or nine times their face value. The success of the T & M owed much to its monopoly of the routes from Liverpool and Manchester via the Potteries to Hull and London. Jean Lindsay, however, points out that the company was also

The northern portal of Brindley's Harecastle Tunnel in 1905. The narrowboat *Westwood* and butty is preparing to go through the tunnel while the horse heads off over the hill. Northern-bound traffic, at that time, used Telford's Tunnel which is to the left of this picture

The Boat Museum archive

extremely well run, paying attention to waybills and the employment of 'walking surveyors' who kept a keen eye on the operation of the line.

By the 1820s, the company had come in for some criticism for the slightly antiquated nature of its canal. Modern waterways were broader, straighter and quicker. To offset some of this criticism and to combat some of the growing competition, the company set about a modernization programme that included doubling up some of the locks and, most expensively, building a second Harecastle Tunnel. These improvements were not enough to prevent the grievances of the canal users – who mostly complained about the small size of the boats and the lack of an alternative – or the incursion of competition. In the 1830s, that competition arrived in two forms: the Birmingham & Liverpool Junction Canal (now called the Shropshire Union) which opened in 1835 and, perhaps more importantly, the Grand Junction Railway in 1838 (which went from Birmingham to join the Liverpool to Manchester line at Newton-Le-Willows).

The T & M responded to the threats from the railway in a canal company's usual fashion. It cut tolls. But in 1845, the routes operated by the T & M were targeted directly by a proposal for four railway lines that were to become the North Staffordshire Railway (NSR). With this blanket attack, the T & M was quicker than most of the canal companies in concluding that its only option was to seek to be taken over. This it duly was on 15 January 1847, following an Act passed in June 1846.

Although one of the first actions of the NSR was to close the Uttoxeter extension of the Caldon Canal, in general the railway company started its management with some good intentions. The Act had guaranteed a certain standard of maintenance and the company reduced tolls to maintain traffic levels. It was impossible, however, to prevent the gradual loss of trade to the railway, so that although the tonnage moved by the canal stayed the same, there was a clear reduction in toll receipts. By the 1870s, receipts were under half what they had been in the 1840s and the NSR decided to concentrate its efforts on its railway business at the expense of the T & M.

This decision cannot have been helped by the high maintenance costs of certain parts of the line. Along a number of stretches in mid-Cheshire, such as around Thurlwood and Rode Heath, the vast salt mines that extend under the canal had caused a series of earth movements and canal bed collapses. This meant that the embankments, locks and bridges needed constant attention and, sometimes, complete rebuilding. The lock and bridge at Thurlwood, for example, have both been rebuilt.

Although by the second half of the nineteenth century, the days of profitable waterway transport were numbered, the trustees of the River Weaver still saw an opportunity to maintain its own traffic levels (primarily in the export of salt from the Cheshire area and the importation of coal) by a programme of improvements. One of those improvements was to develop a permanent junction with the T & M, a move which had previously been resisted by the canal company. In 1872, the trustees prepared a bill to authorise the connection via a barge lift, subsequently known as the Anderton lift. This exuberant structure was completed in 1875 and allowed movement of cargo on to the Weaver, a move which led to further loss of trade along the T & M.

By the early years of the twentieth century, the tonnage carried on the T & M was still in excess of a million tons p.a. In this respect, it remained a successful waterway and many canal-side potteries still found it cheaper to use than the railway. However, the toll receipts had declined to £45–52,000 p.a. and the company's own carrying business had been closed since 1895. There was no suggestion that the railway company had any enthusiasm for its canal business and, when in 1921 the NSR became part of the London Midland & Scottish Railway, the end of the canal as a commercial concern was definitely in sight. Toll receipts now dropped dramatically as the LMS actively sought to move traffic on to its more profitable rails. Traffic in 1940 was less than a quarter of the level thirty years before with receipts down by half despite increased tolls.

Nationalization in 1948 couldn't prevent the continuing fall off of traffic which was now under 100,000 tons p.a. In 1955, the Board of Survey on Canals and Inland Waterways reported that major improvements were needed if the T & M was to compete with other forms of transport. With increasing problems of subsidence, the maintenance of the canal was becoming more and more of a problem. The 1950s were a period of considerable losses not helped by the cessation of the gravel trade in 1958 and the salt trade in 1960. Business in the 1960s consisted mainly of the carriage of coal and of clay to the potteries. With trade disappearing completely during the late 1960s, the canal was effectively rescued by the Transport Act of 1968 when the line was designated as a cruiseway for recreational use. This it remains to this day: an important and highly popular part of the four counties ring with Harecastle Tunnel an exciting venture into the unknown.

The Walk

Start:	Stoke-on-Trent railway station (OS ref: SJ 879456)
Finish:	Kidsgrove railway station (OS ref: SJ 837544)
Distance:	8¾ miles/14 km
Map:	OS Landranger 118 (Stoke-on-Trent & Macclesfield)
Return:	Train Kidsgrove to Stoke (scarce on Sundays). Telephone: (0782) 411411. Buses also run regularly between the two towns. Telephone: (0782) 747000
Car park:	At either station
Public transport:	Stoke and Kidsgrove are both on BR main line

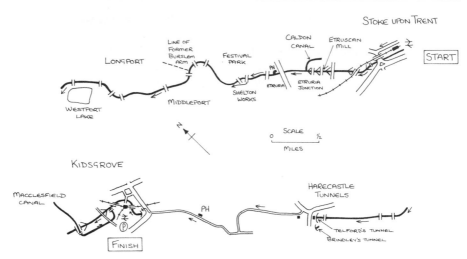

The Trent & Mersey Canal

Leave Stoke station by the main entrance and turn left and left again to pass under the railway. This leads to a superficially impenetrable roundabout. Bear left to go down an underpass which leads directly to the canal. To reach the towpath go over a bridge to the far side of the canal and turn right.

The path immediately passes under the bridges of the roundabout to reach a lock: the bottom of the Stoke flight which takes the canal up 50 ft. The lock is a typical narrow lock (7 ft in width) with a single gate at the upper pound end and a pair of mitred gates at the lower end. The second lock, with an impressive cast-iron milepost, is reached after passing under the railway.

The canal now runs adjacent to a cemetery and on to a road bridge. Here on the left at the Cliffe Vale Pottery is a once characteristic sight of the potteries: two splendid beehive kilns. Although most have been demolished or allowed to fall down, there are still a number throughout the area. The kiln is basically just a large oven in which pottery was baked in a batch process. They were coal-fired and, notoriously, belched thick smoke. Since the Second World War, the beehive has been gradually replaced by gas-fired continuous-flow ovens and these highly individual features of the landscape are steadily disappearing – much to the relief of the local environment.

Although now an area of mixed activity and dereliction, this entire length, between the Harecastle Tunnels and the southern edge of Stoke, was one that not only used the canal but whose existence was firmly based upon it. All the factories along here had their own wharf where raw materials were delivered and packed products shipped away to markets around the world. This is a part of the country therefore that didn't simply play host to a canal but which depended upon it for its industrial lifeblood. At the height of the canal age, this must have been one of the busiest sections of any inland waterway in the country. Now only a barely recognizable wharf or an overgrown factory arm remain to remind us of the role it played.

After the next lock is the tall chimney and low buildings of the Etruscan Flint and Bone Mill. This was built by Jesse Shirley in 1857 for roasting and crushing flint and bone, both of which are used as whiteners and strengtheners in ceramic pottery. The mill is served by a short arm from the canal and its raw materials and products were both shipped along the line. The mill has a museum open to visitors.

The towpath creeps up the side of the fourth lock and then shortly on to Stoke Top Lock (No. 5). Here is a British Waterways maintenance yard and, a little further on, the entrance to the Caldon Canal, a fine, almost unbelievably rural, route to Leek and Froghall (another day, another chapter).

The Shelton steelworks engulfs the Trent & Mersey Canal just north of Etruria

Continue on to a busy road bridge and the site of Wedgwood's first factory, Etruria, opened in 1769 and closed in 1950. Sadly, only two of the original buildings still stand. On the left is a building with a domed roof, now a preserved monument, even though nobody is quite sure what it was for. The bulk of the factory site is now home to a local newspaper. The site was originally at the same height as the canal but has sunk to its present level. For the other survivor of Wedgwood's time, cross the bridge and search among the newer housing for Etruria Hall, which was built in 1770 as a home for the great man within sight of his factory.

From the road bridge take the right-hand bank to pass by the Festival Basin that was built at the time of the International Garden Festival held here in 1986. The festival occupied the right-hand bank of the canal on a site that was formerly covered with blast furnaces. At the basin entrance is The China Garden pub and grill.

The vast Shelton iron- and steelworks now dominate the canal. After a railway bridge, the towpath walker becomes incorporated into the works as vast sheds overhang the entire breadth of the waterway. The complex has long lost its own production facilities (closed in 1978) but is still active as a rolling mill, as the alarming bangs and crashes signify.

Cross the second bridge (a kind of turnover bridge) after the Festival Basin to resume the walk along the left-hand bank. The right-hand side is the site of the garden festival and has been extensively landscaped. On the hill is a ski-lift for an artificial slope. The canal now bends left to go under a bridge and alongside some old tips to continue its sojourn by the steelworks. The scenery opens out as more landscape has been cleaned up, evidenced by the unexpected appearance of greenhouse tunnels on the right-hand bank.

After passing around a left-hand loop and under two bridges, we pass by a milepost and another factory to reach bridge 123. Just before the factory on the far bank is a small valley which was the course of the former Burslem Arm. More old wharves and redundant pottery factories now follow, interspersed, curiously, with a pleasant range of garden shrubs. A little further on the canal passes a splendid covered wharf, a beehive oven, a quaint yard and, when I was here, a man throwing (presumably reject) plates into a skip – all part of the famous Middleport Pottery. On the left-hand side, luckily hidden behind a high fence, the sounds of a vast tip can be heard.

Go under bridge 125 to reach Longport Wharf and bridge 126, which carries the A527 road to Burslem and Tunstall. Here are a small clutch of pubs: The Duke of Bridgwater, The Pack Horse and The Railway. If you turn right across the bridge, the Travellers Rest in Newcastle Street brew their own beer!

Just across the bridge on the right-hand bank is another beehive kiln (part of the Price & Kensington works). A few hundred yards further on, after bridge 127, the land to the left has been landscaped into the Westport Nature Reserve. A fine lake is home to a range of bird species. Interestingly, this was once home to Port Vale Football Club until subsidence rendered the ground unplayable (history does not recall how many spectators noticed).

The canal now passes under bridge 128 and bends gently left and then right in between Tunstall waterworks to the left and the gasworks to the right. There now follows a stretch yet to be reached by the restorers with some overgrown canal arms and a variety of derelict buildings and wharves. After bridge 129, the straightness of the next $\frac{1}{2}$ mile means that those with good eyesight should be able to make out the portals of the old Harecastle Tunnel. Just before reaching the tunnels, cross the canal at the turnover bridge (No. 130).

There are two canal tunnels at Harecastle: Brindley's original on the left and Telford's 'new' tunnel on the right. Brindley's, 2,880 yd long, was started on 27 July 1766 but wasn't open for another 11 years. There were problems with

The southern portal and extractor-fan building of Telford's Harecastle Tunnel on the Trent & Mersey Canal

quicksand, hard rocks, springs of water and methane gas. Progress was only possible with the aid of a windmill for ventilation and an improved Newcomen engine for pumping out the flood waters. When it was finally opened, it was hailed as the eighth wonder of the world. Pleasure-boats took sightseers to admire it, often accompanied by bands playing music. For most of its way the tunnel is lined and arched with brick; the only exception being a 500 yd stretch of solid rock near the middle. As on the Bridgewater Canal, the length is littered with side-tunnels which burrow their way into the coal seams that were discovered during the course of the excavation.

Narrowboats were taken through the tunnel by the cumbersome process of 'legging'. The system was worked by groups of men who occupied a hut at the entrance. To push their way through, the leggers lay on their backs on the top of the narrowboat with their feet pressed against the sides of the tunnel. They then simply walked the boat to the other end, a process that could, apparently, take two to three hours. For this, the leggers received 1s. 6d. and some very tired legs.

From the outset there were problems with the quality of the tunnel and constant grievance at the time taken to pass through it. By the 1820s the situation was so bad that John Rennie was called in to try and improve matters. He agreed that the brickwork and mortar were in a poor state and that the line was crooked and difficult to navigate. Although he made a number of alternative suggestions, his recommendation was to build a new tunnel.

Sadly, Rennie died before he could start the work and Thomas Telford was employed in his place. Telford's new, 2,926 yd long, tunnel was built to the east of the old tunnel (i.e. to its right as it is looked at) and about 25 yd away from it. The intention was to pass through undisturbed ground where possible. Telford's tunnel is taller and wider than Brindley's and originally included a towpath. It took just three years to build and was opened on 30 April 1827. Telford was very pleased with it and described it as 'quite perfect'. One boatman, asked what he thought of the new tunnel, is said to have replied 'I only wish that it reached all the way to Manchester'. With two tunnels in operation, movement along the canal was much faster with each tunnel becoming one-way: north through the Telford and south through the Brindley.

In 1891, the North Staffordshire Railway provided steam tugs to tow boats through the tunnel. In 1914, the steam tug was replaced by an electric one (firstly it hauled a battery boat and then it was run by picking up current from overhead wires). This remained in operation until 1954. Brindley's tunnel, however, had sunk so much because of the various mine workings that by the First World War it was abandoned and has been shut ever since.

Nowadays a keeper is on duty at each end of the Telford Tunnel to manage the passage of leisure traffic. An average time for travelling the entire length is about fifty minutes. The brick-built, box-like structure around the entrance houses extractor fans. It was installed in 1954 to help keep the line clear of diesel fumes. The loud whirring sound of the fans only ceases when the apparatus is turned off to allow boats to enter or leave. As Telford's tunnel has also sunk, the head room along some stretches has been reduced to about 6 ft. In fact, the tunnel needs a considerable amount of maintenance and remains liable to subsidence or falls of roof lining. It was closed for four years in the 1970s (during which time the towpath was removed to make more room) and although now in good condition, it must remain a concern to British Waterways.

To reach the northern end of the tunnels, walkers have to follow the route previously taken by the barge horses while their boats were legged through below. It's not an exciting route so if someone offers you a lift, take it. Who knows you may get a chance to see the ghost of the woman who is said to have been murdered halfway along.

Those without a lift should take the steps on the right up to Chatterley Road and turn left. As the road bends sharp left, cross to Holywall Lane and then take the small road left. This passes a terrace of houses (with some very 'yappy' dogs) and steadily climbs for about a $\frac{1}{2}$ mile before bending left to pass by a caravan site. An unmade road then continues to a farmyard. At what amounts to a T-junction, turn right along a metalled lane. This is Boathorse Road which bends and curves to reach a small group of houses and the Rifleman pub. From there the road gradually descends into the outskirts of Kidsgrove. Keep on this course, past some pleasant housing to reach a bend in a wide and busy road. Cross this road and bear right. Within 20 yd, a signpost points left towards the Harecastle Tunnels. After five to ten yards, turn right to go down a tree-lined track. This eventually leads to the northern entrance of Brindley's tunnel and then around to Telford's where the bright orange colour of the canal water comes as a bit of a shock. This isn't some new kind of pollution. It's caused by the presence of iron salts that have been leached from the rocks exposed along the length of the tunnel.

The towpath shortly passes under a turnover bridge and a rail bridge. Kidsgrove station can be reached by going up some steps to the left. Before finishing, however, there are still sights to see. Carry on along the path for about a $\frac{1}{4}$ mile to Hardings Wood Junction, the entrance to the Macclesfield Canal. Go over the roving bridge to pass a double lock, the first of the Red Bull Locks. This and the other double locks to be seen along this stretch were originally single but were doubled up in the 1820s as part of the attempt to speed up the flow of traffic. Cross both channels at the downstream end towards the Tavern, to pass under the road bridge and onto the second Red Bull Lock which is dominated by the sight of Pool Lock Aqueduct. This carries the Macclesfield Canal over the T & M and on to Macclesfield, Congleton and Marple where it meets the Peak Forest Canal. The entire length of this canal is open to walkers and is covered in *Canal Walks: North*.

Cross both channels using the lock's cantilevered bridges and take the thin track which passes up to the Macclesfield Canal. Turn left to pass a boatyard and then around a gentle bend to reach the roving bridge at Hardings Wood Junction. From there turn right, across the bridge, to return to Kidsgrove station.

Further Explorations

The Trent & Mersey is walkable along its entire length and thus holds out the possibility of a 93 mile long-distance footpath which could be tackled over a period of a week. The walk might start at Shardlow (on the A6 south-east of Derby) with nights at: Burton-on-Trent, Rugeley, Stone, Kidsgrove, Middlewich, Anderton/Northwich and then on to Runcorn (along the Bridgewater Canal).

However, those who think of a long-distance walk as being under a mile, should head for the small village of Anderton, just a couple of miles to the north of Northwich in Cheshire. Anderton can be found on OS Landranger 118 at ref SJ 647752. If you take the small side road off Anderton main street (opposite the general stores and post office), you will shortly arrive at a car park near the Stanley Arms pub. Here is the Trent & Mersey Canal quietly and unobtrusively going on its way to Preston Brook. But beyond is the loud noise of the Anderton lift.

There is nothing else like it on the British canal system. The rusting bones that constitute the Anderton lift was once the connecting link between the T & M and the Weaver Navigation some 50 ft below. If you cross the T & M you will reach a small basin and the top of the lift. Here are two 162 ft long aqueducts that pass from the basin over to the lift.

In the 1860s, there was still a relatively large amount of trade between the Weaver Navigation and the Potteries via the T & M. As there was no physical link at that time, cargo was transhipped using chutes and tramways, the remnants of which can still be seen as you walk down the footpath to the lower basin. Edward Leader Williams, engineer to the Weaver, proposed a lift and Edwin Clark designed it. It was opened in 1875.

Having traversed the aqueduct, boats were loaded into a caisson, effectively a water-tight tank measuring 75 ft by 15 ft by 5 ft capable of holding two narrowboats. The tanks were originally supported on hydraulic rams and connected so that when one tank was down the other was up. To get the system moving, the opened ends of the caissons were sealed and 15 tons of water (6 in depth) was removed from the bottom tank. As the weight of the upper caisson forced it downwards, so hydraulic pressure pushed the lower tank upwards. The final push was powered by a steam engine. At the top, the 6 in of water were then restored to the caisson and the gates opened.

In a series of alterations in the early part of the twentieth century, steam power was replaced by electric, and the hydraulic system was replaced by a series of counterbalancing weights. This meant that each caisson could be worked independently.

To see the bottom basin, take the steep path that runs down to the right of the top basin. Half-way down, the remains of one of the chutes can be clearly seen. At the bottom, ICI's Winnington Works belches steam and noise and the forlorn-looking lift looms like an inmate of the Natural History Museum.

The lift, which is a designated ancient monument, is currently in the process of restoration and funds are being raised by the Anderton Boat Lift Development Group in order to undertake the process. This will be the second time since the Second World War that restoration has been needed. In the mid-1970s, the lift was closed for two years while urgent repairs were undertaken. It closed again in 1983 and has been becoming increasingly derelict ever since. At the time of my visit, the counterbalance system had been stripped off and was lying in the nearby field. Maybe you'll have more luck.

Further Information

The Trent & Mersey Canal Society Limited was founded in 1974, incorporated in 1984 and has a Wedgwood (John) as its patron. They can be contacted at:

    c/o Michael Mitchell,
    34 Kennedy Avenue,
    New Sawley,
    Long Eaton,
    Nottingham,
    NG10 3GF.

For those interested in the history of the canal, there are none better than:
Lindsay, Jean, *The Trent & Mersey Canal*. David & Charles, 1979.
Hadfield, Charles, *The Canals of the West Midlands*. David & Charles, 1969.

12
THE WORCESTER & BIRMINGHAM CANAL

The Tardebigge Locks

Introduction

The Worcester & Birmingham Canal was a late entrant into the canal race and seems to have had a problematical birth and a difficult life. The intention was to produce a short cut from the River Severn at Worcester into Birmingham. This it does by a head-on assault on the south-west face of the city via a spectacular series of locks that instill themselves on the memory and lock-gear turning arms of those that attempt it. Never given the same respect as the Staffordshire & Worcestershire, the W & B offers some fine country and some good walking.

The 30 miles of the W & B start at the Gas Street Basin in central Birmingham. Here for many years was the Worcester Bar, a barrier set up between the W & B and the Birmingham Canal that prevented traffic between the two canals and which wasn't finally removed until 1815. The first stretches of the W & B pass through the leafy reaches of Edgbaston and alongside Birmingham University to Selly Oak. Here, at one time, the Dudley No. 2 Canal intersected to form a major bypass route to the Black Country, much to the annoyance of the Birmingham Canal Navigations Company. The W & B, meanwhile, continues on through the outskirts of Birmingham, via Bournville, the home of Cadbury's chocolate, to King's Norton. Here the Stratford Canal leaves the W & B to pass through an elaborate guillotine stop lock and on to Lapworth, Stratford and, eventually, London. The W & B, however, heads straight into the 2,726 yd long King's Norton or West Hill Tunnel and then around the edge of Alvechurch and into Shortwood Tunnel. Tardebigge Old Wharf was, from 1807 to 1815, the terminus of the canal, still only half-way to Worcester. But with further funds the company was able to push its way through Tardebigge Tunnel and down the famous Tardebigge flight of locks where thirty locks take the canal down 217 ft to Stoke Prior. After Stoke Works, the former site of a massive salt mine, the canal passes down the Astwood flight to the junction with the Droitwich Junction Canal at Hanbury Wharf, some 9¼ miles from the Severn at Worcester. The canal now heads through Dunhampstead Tunnel and on to the small village of Tibberton. From here, the line crawls its way through the suburbs of Worcester and on to Diglis Basin and the River Severn.

You'll talk to more people on the towpath at Tardebigge in half an hour than you will in an entire day at Gas Street. It's that kind of place. The pound between the top lock and the tunnel has become a site of pilgrimage – a place to remember that the restoration of the canal network wasn't an inevitability but was only achieved by the work of far-sighted individuals who saw the inland waterways as more than just a place to put redundant Ford Escorts. It's worth going to just for that.

History

In the 1780s the only route for canal traffic from Worcester (and hence Bristol) to Birmingham was along the, often unnavigable, River Severn to Stourport, up the Staffordshire & Worcestershire Canal to Aldersley and then, after an approximately 120° turn towards the south-east, along the Birmingham Canal and into town. Although a short cut via the Stourbridge and Dudley Canals was conceived, the shallows of the Severn would remain and the line was still not as direct as perhaps it could be.

By 1789, this fact had reached the minds of those who were preparing their wallets for canal mania. A proposal was made for a direct line to Birmingham from the Severn at Worcester via Tardebigge. It was promoted in a pamphlet, published in 1790, which pointed out that the new canal would be 30 miles shorter than the Aldersley route and 15 miles shorter than the still to be finished Stourbridge/Dudley route. It was clear, the authors claimed, that the proposed line would more than halve the cost of tolls for goods being moved from Worcester to Birmingham.

Predictably, the other canal companies, notably the Staffordshire & Worcestershire and the Birmingham, weren't supportive of the new canal and successfully fought and defeated the first bill of 1790. A second bill followed with more solid support, including active participation from Birmingham businessmen who objected to the protectionist antics of the other canal companies. As a consequence, the 1791 Act enabling the building of the Worcester & Birmingham Canal was passed, a result that was celebrated in Worcester with the ringing of bells and the lighting of bonfires.

Although the Act was successful, the rival companies obtained their pound of flesh. The Birmingham, set against a junction between the W & B and its own

In times of freezing weather boatmen were unable to move their craft and were thus unable to earn a living. In 1895, the boatmen who usually plied the W & B were forced to make collections in Worcester on behalf of the Seamen and Boatmen's Friends Society

British Waterways

canal, insisted that the two be at least 7 ft apart at Gas Street Basin, a barrier that became known as the Worcester Bar. This bit of petulance on behalf of the Birmingham was reputedly to prevent water loss from its line but in practice it forced a costly transhipment of cargo and, coincidentally of course, the payment of an extra toll. The W & B was also forced by the Act to placate the Stourbridge and Dudley Companies, the proprietor of the Lower Avon Navigation, the Droitwich Canal Company and even the Worcester Corporation, by various forms of payment or financial guarantee.

In the Act, £180,000 was deemed to be sufficient to build the new line with a further £70,000 if needed. Despite the apparently onerous commitments to its rivals, the shares were fully subscribed by July 1791. In fact, following the financial success of canals like the S & W, many W & B shares changed hands at a profit even before any work was started.

The preliminary survey for the W & B was carried out by John Snape and Josiah Clowes, with Thomas Cartwright employed as engineer to build the line. The original plan was for a broad canal so that barges from the Severn could reach Birmingham without transhipment. But by mid-1794, partly due to cost and partly because of the planned junctions with the narrow Dudley and Stratford Canals, they had decided to opt for the odd combination of broad tunnels and bridges but narrow locks.

By 30 October 1795, 3 miles of canal between the Worcester Bar and Selly Oak were open, and work on King's Norton Tunnel was underway. However, in order to ensure that mill streams to the south were not starved of water, the company had to build reservoirs to supply the summit level. This additional engineering work meant that by the end of 1795, £154,067 had already been spent and the company urgently sought to access the additional £70,000. Despite these problems, the canal reached its junction with the Stratford Canal at King's Norton in May 1796 and opened the 2,760 yd long King's Norton Tunnel on 27 March 1797. Barges of 60 and 70 tons were now able to navigate from central Birmingham to Hopwood Wharf – a distance of about 8½ miles.

The smouldering financial difficulties now began to flare and building work was seriously affected. The situation was not improved by the fact that the treasurer, Thomas Hooper, had secreted away an estimated £13,800 of the company's funds. His resignation didn't remedy the situation and the construction work slowed to such an extent that five years on, in May 1802, the southern terminus remained at Hopwood.

In an attempt to improve matters, an Act was passed in 1804 to raise more money from existing shareholders. This action improved the short term prospects and progress was once again possible. In 1805, Cartwright was re-engaged to take the canal to Tardebigge. By March 1807, he had moved on 6 miles to a point between the Shortwood and Tardebigge Tunnels where a wharf (Tardebigge Old Wharf) was built to off-load cargo for land shipment south. This proved to be quite a success and trade was brisk. There was even a regular passenger boat running between Alvechurch and Birmingham. Spirits in the W & B had now revived and yet another Act was passed in 1808. This offered attractive terms for new money and some £168,000 was, perhaps surprisingly, raised by January 1809 with shares trading at a premium. Activity on site also appeared to be more in evidence with the next big challenge: to take the canal down the 217 ft from Tardebigge to Stoke Prior.

The original proposal for seventy six locks at Tardebigge was daunting, both in terms of the cost and in the amount of time and water that would be needed to use them. The canal engineer, William Jessop, acting as consultant, agreed that there was a potential problem with water supply at the summit. So, following a proposal from the new engineer, John Woodhouse, the company agreed to experiment with a lift system in which craft would be raised and lowered in massive tanks. By the time Tardebigge Tunnel was finished in 1810, a test rig was ready and, in the winter of 1811, the lift was run for nearly three weeks, showing itself to be quick, efficient and water saving. However, by this time the committee had grown sceptical about the robustness of lifts generally and, as

Tom Rolt spent 1,800 days at Tardebigge during the Second World War when he inspected factories for the government. Here he first met Robert Aickman and from that meeting was born the Inland Waterways Association. This picture shows Rolt on board his narrowboat, *Cressy*, at Tardebigge

Tom Rolt Archive, British Waterways

another eminent engineer, John Rennie, agreed with them, they decided to build a reduced number of locks instead.

So, from 1812, a major effort on the Tardebigge Locks was under way. Progress was encouraging and with this new investors came in. By 1814, the reservoirs at Crofton and King's Norton were ready and it had been agreed to remove the Worcester Bar (replaced with a lock in July 1815). After raising £36,000 by annuities in early 1815, the company was able to complete the construction. The basin at Diglis on the Severn below Worcester was built and the whole line of 30 miles, fifty-eight locks and five tunnels was opened on 4 December 1815. Charles Hadfield suggests that the total cost was £610,000, three-and-a-half times the original estimate.

The financial position of the company was difficult from the start. It owed large sums on annuities and loans, and it was forced to raise more from shareholders. It needed to get the traffic moving as quickly as possible and promptly reduced its tolls – a tricky matter as the S & W, Stourbridge and Dudley Canals were all doing likewise to maintain their own trade. However, things improved once the canal was fully open and trade started flowing:

industrial goods and coal to Worcester, grain, timber and agricultural products to Birmingham. In 1820, toll receipts reached £14,625 and the princely dividend of £1 per share was paid in 1821.

Receipts continued to rise as traffic, especially in salt, improved. In 1825, the company expanded its activity by leasing the Lower Avon (Tewkesbury to Evesham) in order to develop markets and look for a new route to the Midlands. By 1830, rock salt from Stoke Works had became a major cargo and receipts rose sharply. This meant that by 1836–8, average receipts reached new highs at £43,488, with dividends of £4 a share. However, the potential impact of the railway age became apparent when, with the mere threat of a line between Gloucester & Birmingham, the W & B share price fell from £105 to £75. This threat became reality in 1836 when the Birmingham & Gloucester Railway Act was passed. By October 1841, the line was open and, virtually immediately, it started taking trade from the canal. The W & B and the Birmingham were forced to reduce shipping tolls and to do away with those on coal passing over the Worcester Bar. This had its inevitable effect on dividends.

In 1847, the railway dealt a body blow to the canal by taking the lion's share of the Stoke Works salt trade. The W & B tried to respond by starting its own carrying company and by promoting the Droitwich Junction Canal to link it with Droitwich. Neither gambit was to be a success. After another toll cut in 1857, the company opened negotiations with the Oxford, Worcester & Wolverhampton Railway. The talks resulted in a proposal that the railway should lease the canal for a period of twenty-one years at a rent that would yield a dividend of 1 to 2 per cent. This move was fiercely opposed by the other canal companies who feared closure and loss of a through route for their own cargoes. Following an appeal to the Board of Trade, the arrangement was indeed rendered illegal by an Act of Parliament.

Having the lease option closed to it and unable to pay dividends, the W & B then sought, in 1864, to build a railway of its own from New Street Birmingham to King's Norton. The necessary finance was not forthcoming and the plan was dropped. It was then proposed that the canal be sold to a group of railway contractors who would themselves convert it to a railway. This idea also evaporated in the face of more fierce opposition from the other canal companies.

Trade had now reached the point where it wasn't possible to pay dividends. In 1868, with some £100,000 still owing and little or no chance to raise any further money, the receiver was called in. After further opposition to another railway proposal, the Sharpness New Docks Company offered to take over the canal, including its liabilities for the Droitwich Canals, for £6,000 p.a. or £1 per share, after converted to £150,000 of 4 per cent debentures, and the liquidation of the debt of £100,473. This offer was accepted gratefully and authorized by an Act of 1874. The new company was called, rather cumbersomely, the Sharpness New Docks & Gloucester & Birmingham Navigation Company.

Although the new company started optimistically, after only four years it was losing money and only remained solvent because of the profitability of the Gloucester & Sharpness Ship Canal. Without this, the W & B would have closed. Some additional help came to hand in 1926, when George Cadbury inaugurated a fund to provide annual subsidies against losses. In return, the Docks company agreed not to close the line for as long as the subsidies were forthcoming. The trading loss at the time was £3,605 p.a. with a subsidy averaging £1,589.

In 1939, the Droitwich lines which had been disused for many years were officially abandoned. A similar fate was proposed for the W & B after it was nationalized in 1948. In the Report of the Board of Survey for the British Transport Commission in 1955, it was proposed that either the W & B or the Staffordshire & Worcestershire be retained but not both. Somehow both survived even though the remaining commercial traffic on the W & B (coal from Cannock to Worcester, and chocolate crumb from Worcester to Bourneville) ceased in 1960 and 1961 respectively. Fortunately, the W & B was never abandoned. As we all know, the 1960s saw a great resurgence in interest in canals, not from the nation's industry but from boating enthusiasts in the age of increased leisure time. The Transport Act of 1968 listed both the W & B and the S & W as cruising waterways to be retained and maintained into the future for leisure traffic. The line is now part of the increasingly popular cruising circuit and one link of the Avon Ring: a route which includes the Lower & Upper Avon Navigations, the Stratford Canal and the River Severn.

Walk

Start and finish:	Tardebigge (OS ref: SO 997694)
Distance:	5 miles/8 km
Maps:	OS Landranger 139 (Birmingham) and 150 (Worcester & The Malverns)
Car park:	At Tardebigge room for about four cars on verge on the Redditch side of the British Waterways depot on the (minor) Finstall road
Public transport:	Regular Midland Red West buses journeying between Bromsgrove and Redditch stop at Tardebigge. They do not run on Sundays. Telephone: (0527) 72265

This walk goes from Tardebrigge to the Queen's Head pub and back again. If you prefer not to have to return along the same route, it is suggested that you start from Bromsgrove where you can take the bus to Tardebigge. From the Queen's Head continue on to Stoke Works (bridge 42 with the Boat & Railway pub)

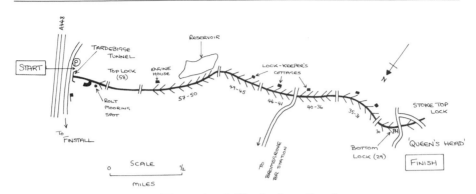

The Worcester & Birmingham Canal

from where another Midland Red West bus can convey you back to Bromsgrove. This would yield a walk of about 4 miles. However, it is clearly vital to check the times of the buses from Stoke Works using the telephone number above.

If arriving by bus at Tardebigge, continue on the road towards Redditch for approximately 50 yd to what appears to be a bridge over the canal but which is in fact the entrance to Tardebigge Tunnel. Here you'll meet those who came by car. A path leads down to the left-hand bank of the canal where, to the right, the

The southern portal of Tardebigge Tunnel

entrance to the 580 yd long tunnel can be seen. It's worth noting that this is a relatively wide tunnel for a narrow-locked canal. This wasn't a deliberate design to facilitate passage both ways but is a remnant of the original plan to make the line a barge canal. There is no towpath through the tunnel and barge horses were forced to go up to the road where they turned left and then right along a footpath which leads to the northern portal. Originally boats were taken through the tunnel by 'leggers', men who lay on their backs to push the boats through by walking against the roof. This practice was abandoned in the 1870s when company tugs were employed to pull boats through. This, in turn, was abandoned during the early part of the twentieth century. There are three more tunnels (Edgbaston, West Hill and Shortwood) between here and central Birmingham, some 14 miles to the north-east.

A few yards further on, the British Waterways' yard situated at what was once called New Wharf (Old Wharf is on the northern side of the tunnel) offers comfort to passing boaters.

The scene around Tardebigge New Wharf might seem quiet and unspectacular but it has a unique place in waterways restoration history. During most of the Second World War (for 1800 days to be exact), the author Tom Rolt and his wife Angela moored their boat *Cressy* at a point about 50 yd beyond the BW basin. Here they lived while Tom was employed by the government during the day and started on a fledgling writing career by night. In doing the latter, he had written a book that was to be a landmark in waterways restoration. *Narrow Boat* was originally published in 1944 and had almost immediately attracted the attention of Robert Aickman, a man who felt deeply for the country's waterways but who otherwise was a novice in the ways of the canals. Aickman was moved to write to Rolt to express his admiration for both the book and the motives. In response, Rolt invited Aickman and his wife, Ray, to visit *Cressy* and so, in August 1945, the four met for the first time. From that meeting the Inland Waterways Association, the premier force in the fight against the abandonment of the British canal network, was formed. To mark the spot, the Worcester & Birmingham Canal Society have erected a plinth just on the other side of the canal. Interestingly, this notes the first meeting between Rolt and Aickman as being in 1946. To reach the plinth, cross the canal at the lock.

Although the start of the IWA was inauspicious, and some of its early days a little shaky, the association has gone from strength to strength. The IWA campaigns for the restoration, retention and development of inland waterways in the British Isles and for their fullest commercial and recreational use. There are branches up and down the country with literally thousands of members contributing both directly and indirectly to the aims. New members are always welcome and the address of the association is included at the end of this book.

Even without the birth of the IWA, Rolt's descriptions of life here at Tardebigge during the Second World War would still make the place special. The goings on are wonderfully told in his autobiography *Landscape with Canals* which contains portraits of the characters who staffed the boatyard and worked the locks of what was then still, officially at least, a working waterway.

About 50 yd beyond the plinth is the start of the other reason why Tardebigge is so famous throughout the canal world. The view to the south should give a hint as to what is to come. The land in front of us and to the right falls rapidly towards the Severn Valley to offer views of the Welsh border and even the Malvern hills away in the distance. Thus Tardebigge plays host to the biggest flight of locks in the country. There are thirty of them in the Tardebigge flight proper with a total rise/fall of some 217 ft. If a further six at Stoke are included, the total rise/fall comes to 259 ft. All these locks take the W & B from what is known as the Birmingham Level (453 ft above sea level – the same height as the Birmingham Canal at Gas Street) down, through a further twenty-two locks, to the River Severn at Worcester.

Tardebigge Top Lock (No. 58), apart from being the first of the flight, also has the honour of being one of the deepest narrow locks on the system with a fall of some 14 ft. The reason for this is one of those peculiar accidents of history. When the canal was being built in the early 1790s, it was thought that the company would have to pump water for their part of the Birmingham level all the way from the Severn. To avoid this, radical answers were sought. John Woodhouse, an engineer on the Grand Junction Canal (now the Grand Union), offered to build an experimental vertical lift at Tardebigge at his own expense if the company agreed to pay for the excavation work and the masonry. The objective was to reduce the number of locks planned for the flight and to avoid excessive loss of water down the line. As a lift would use hardly any water in operation, the W & B company agreed to the experiment and so, while they still continued with the construction of reservoir capacity at the summit, Woodhouse built his lift.

The test rig was ready in June 1808. It was a 12 ft lift and was situated where the top lock is now. It consisted of a wooden tank, big enough to carry a narrowboat and weighing 64 tons fully loaded, counterbalanced by a platform full of bricks. The two sides were connected by chains passing over a set of cast-iron wheels. The canal was sealed using wooden paddles at each end of the tank which were raised or lowered to allow boats to come and go. The lift was moved by two men who wound the tank up and down. From 25 February to 16 March 1811, the lift worked solidly with a peak on 15 March, when 110 boats were raised or lowered in just twelve hours.

This performance failed to impress the W & B committee, however, who invited comments from the well-known canal engineer, John Rennie. It is true to say that such lifts did not have a good reputation for their robustness and whatever Woodhouse had achieved was almost irrelevent. Locks were a slower and more water-costly system but they were at least reliable and a known quantity. On this basis, both Rennie and the committee abandoned the idea of using lifts at Tardebigge and the test lift was dismantled. The actual site of the test rig was a slightly larger than normal drop but was converted into a single fall lock in 1815. Thus the deeper than average fall.

Our walk now passes under a neat accommodation bridge and on to bridge 55. Just beyond, by lock 57, is the Engine House, a former pumping house that is now a restaurant and night club. Originally the house contained a steam engine which pumped water from the reservoir, soon to be visible to the left of the canal, up into the Birmingham level.

As you walk down the hill, the long line of locks to the next bridge appears before you, two with neighbouring cottages. At one point there is a sudden and unexpected view over to the reservoir on the left. Just before the bridge (No. 54) a number of rough-hewn paths up the bank to the left invite the curious for a closer look at the reservoir as well as a less obstructed view to the south. In summer months, an endless string of boats with their, by now, weary crews can be seen passing up the locks, no doubt wondering when their toils will be over and dreading the return trip when the whole procedure will have to be gone through all over again.

There is now a mile of steadily changing scenery where locks, cottages and boats intermix to make a peculiarly bustling foreground in what is otherwise a rather remote and peaceful part of the world. The toil involved in negotiating the flight brings out a camaraderie among the boaters who share their experience with anybody and everybody in reach. From all the activity, it isn't

Working the Tardebigge flight of thirty locks on the Worcester & Birmingham Canal is often better to watch than to participate in!

too hard to imagine the scene here when the canal was a working one. Interestingly, boats on the W & B were often towed by donkeys, working in pairs, rather than horses. Why this should be the case isn't known.

Before long you reach Tardebigge Bottom Lock and realize that there are no more locks in front of you. It makes you feel somewhat deprived like a toddler who's had a favourite toy taken away. The line bends around slightly to the left and the canalside gardens of the Queens Head pub are seen over to the right. By taking the leafy track to the left of the next bridge (No. 48) you will reach the road. Turn right for the pub.

Here you can either meet someone kind enough to take you back to Tardebigge or you can simply about face to re-meet with all those boat crews who are still struggling up the line.

If you plan to go on to Stoke Works for the Bromsgrove bus, continue along the left-hand bank of the canal, past six locks to Stoke Wharf, a pleasant spot with a warehouse, some cottages, a crane and the Navigation pub. Another ½ mile will take you past the remnants of an enormous salt works (now a pharmaceutical factory) to Stoke Works. The salt mines here once brought a considerable amount of trade to the canal but are now long gone. To reach the bus stop, you should turn off right at bridge 42 near the Boat & Railway pub (OS map 150, ref: SO 944663).

Further Explorations

The richness of public transport between Worcester and Droitwich should tempt some to devise a walk between the two towns, a distance of approximately 12 miles. Use OS Landranger 150 (Worcester & The Malverns).

From the Market Hall car park in the centre of Droitwich, cross the road to find the fine barge lock which actually has a swing bridge halfway along it. Those interested in this particular kind of oddity will find a similar structure at Hungerford Marsh Lock on the Kennet & Avon Canal. This lock marks the beginning of the Droitwich Barge Canal that goes south-west from here to join the River Severn near Hawford. This in itself is worthy of further exploration and those with the time and inclination can turn left to follow the public footpath for 6 miles to the River Severn. The barge canal was opened in 1771 having been surveyed by James Brindley no less. The route was an important route for Droitwich salt. By the Second World War, the line was officially abandoned but, with the formation of the Droitwich Canals Trust in 1973, it is being gradually improved throughout.

Towpathers to Worcester will, however, not be so distracted and should turn right at the barge lock to join the River Salwarpe and on to a small, ornate road bridge. Go up to the road and bear right along the A38 to traffic lights and turn left on to the B4090. This is the Salt Way.

The road continues on to a point where it passes under the M5 motorway. Just before the motorway, the River Salwarpe, which has been on the left-hand side of the road, deviates northwards. From this point, the Droitwich Junction Canal was built with seven locks in 1851 to provide the salt town of Droitwich with a direct link to Birmingham via the W & B. The DJC fell into disuse in the 1920s and was officially abandoned in July 1939. However, in recent years there have been moves to restore the line and it is now owned by the Droitwich Borough Council. By dodging right along lanes off the Salt Way, traces of the old line can be seen running parallel with the road. After a while, you need dodge no longer as various moored boats indicate that the line is intact and in water.

The Salt Way crosses the W & B at Hanbury Wharf where the Eagle & Sun offers refreshment. Towpathers to Worcester should cross the canal and turn right to pass along the left-hand bank of the line. After nearly 2 miles, the canal goes through Dunhampstead Tunnel where walkers have to deviate off to the left and over the hill, returning to the canal at bridge 30. From there the canal winds through Tibberton (opportunity here for lunch and/or some shopping) before going under the M5 and down the six Offerton Locks.

By now the outskirts of Worcester are making themselves evident and the canal goes down a further six locks to skirt Worcester City football ground where the towpath changes sides. From there, the canal slips quietly through the town. Bridge 10 is a splendidly arched and extrovertly 'holed' railway bridge just before Lowesmoor Wharf. Re-cross the canal at bridge 5 and continue on past two more locks and into Diglis Basin.

Diglis isn't as grand as Stourport but has similar intent. Beyond the wide basin, the dry dock and the two massive barge locks, lies the River Severn. Turn right here to take the river path round to the main river bridge. If you turn right here you will reach firstly the bus station (both Midlands Red West (No. 144) and Citibus (No. 22) offer services to Droitwich) or Foregate Street station from where trains compete for your return trip.

The proximity of train and canal at the Birmingham end also allows an easy one-way walk between New Street station and Bournville; a distance of about 5 miles. To reach Gas Street Basin from New Street follow the instructions given under Further Explorations of the Birmingham Canal Navigations. Cross the Worcester Bar (where the W & B joins the BCN) and turn left to take the right-hand side of the canal. Within a short distance the line crosses a small aqueduct and turns abruptly right. Our course is now due south and we pass through the leafy suburbs of Edgbaston including, for a time on the right, the Botanical Gardens and, on the left, the campus of Birmingham University. After passing under a railway bridge, the canal reaches Selly Oak. At bridge 79A, the railway re-crosses. Three bridges further on, pass up to the road and turn right to reach Bournville station. Before returning to New Street, you might consider visiting the Cadbury World visitor centre in the nearby chocolate factory. How can you resist?

Further Information

The address of the Worcester & Birmingham Canal Society is:
  1 Southwold Close,
  St Peter the Great,
  Worcester,
  Worcestershire,
  WR5 3RD.

The history of the W&B can be traced through the pages of:
Hadfield, Charles, *The Canals of the West Midlands*. David & Charles, 1969.

Tom Rolt's account of Tardebigge during the Second World War and his account of the birth of the Inland Waterways Association can be found in:
Rolt, L.T.C., *Landscape with Canals*. Alan Sutton Publishing, 1977.

A more independent history of the IWA and a fascinating account of how the country's waterways were saved can be found in:
Bolton, David, *Race Against Time*. Mandarin Paperbacks, 1990.

APPENDIX A

General Reading

This book can, of course, only provide you with a brief glimpse of the history and workings of the waterway network. Other authors are far more qualified than I to fill the gaps and the following reading matter may help those who wish to know more.

Magazines

There are two monthly canal magazines that are available in most newsagents: *Canal & Riverboat* and *Waterways World*. Towpathers may be particularly interested in *Waterways World* as it has a monthly canal walks column.

Books

There are a wide range of canal books available, varying between guides for specific waterways to learned historical texts. There should be something for everyone's level of interest, taste and ability to pay. Libraries also carry a good stock of the more expensive works and are well worth a visit.

All the books listed here are available in paperback unless marked with an asterisk.

For a good general introduction to canals that won't stretch the intellect, or the pocket, too far:
Smith, P.L., *Discovering Canals in Britain*. Shire Books, 1984.
Burton, A. and Platt, D., *Canal*. David & Charles, 1980.
Hadfield, C., *Waterways sights to see*. David & Charles, 1976.*
Rolt, L.T.C., *Narrowboat*. Methuen, 1944.

This can be taken a few steps further with the more learned:
Hadfield, C., *British Canals*. David & Charles, 1984; new edition Alan Sutton, 1993.

There are a number of books that are predominantly collections of old photographs. Two examples are:
Ware, M., *Canals and Waterways*, History in Camera Series. Shire Books, 1987.
Gladwin, D., *Building Britain's Canals*. K.A.F. Brewin Books, 1988.

At least three companies publish boating guides:
Nicholson's Guides to the Waterways. Three volumes.
Pearson's Canal & River Companions. Eight volumes (so far).
Waterways World. Eight volumes (so far).
Of the three, Pearson's guides are the most useful for towpathers, with their volume on the Birmingham Canal Navigations heading the list. The only problem for users is the duplication between volumes. Other Pearson volumes applicable to Midlands canals are:
Four Counties Ring, Midlands Ring, Severn & Avon (Avon Ring), *Shropshire Union & Llangollen Canals, South Midlands & Warwickshire Ring*.

APPENDIX B

Useful Addresses

British Waterways

BW are the guardians of the vast majority of the canal network and deserve our support. There are offices all over the country but their customer services department can be found at:

British Waterways,
Greycaine Road,
Watford,
WD2 4JR.
Telephone: (0923) 226422

Inland Waterways Association

The IWA was the first, and is still the premier, society that campaigns for Britain's waterways. They publish a member's magazine, *Waterways*, and provide various services. There are numerous local groups which each hold meetings, outings, rallies, etc. Head office is at:

Inland Waterways Association,
114 Regent's Park Road,
London,
NW1 8UQ.
Telephone: (071) 5862556

Towpath Action Group

The Towpath Action Group campaigns for access to and maintenance of the towpaths of Britain and publish a regular newsletter. They are thus the natural home of all keen towpathers.

Towpath Action Group,
23 Hague Bar Road,
New Mills,
Stockport,
SK12 3AT.

APPENDIX C

Museums

A number of canal museums are springing up all over the country. The following are within reach of the area covered within this book and are wholly devoted to canals or have sections of interest:

MUSEUM OF SCIENCE AND
INDUSTRY
Newhall Street,
Birmingham,
B3 1RZ.
Telephone: (021) 2351661

BLACK COUNTRY MUSEUM
Tipton Road,
Dudley,
West Midlands,
DY1 4SQ
Telephone: (021) 5579643

THE BOAT MUSEUM
Dockyard Road,
Ellesmere Port,
South Wirral,
L65 4EF
Telephone: (051) 3555017

THE NATIONAL WATERWAYS
MUSEUM
Llanthony Warehouse,
Gloucester Docks,
Gloucester,
GL1 2EH
Telephone: (0452) 307009

THE CANAL MUSEUM
Stoke Bruerne,
Towcester,
Northamptonshire,
NN12 7SE
Telephone: (0604) 862229

INDEX

CANAL
WALKS

SOUTH

To Mum and Dad
who have never had
a book
dedicated to them
before.

CONTENTS

ACKNOWLEDGEMENTS

This book would have been impossible without the splendid resources of various libraries: communal ownership in practice. Despite chronic under-funding, the information and help received was substantial.

Help and advice on the routes came from: Mrs Audrey Wheatley of the Bude Canal Society; W.J. Spall of the Chelmer & Blackwater Navigation Ltd; staff of the National Rivers Authority; David Jowett of the Cotswolds Canals Trust; Brian Forder of the Monmouthshire, Brecon & Abergavenny Canals Trust Ltd; and John Wood of the Wey & Arun Canal Trust. Thanks also to the various employees of British Waterways who, as ever, have been both helpful and cooperative.

Assistance with archive photographs came from Roy Jamieson of British Waterways; Lynn Doylerush of the Boat Museum; Ms J.T. Smith of Essex Record Office; R. Bowder of Marylebone Library; Victoria Williams of Hastings Museum; Mr P.A.L. Vine; and the staff of the National Monuments Record. Thanks also to members of the Sussex Archaeological Trust. Assistance with, though no responsibility for, the author's pictures came from Mr Ilford and Mr Fuji and two old and steadily deteriorating Olympuses fitted with 28 mm and 75–150 mm lenses.

For assistance during walking trips, thanks go to Mike and Val of Berry Park, Welcombe, and to Kath (who also helped with some inter-library loans!). Thanks also to Tessa and Miles for helping me with at least one 'two-car trick'. As before, much appreciation to Taffy for most of the trans-portation. And, of course, thanks to Mary who continued to provide her wonderful support throughout.

LOCATION OF WALKS

1. Basingstoke Canal at Woking
2. Brecon & Abergavenny Canal at Brecon
3. Bridgwater & Taunton Canal at Taunton
4. Bude Canal at Bude
5. Chelmer & Blackwater Navigation at Heybridge
6. Grand Junction Canal at Tring
7. Kennet & Avon Canal at Bradford-on-Avon
8. Oxford Canal at Oxford
9. Regent's Canal at Camden
10. Royal Military Canal at Rye
11. Thames & Severn Canal in the Golden Valley
12. Wey & Arun Junction Canal at Loxwood

KEY TO MAPS

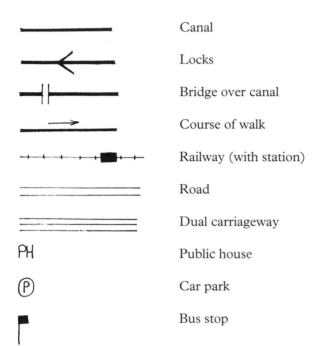

Canal

Locks

Bridge over canal

Course of walk

Railway (with station)

Road

Dual carriageway

Public house

Car park

Bus stop

INTRODUCTION

On a warm, sunny Friday in May 1991, a crowd of enthusiasts at Frimley Lodge Park, near Aldershot, cheered and waved as the Duke of Kent proclaimed the Basingstoke Canal officially re-opened. The canal, from its junction with the River Wey near Byfleet in Surrey to the closed tunnel at Greywell in Hampshire, was once thought to have been lost in a mass of scrubby vegetation, old mattresses and discarded supermarket trolleys. The commitment of 2,000 local canal society members and the vision of two county councils has turned a public embarrassment into a magnificent amenity. The Basingstoke Canal of the 1960s bears no resemblance to that of the 1990s. The 30 or so miles of restored waterway are now the pride of the area and a great attraction to boaters, naturalists, anglers and walkers alike. The Basingstoke is a modern-day success story and it typifies the change in attitude towards the nation's canals. Whereas twenty to thirty years ago they were objects of derision, now we all, it seems, appreciate them for the fine resource that they are.

Although many of the country's waterways have been restored in this way, only recently has their value for walking been fully appreciated. Even though British Waterways produces leaflets which proudly proclaim that it owns 1,500 miles of walking (more, as it puts it, than the Pennine Way and the Offa's Dyke Path put together), to date towpath walkers have mostly been tolerated rather than encouraged. Now, apparently, things are to change. The publication of this book coincides with British Waterways' announcement of the first National Waterway Walk (a title designed to be equivalent to a National Trail, like the aforementioned Pennine Way and Offa's Dyke). To commemorate the two-hundredth anniversary of canal mania, the 140 miles from the Gas Street Basin in central Birmingham along the Grand Union Canal to Paddington in London are set to become the bee's knees of canal walking. The towpath is to be upgraded to render it usable by the less able-bodied; there is to be better access; waymarks and information notices are to be erected; and, in that ultimate accolade which tells us that it's all kosher, there is to be an official walking guide.

Is this the answer to all those critics who say that towpaths are an ignored and abused resource? Well, nearly. The route will never be directly equivalent to a National Trail. National Trails are public rights of way on which walkers have the legal right of access at all times. British Waterways, however, is very firm that it will preserve its rights over the towpath. Unless they are already so dedicated, British Waterways' towpaths will not become public rights of way. British Waterways will retain the privilege of being able to withdraw access just as quickly as it gave it.

It is, of course, easy to carp. After all, on non-British Waterways canals (in this volume the Basingstoke, the Bude, the Chelmer & Blackwater, the Royal Military, the Thames & Severn and the Wey & Arun are all non-British Waterways-owned canals), walkers are restricted to public rights of way only and many miles remain closed. The establishment of the National Waterway Walk is therefore a great first step to the full appreciation of the nation's canals as long-distance walking routes. The Kennet & Avon is already acknowledged as one (if not officially). The Staffordshire & Worcestershire would make a splendid one if only some of the towpath was restored, and the Leeds & Liverpool could be become a fine, relatively flat, alternative to Wainwright's Coast to Coast. All towpathers should therefore welcome this move by British Waterways and encourage it to do more of the same. Furthermore, we might encourage them to convert all the nation's towpaths into the public rights of way that they should surely be.

The Southern Canals

At first glance, the southern canals appear to have less to offer than their Midlands counterparts. How can any line have the richness of history and scenery of a Staffordshire & Worcestershire or a Trent & Mersey? Yet there's no busier canal in the country than the Grand Junction and few lines anywhere share the variety of the Kennet & Avon. And none, simply none, can compare with the wonderful mountain setting of the Brecon & Abergavenny.

Mostly, however, the South's canals take on a much more somnolent air. The Basingstoke is the archetypal green finger through the suburbs of Surrey, and the Bridgwater & Taunton is so peaceful it almost snoozes its way between the two Somerset towns. The Chelmer & Blackwater is undoubtedly the acceptable face of Essex Man, while the Oxford Canal, despite the constant fear of finding one of Morse's corpses, is nothing short of a rural retreat. Even in their most urban form, and the Regent's Canal through central London cannot be more so, the southern canals have plenty to offer the towpather.

It must also be said that some of the South's canals have become so rural that they've all but disappeared into the undergrowth. The Kennet & Avon and the Basingstoke have only just survived to be re-opened within the last 2–3 years. Others, such as the Wey & Arun, the Thames & Severn and the

Bude, are in various states of disrepair and still have a somewhat doubtful future. And finally, what can be said of that most peculiar of defensive fortifications, the Royal Military Canal? Is it a canal? Is it a drainage ditch? Or is it simply a super place for an afternoon stroll?

Here is just a selection of the canals of the South and, although by no means arbitrary, it is strictly a personal choice. Walkers should not ignore the fine Wey Navigation (through Guildford), the Lee & Stort, the Gloucester & Sharpness, or the long-lost lines such as the Wiltshire & Berkshire, the Gloucester & Hereford, the Somerset Coal Canal, the Grand Western or the many other small canals of south Wales or south-west England. All these are worthy of further investigation and, with the aid of an Ordnance Survey map and some keen legs, can be traced both on paper and, often, on the ground.

As with all personal choices, some people's favourite lines or stretches of waterway may have been omitted. There is also the problem of what is a southern canal. Both the Oxford and the Grand Junction canals go just as far north as the Stratford and yet the latter was included in *Canal Walks: Midlands*. Furthermore, what exactly constitutes a canal? Could it not be argued that the best waterway walk in the South is the Thames 'Navigation'? In all these matters, only the author is to blame. I have not tried to be a sage on these matters, merely a stimulus. Walking the canals of the South should be an adventure with plenty to see and discover for yourselves. And it is quite likely that you will see even more than I did and enjoy them every bit as much.

Walking the Towpaths

The walks in this book are all straightforward and require no special feats of strength or navigation. Towpath walks have two great virtues: they are mostly on the flat; and they have a ready-made, unmistakable course to follow. Getting lost should therefore, in theory at least, be relatively hard. The key problem with towpath walks is that if you want to spend most of the day by the canal, circular routes to and from a vehicle or a particular station or bus-stop become difficult. Many of the walks in this volume involve walking one way and returning by public transport. This means that you must check the availability of the bus or train before travelling. Telephone numbers are provided and your local library should have the main British Rail timetable.

Walkers should generally plan for 2 to 2¹/₂ miles an hour so that stops can be made for sightseeing or a break. Head-down speedsters should easily manage three miles an hour on a good track. You should, of course, add a little time for stoppages for refreshment and add a lot of time if you are accompanied by photographers or bird-watchers.

No special equipment or provisions are needed to walk the towpaths of Britain. On a good day and on a good path, any comfortable footwear and clothing will do, and you'll be able to leave the laden rucksack at home. However, for longer walks through more remote country you should be more prudent. Even in a drought, towpaths can be extremely muddy and, from experience, it can not only rain virtually any time but usually does. Boots and a raincoat of some sort are therefore advisable. Similarly, although pubs and small shops are often fairly common along the way, it may be useful to carry some kind of snack and drink with you.

This book includes sketch maps that show the route to be taken. However, the local Ordnance Survey map will always be useful and the appropriate map numbers and references are provided in each chapter. Again, your local library may well have them for loan.

Finally, the dangers inherent in walking along a waterway are often not fully appreciated. Over the 1990 Christmas holiday, three children died after falling into a lock on the Kennet & Avon at Burghfield. A year later their mother committed suicide, having been unable to endure her loss. Locks are deep, often have silt-laden bottoms, and are very difficult to get out of. Everybody, especially children, should be made aware of this. If somebody does fall in, you should not go into the water except as a last resort. You should LIE on the bank and use something like a coat for the person to grab so that you can then pull them in. Better still, keep children away from the edge.

Otherwise, please enjoy.

1
THE BASINGSTOKE CANAL
West Byfleet to Ash Vale

Introduction

The commuter-land of north-west Surrey is an unpromising area for a canal, let alone an agricultural one, but herein lies a gem. Although it's often given the somewhat over-used epithet of 'green finger', the Basingstoke Canal is as close as you'll come to such a thing.

Starting where the London-Southampton railway and the M25 compete for attention at West Byfleet, the canal leaves the River Wey to follow a leafy course through Woking, Brookwood and Frimley before entering Hampshire at Aldershot. After skirting Farnborough airfield, the line goes through Fleet and on to the villages of Crookham and Odiham. Contrary to expectations, the current route of the Basingstoke Canal does not go to Basingstoke. Instead it stops 4 miles to the east at one of the most notorious sites in canal restoration: the Greywell Tunnel. Boat enthusiasts argue in favour of re-opening the tunnel which was closed in 1932. However, the cave-like interior is an important roosting site for that most beleaguered of animals, the bat. Nature lovers, including English Nature, argue that restoration would disturb the bats in one of their last major refuges in this part of the country. At the time of writing, the debate continues and the tunnel remains closed, a situation no doubt welcomed by the bats. Whatever the arguments, it is true that re-opening the tunnel would not add much to the navigable length of the waterway. Beyond Greywell Hill, the canal runs for just another mile before coming to a complete stop. Although it is still possible to trace the course of the line into central Basingstoke, beyond Up Nately the land has been sold and a lot of it redeveloped.

Despite all of this, the Basingstoke is a breath of fresh air for the area and justifiably a popular spot. But if you take your wildlife identification guides with you, it may be tactful to keep them hidden from passing boaters!

History

Whereas most canal sponsors saw their main source of revenue in the shipment of minerals, coal or manufactured goods, the prime purpose of the investors of north Hampshire was to give a boost to the local farming community. They believed that the benefits of the canal would be seen through the cheap importation of fertilizer and in the reduced cost of shipping produce to the great markets in London.

The first scheme to link Basingstoke with the Thames (and hence London) was suggested in 1770. Benjamin Davis, who surveyed the route, planned to follow the Loddon Valley as far as Twyford and then to bear east via White Waltham to Monkey Island near Bray on the Thames. This proposal was later changed to a route from Basingstoke to the navigable River Wey at Byfleet and was surveyed for a group of enthusiastic investors by Joseph Parker in 1776. The sponsors of the canal claimed that it would 'furnish timber for the Navy' (a politically expedient claim at the time) 'and also supply the London markets with flour and grain at a cheaper rate' (politically expedient at any time). Parliament was obviously convinced by the argument. An Act authorizing the route and providing the powers to raise £86,000 (with £40,000 more if necessary) followed in 1778. This was despite opposition from the people of Reading who feared loss of their country trade to the new, possibly quicker, route to London.

Parker's surveyed course broadly followed the route we see today. The only significant difference occurs between Odiham and Basingstoke. Here Parker proposed a 6 mile loop to Turgis Green. This idea was dropped following the protestations of the local landowner, Lord Tylney, who felt that he was being encircled by the waterway. As a result, the Greywell Tunnel was built. This cut the total length of the canal from the original 44 miles to the built 37$\frac{1}{2}$. The tunnel had the additional benefit of tapping an underground water source and remains to this day an important supply, with the long summit pound from Aldershot to Greywell acting as a kind of linear reservoir for the rest of the canal.

The uncertainty caused by the American War of Independence delayed construction work and it wasn't until 1788 that the Basingstoke Canal Navigation Company appointed William Jessop as engineer. He promptly carried out the final survey, engaged John Pinkerton as contractor and building work began. The first tolls were collected in 1791 and by August 1792 some 32 miles (from the Wey to Greywell), including twenty-four locks, were navigable. However, the cost involved in doing so meant that a new Act was needed in 1793 to raise a further £60,000. Greywell Tunnel was finally ready and the entire length (37$\frac{1}{2}$ miles and twenty-nine locks) opened on 4 September 1794. The canal could handle barges 82½ ft long by 14½ ft wide, each carrying 50 ton loads and able to travel the 71 miles from London to Basingstoke in three or four days. In the early years, barges to London carried malt, flour and timber, while those from London carried coal and groceries.

To cope with the expected rush, company wharves were built at Horsell, Frimley, Ash, Farnham Road (west Aldershot), Crookham, Winchfield, Odiham, Basing and Basingstoke. The company also maintained its own fleet of boats. At Basingstoke, road transport was available to carry goods on to Winchester and Southampton. Later, it was hoped, the Itchen Navigation, which ran from Southampton to Winchester, would be extended so that a continuous water line from London to Salisbury and Southampton would be available. Sadly, this hope was not to be fulfilled and the canal was never a financial success. With a cost of about £153,000 (against the original estimate of £87,000), the line was always in debt and the company found it hard to keep up the payments of its loan interest, a factor which added a further £37,000 to the overall bill. As a result, the shareholders never received a dividend. The reason for the failure was that the canal was simply not as busy as forecast. Whereas the company had predicted business of 30,700 tons p.a. carried for 5s. per ton, the actual figures were considerably lower. In 1801–2, for example, 18,737 tons were carried at about 4s. per ton. Thus, the predicted annual revenue of £7,783 was never realized and annual income was usually around £2,000. The end of the war with France in 1815 didn't help. A lot of traffic from the Isle of Wight and the Channel Islands, which formerly used the inland route, could now risk the sea passage to London and did so, as it was both quicker and cheaper. By 1818, the canal also faced competition from road transport which, although marginally more expensive, was both faster and more direct.

Undoubtedly the key disadvantage of the canal was that it stopped at Basingstoke. Although proposals were made to take the line to Southampton, the company could never raise either interest or funds to build the extension. One scheme, proposed in 1807, was the Portsmouth, Southampton & London Junction Canal, which would have linked the Grand Surrey Canal (which joined the Thames at Rotherhithe) with the Basingstoke and (via Aldershot, Farnham and Winchester) the Itchen Navigation. This idea floundered on an unfavourable report from John Rennie and opposition from land and mill owners. A link with the Kennet & Avon to produce an alternative London–Bristol route was proposed in 1825 but never started. The Hampshire & Berkshire Junction Canal, as it was called, would have linked Basing, a village to the east of Basingstoke, to Newbury. The proposal was opposed by the Thames Commissioners who were successful in having the idea scrapped despite the enthusiasm (and the cash) of both the Basingstoke and K&A companies.

When the London & Southampton Railway (later renamed the London & South Western) was built, the decline was almost complete. Ironically, things improved slightly while the new railway was under construction as the canal was used to ship building materials for the new line. But in June 1839 the railway reached Basingstoke and in May 1840 it was open all the way to Southampton. The end of the canal as a going concern was inevitable. In the first year after the railway opened, canal toll receipts fell by 30 per cent. Business improved when the army camp at Aldershot was being built in the 1850s, but by 1865 income had dropped by nearly 80 per cent. As a result, the company went into liquidation in 1866.

There now followed a series of speculative and dubious financial deals:

1866   Original company liquidated – in hands of receiver
1874   William St Aubyn forms the Surrey & Hants Canal Company
1878   Canal in hands of receiver
1880   Canal bought by Messrs Dixon & Ward (for £14,800)
1880   Canal bought by J.B. Smith
1880   Surrey & Hampshire Canal Corporation formed nominally to sell water to London but more probably as a way of extracting funds from foolhardy investors
1882   Canal in hands of receiver
1883   London & Hampshire Canal & Water Company formed by some of the creditors of the Surrey & Hants
1887   Canal in hands of receiver
1895   Canal bought by Sir Frederick Hunt
1896   Woking, Aldershot & Basingstoke Canal and Navigation Company formed
1900   Canal in hands of receiver
1905   Canal bought by William Carter and sold to Horatio Bottomley's Joint Stock Trust & Finance Corporation
1908   London & South-Western Canal Company formed which again successfully invited investors to make the proprietors rich
1909   Canal rebought by William Carter
1914   Basingstoke Canal Syndicate bought the canal for £15,000
1921   Canal in hands of receiver and bought by William Carter
1923   Canal bought by A.J. Harmsworth
1937   Weybridge, Woking & Aldershot Canal Company formed by Harmsworth
1949   Canal in hands of receiver
1950   Canal bought for £6,000 by the New Basingstoke Canal Company

Despite the many dubious business transactions and the numerous name changes, by 1901 commercial traffic on the canal had virtually stopped after

A.J. Harmsworth, standing on the cabin of *Basingstoke*, attempts to pass through the canal near Up Nately in late 1913. This, the last attempt to go along the entire canal, was not successful but did enable the line to be kept open as a commercial concern

British Waterways

the brickworks at Up Nately had closed. In 1913, Alec Harmsworth took the narrow boat *Basingstoke* along the whole length of the canal in order to prove that it could still be done. It took three months. The First World War revived fortunes, but only temporarily, and for a while the canal was used as a reservoir. By 1923, when Alec Harmsworth bought the canal, there was no traffic west of Woking. The potentially disastrous collapse of Greywell Tunnel in 1932, which finally sealed the route to Basingstoke, therefore had no practical effect on trade. The years of Harmsworth's ownership, how-ever, were good for the canal. Here at last was a man who was actually interested in the line as a working waterway. Because of the traffic to the gasworks at Woking, trade peaked in 1935 at 31,577 tons; a period of pros-perity second only to that of 1838–9 when the railway was being built. This Indian summer came to an end in 1936 when the Woking District Gas Company ceased to make its own gas. After Harmsworth died in 1947, some timber traffic continued but on 15 March 1949 the final load was delivered to Spanton's Yard beside Chertsey Road, Woking. This was quickly followed by yet another company liquidation. By the 1960s the whole line was derelict, the locks inoperable and the canal itself mostly overgrown with weed or filled with rubbish and silt.

The revival of the Basingstoke Canal began in 1966 when a group of enthusiasts formed the Surrey & Hampshire Canal Society. Their first major success was in persuading the relevant county councils to recognize the value of the navigation as an amenity and to purchase it. The Hampshire stretch was bought in 1973 and the Surrey section in 1975. Since then, this consortium of interests, fronted by the Basingstoke Canal Authority, has dragged the canal back to health by clearing the waterway, repairing the locks and improving the towpath and other facilities. It has cost about £3–4 million. This effort was rewarded on 10 May 1991 when the Duke of Kent declared the line re-opened. Today, given adequate water, the entire length from the Wey through to Greywell is theoretically open to leisure cruising.

The Walk

Start:	West Byfleet BR station (OS ref: SU 042611)
Finish:	Ash Vale BR station (OS ref: SU 893534). Shorter routes stop at Woking or Brookwood
Distance:	14 miles/22^1/2 km (or 4 miles to Woking; 87 miles to Brookwood)
Maps:	OS Landranger 186 (Aldershot & Guildford) plus 1/4 mile on 176
Return:	BR Ash Vale to West Byfleet. Short walks return from Woking or Brookwood stations (enquiries: 0483-755905)
Car park:	At West Byfleet station or by the canal off Camphill Road (OS ref: SU 046616). It is also possible to park at Ash Vale station
Public transport:	The London to Southampton BR line at West Byfleet

The walk starts at West Byfleet station. There is a large car park to the east of the station which is free at weekends. From the station, turn left to a minor crossroads and then left along Madeira Road. At a T-junction, turn left along Camphill Road past a school and under a railway bridge. The road bends right and then left to reach a bridge (Scotland Bridge) over the canal. The main walk now turns left. However, if you wish to see the junction of the canal with the River Wey Navigation, turn right to go down this (southern) side, passing firstly some houseboats and then lock 1. Before long the M25 can be seen perched on stilts in the distance. This detour (including the return) is just over 1 mile.

The Wey Navigation runs from the Thames at Weybridge to Guildford

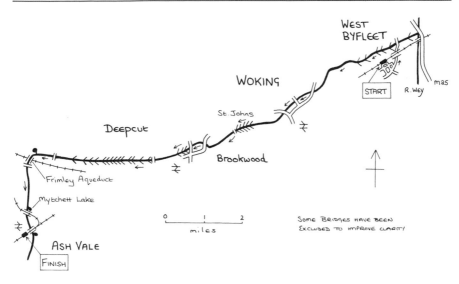

The Basingstoke Canal

and then on to Godalming. At one time it was possible to navigate from London to Portsmouth along this route (via the Wey & Arun Canal), but alas no longer. The junction here at Byfleet isn't spectacular scenery but offers the interesting sight of three ages of transport overlaying each other in a kind of concrete sandwich. The Wey Navigation of the mid-seventeenth century is bridged by the London and South-Western Railway of the mid-nineteenth century, which in turn is bridged by the M25 of the mid-twentieth. It is, of course, a matter of taste as to which is preferred.

From the Wey, return along the canal to Scotland Bridge. The main walk starts on the southern bank. From the bridge turn left so that the canal is on your right. Within a short distance you pass a group of locks (2–6) known as the Woodham Locks. It was this section of the canal that was started first and these locks were open in mid-1791. The first toll was collected from a boat carrying 'merchandise' to Horsell. Tolls were collected at the lock-house at lock 3. At the time of my visits, the pound between locks 4 and 5 was completely dry. Basingstoke Canal has no artificial water supply and relies entirely on natural springs, such as those in Greywell Tunnel, and the wash-off from roads, soil, etc. During periods of drought, therefore, levels can become very low, leading to the closure of the navigation.

After a road bridge, the towpath passes the last Woodham Lock and loops in between houses, some of which have elaborate breeze block mooring spaces. Eventually, the housing gives way to some typical Surrey heathland with birch, pines and gorse. The canal follows the contours to reach a new, ornamental office building at Britannia Wharf. If you go under the road bridge you will reach the remains of a small wharf crane, almost the last remnant of the wood trade that continued into the 1940s. The canal now enters a straight section to the skewed Chertsey Road Bridge. The area to the left here was the site of Spanton's Wood Yard, one of the major and last users of the canal. Beyond is Boundary Road, formerly the home of the Woking Gasworks, the single biggest customer of the canal in the twentieth century. At one stage the trade was so brisk (at about 14,000 tons p.a.) that a tramway was built to carry coal from the canal to the works and, presumably, ashes and tar back to the canal. The works were opened in 1892 and closed in 1936. The original Chertsey Road Bridge was built in 1792 but by the turn of this century this and a number of the others around Woking were falling down, a situation that led to some friction between the council and the canal company. The original bridge was replaced by a temporary wooden structure in 1910 while the situation resolved itself. It didn't and the council resorted to an Act to enable it to build new bridges and then charge the company. So in 1922 the present bridge was built. However, as the canal company never had any money, the council was never paid.

Go under the bridge and continue on to the next which carries the Chobham Road. This was also rebuilt in the 1920s at the expense of the council as the original Wheatsheaf Bridge was both falling down and very narrow. To the left are the office blocks, multistorey car parks and shops of Woking. It's said that when the railway opened in 1838, Woking comprised a few houses, a pub and bare heath! If completing your walk here, turn left and follow the signs for the BR station. If continuing to Brookwood, cross both the road and the canal to continue the walk on the right-hand, northern bank.

The walk continues with the canal and Woking to the left. Within 100 yd, the path passes a large car park and goes under a skew bridge. For some distance the canal is bordered by a noisy road. This eventually recedes and the line passes under a small footbridge, Arthur's Bridge. If thirsty, cross the canal for the Bridge Barn pub and restaurant, a converted sixteenth-century barn. Otherwise continue on this side of the canal to pass the front doors of a row of houseboats. After a road bridge, the towpath continues to a brick-built farm accommodation bridge typical of the original style. This is now used solely as a footbridge. The area to the north is Horsell. The canal along here collapsed when a new sewer system was installed in September 1899 and the navigation was closed for six months at a critical time for the survival of the line as a commercial concern beyond Woking. In fact, it never really recovered.

After a set of locks (7–11) collectively known as the Goldsworth Locks, you reach St Johns, formerly the site of the kilns that were used to produce bricks needed at this end of the canal. Here you cross the road and the canal to continue on the left-hand bank. After an iron footbridge, the high-walled railway embankment dominates to the left. The canal swings away from the railway after a couple of hundred yards and broadens to Hermitage Road Bridge. Go through the narrow tunnel under the bridge to a lagoon, home, when I passed, to a group of houseboats. The route now enters countryside with woods on the left and fields to the right. Go under the next bridge (which carries the A322 Bracknell–Guildford road) and up the steps. Cross the canal and turn left to continue the walk along the northern bank. This is Brookwood.

At Brookwood is a 400 acre cemetery, the largest in western Europe. It was built in the 1840s when London's cemeteries were literally overflowing with cholera victims. The London Necropolis & National Mausoleum Company chose the wilds of Woking as the site and on the government's nod bought 2,300 acres of Surrey commonland, most of which was promptly sold off at a vast profit. Bodies and mourners were shipped in from London in huge numbers by the recently built railway line. If you wish to see this particular example of Victorian splendour, turn left at the bridge, go under the railway and the entrance is on the right. However, a good view can be had from the train on the return from Ash Vale.

Brookwood Locks 12–14 now follow in quick succession and within a short distance is an iron road bridge (Sheets Heath Bridge). If finishing the walk at Brookwood you should cross the bridge and go along the road to Brookwood station for trains back to West Byfleet. You are now about 5^1/$_2$ miles from Ash Vale station.

If continuing the walk, go under the bridge. At the next bridge (Pirbright Bridge), cross the road and the canal to continue on the left-hand bank past a house and lock 15. This is the first of the Deepcut Locks, a flight of fourteen over 2 miles which raises the canal 100 ft onto the Surrey Heath. Daniel Defoe visited here in the eighteenth century (before the canal) and described it as 'not good for little but good for nothing'. Defoe may have been aware that the heath was alive with highwaymen. The most famous of all, Dick Turpin, had a hide-out near here from where he made forays on the London to Winchester and Portsmouth roads. Interestingly, I've not seen any references to highwaymen attacking canal boats; maybe the bargees were simply too tough a target?

The canal is now in a deep, quiet slumber. In winter there's a sombre feel to the place. The railway embankment is high to the left and there are extensive woodlands to the right. The canal follows a dank and echoing chasm. Curiously, the railway's amendment Act of 1837 compelled the railway company to build these high banks and walls so that the trains didn't frighten the barge horses. After lock 16, go under Cowshott Bridge to reach locks 17–19. On the other side of the fence to the right is Pirbright Camp, just one part of the extensive area of army encampment in this part of Surrey. The army settled here in 1855 when the first camp was built on the banks of the canal at the Farnborough Road Bridge.

Our way winds through dense woodland to pass locks 20–25. An army swimming pool was built into the canal between locks 22 and 23 and the

A train for Southampton goes under the Frimley Aqueduct

brick walls can still be seen. In 1957 some troops returning from an exercise decided to blow up lock 22, an action which lead to the draining of the pound and deterioration of this part of the canal generally. After a new bridge the canal opens out to form a lagoon with an island (Fred's Ait). Locks 26–28 follow in quick succession, after which is a bungalow formerly a carpenter's shop and forge, and a dry dock rebuilt by Youth Training Scheme workers in 1984. Deepcut Bridge was being renovated as I passed. It was pristine white and very clean. After a cottage to the left, the bank on the right is covered in rhododendrons. It then breaks to an entrance to Wharfenden Lake with its extensive leisure developments. The canal then bends left to pass over Frimley Aqueduct. There are now just 2 miles to Ash Vale.

The aqueduct was built by the London & South Western Railway in 1839 and is 139 ft long. To build it the railway company made a huge cutting up to each side of the canal, then bored through and lined the channels; a process more like tunnelling than aqueduct building. To ensure that the structure was watertight, the canal bed was bricked and a clay 'puddling' layed in the classical fashion. The railway company widened the structure in 1900 and two more arches were added. But by 1925 a sag had developed and a leak started that got progressively worse. A 190 ft long wooden aqueduct was put into place while the main canal was repaired so that the, albeit infrequent, canal traffic could continue. The aqueduct meanwhile was dug out, rebricked, puddled, lined with lead sheets with soldered joints, puddled again and then further layered with bricks. Even this apparent solidity didn't last and the aqueduct had to be relined again during the 1980s. It is worth noting the two stop gates at either end of the aqueduct which can be closed in the event of any further leakage! At the end of the aqueduct, take the metalled lane that bears away from the canal and up to the road. Turn right to cross the bridge. To continue the walk, the path down to the right-hand side of the canal can be found about 10 yd along the road on the left. The King's Head is 50 yd along the road to right.

The canal now takes a straight course with Frimley Lodge Park on the right. In the field can be found the plaque unveiled by the Duke of Kent in May 1991 when he declared the canal officially re-opened. A little way further down the canal, there is an information notice and map of the navigation. Continue along the popular path, passing a small pond, Potter's Pool, to the left. Eventually a railway line (the Staines–Aldershot line) joins the canal, with a housing estate (Mytchett) just beyond. Go under a bridge and along to a point where the canal widens to form Mytchett Lake. The path now goes under a railway bridge. After another couple of hundred yards the railway bends round to pass over the canal again. Just after the bridge, turn right up to a road.

The corrugated-iron shed just after the turn-off point is Ash Vale Barge

The barge *Aldershot* is seen here under construction at A.J. Harmsworth's Ash Vale yard in January 1932

British Waterways

Yard. This was the site of the barge building and repair business run by Alec Harmsworth, who owned the canal for twenty-four years until he died in 1947. Barges were built on this side, while repairs were done on the other. It's said that new boats were produced at about one a year and fell into two types: 'Resos' (residential boats) and 'Oddns' (those without accommodation). They were built of English oak with Columbian pine bottoms to Harmsworth's design. The boats were launched broadside after the remaining struts were knocked out. They were then taken across the canal to be fitted out. The yard was open from 1918 to 1949.

The final stretch is now in view. Bear left and Ash Vale station is about 100 yd downhill on the right. Trains leave for West Byfleet three times an hour even on Sundays. If you want a good view of Brookwood cemetery sit on the right, otherwise the left side offers views of the canal.

Further Explorations

The Surrey and Hampshire county councils must receive the highest praise for their efforts. The whole length of the Basingstoke is good walking and there are a number of specially-built car parks along the way, such as at Odiham (Colt Hill), Crookham Wharf and Wharf Bridge, Aldershot (on the A325). Any of these points can be the start of a quiet and easy walk along the lock-less Hampshire section.

The stretch from Greywell (OS ref: SU 718514) to Odiham is particularly popular. This walk (5 miles) starts at Greywell, a small village 1¼ miles south-west of junction 5 of the M3. Almost opposite the Fox and Goose pub, a road is signposted to North Warnborough and Odiham. About 20 yd along the road, a sign points left over a stile. This passes over and then down to the canal. Here, to the right, is the entrance to the 1,230 yd Greywell Tunnel. The tunnel was started in 1788 and opened in 1794. At its deepest it is 140 ft below Greywell Hill. It is said to have been rather jerry-built and accusing fingers have been pointed at the local labour employed by John Pinkerton, the contractor. Others say that it suffered from earth movement. Whatever the case, the collapse in 1932 sealed the route to Basingstoke, possibly for ever. It is now a major roosting site for bats. If you peer into the tunnel you can see the barred gate which prevents human access while allowing bats to fly in and out. The portal was given a face-lift in 1976 as part of the European Architectural Heritage Year. A number of springs are tapped along the length of the tunnel and these are the main source of water for the canal. The top pound of the line down to the first lock (Ash Lock in Aldershot) thus acts as a kind of reservoir for the rest of the line.

The eastern portal of Greywell Tunnel

The next 400 yd are a delight. This short stretch fully lives up to the canal's reputation as a nature lover's paradise. In the space of five minutes I saw mallard ducklings, various warblers, numerous dabchicks, swans and a stoat who, no doubt, was viewing the others as potential lunch. After passing the remains of an old lock, at a point where the canal goes through a brick-built channel, the waterway opens out to a winding hole. Within a few yards a small aqueduct carries both canal and towpath over the River Whitewater. This rather insignificant bridge had a minor but important leak for many years before restoration. Shortly after the aqueduct the remains of Odiham Castle are on the left. The castle, popularly known as King John's Castle, was built of stone and flint in the early thirteenth century as a hunting lodge. It was from here that King John rode to Runnymede in June

1215 to sign the Magna Carta. It has been derelict since the fifteenth century.

Back on the canal, the towpath gently winds round the contours to reach North Warnborough Lift Bridge. On some maps this is still labelled as a swing bridge, though there hasn't been one here since 1954. From here, the path again twists to go under the brick-built Swan Bridge. You now pass into more open countryside with only Odiham bypass, away to the left, to remind you that you're still in the twentieth century. Here the canal straightens to pass Lodge Copse Bridge, a simple accommodation bridge, and goes on through pasture to pass underneath Colt Hill Bridge to Odiham Wharf. At weekends you can buy a cup of tea at a kiosk. Alternatively, on the other side of the bridge, the Water Witch pub serves bar meals and has a canalside garden. On summer weekends the society's boat, *John Pinkerton*, provides short trips from the wharf. This fifty-seater boat is operated by enthusiasts and has provided over £100,000 towards the restoration work.

To return to Greywell, you can either retrace your steps or continue past the Water Witch and into the small town of Odiham with its Georgian main street. A footpath sign alongside the George Hotel takes you through some new buildings. Bear left along a narrow path next to a wall and into a field. Turn left and follow a wall and then a hedge to reach a pond. Cross the stile and walk on to reach a road. A footpath sign points you between houses and back to the canal. Turn left, cross the Swan Bridge and then retrace your steps to Greywell.

Further Information

The canal has a very active society which is worthy of attention:
 The Surrey & Hampshire Canal Society,
 The Spinney,
 Meadow Road,
 Ashtead,
 Surrey KT21 1QR.
 Tel: 03722-72631.
It publishes a number of small booklets about the canal, its history and its restoration:
Cansdale, R. & Jebens, D., *A Guide to the Basingstoke Canal.* 1989.
Crocker, G., *A History of the Basingstoke Canal.* 1977.
Gerry, D., *Towpath Walks by the Basingstoke Canal.* 1987.

There is also a useful map of the canal which can be obtained in local shops or from GEOProjects (UK) Ltd, Henley-on-Thames.

2
THE BRECON & ABERGAVENNY CANAL
Talybont to Brecon

Introduction

To build a canal through the Brecon Beacons was pure genius. Whatever it was for; however it was built; whether or not it was a financial success; all irrelevant. Whoever did it was my kind of canal builder. It's simply terrific.

The Brecon & Abergavenny Canal has lived under a number of different names and guises over the years. Correctly speaking, the B&A is merely a section of the Monmouthshire & Brecon. This latter designation (and

The railway viaduct at Crumlin on the Monmouthshire Canal was built in 1857 by Liddle & Gordon to carry the Newport, Abergavenny & Hereford Railway over Ebbw Vale. It crossed the upper reaches of the Crumlin branch. At the time it was the largest viaduct in the world – 1650 ft long and 210 ft high. It was dismantled in 1965

The Boat Museum Archive

British Waterways calls it the M&B) includes the Monmouthshire Canal, a no-longer navigable route which runs from Newport via Cwmbran to Pontymoile, and a branch called the Crumlin Arm, which leaves the main canal at Malpas to head west and then north to Newbridge and Crumlin. The 33 mile long B&A starts at Pontymoile (just south-east of Pontypool) and runs north to Llanfoist near Abergavenny. The line then turns north-west through Gilwern, Llangattock, Llangynidr and Talybont-on-Usk, and on to Brecon.

Here is an easy walk with plenty of good pubs and picnic spots, and so many great views that it could take you all day.

History

It was in 1791 that the industrialists and colliery owners of Monmouthshire saw the advantage of building a canal from Pontnewynydd, just north of Pontypool, to the mouth of the River Usk at the then small port of Newport. After the passage of an Act in June 1792, the Monmouthshire Canal was built to serve the ironworks, limestone quarries and collieries of the area of Pontnewynydd and of Crumlin, to which a branch line had been built. The main canal was 11 miles long and rose 435 ft by forty-one locks.

With the construction work on the Monmouthshire under way, the people to the north started to consider the potential for a canal of their own. One plan that circulated in August 1792 proposed a line from the River Usk at Newbridge to Abergavenny and on to Llanelly near Gilwern. However, it wasn't too long before it was realized that there were advantages in moving the southern end further east to join the Monmouthshire at Pontymoile, just south of Pontypool. The Monmouthshire's proprietors were so convinced of the advantages that they offered the new canal's promoters a £3,000 inducement to continue. They also promised to supply the water to the lower pound that would run from Pontymoile to Llangynidr. All these matters were discussed and agreed at a meeting held on 15 October 1792, by which time it was also agreed to extend the waterway further north to Brecon.

The proposed line, of what at the time was known as the Abergavenny Canal, was surveyed by Thomas Dadford, Junior, son of the Thomas Dadford who had built the Glamorganshire Canal. At the time, Dadford was employed as engineer to the Monmouthshire Company and it was it which paid him for undertaking the survey. His recommended line was 33 miles in length, from Brecon to a junction with the Monmouthshire at Pontymoile. The plan included three tramroads: Gilwern to Beaufort;

Gilwern to Glangrwyne; and Llanfoist Wharf to Abergavenny Bridge.

Dadford's survey was approved at a meeting in November 1792 and the Act received Royal Assent in March 1793. The authorized capital for the Brecknockshire & Abergavenny Canal was £100,000, with approval to raise a further £50,000 if it was needed. The Act included powers to build tramroads up to a distance of 8 miles from the canal. While the Monmouthshire attracted shareholders from a wide area, the B&A was a much more parochial affair, with most of its support coming from local sources. There were some shareholders, however, such as the Duke of Beaufort and Sir Charles Morgan, who held shares in the Monmouthshire Canal, and the members of the Wilkins family of the The Old Bank, Brecon, had interests in a number of other Welsh canals. Right from the beginning, the B&A was different from the majority of the other canals in south Wales in that it was not intended as an industrial line. Whereas most of the recently constructed waterways were built to carry iron and coal out to seaports, the B&A was primarily built to carry coal, lime, manure and agricultural produce inland to Brecon and other places along the canal.

Construction work started in 1794 when a tramroad from the coal mines at Gelli Felen in the Clydach valley to Glangrwyne was built by John Dadford, Thomas junior's brother. Strangely, and much to the annoyance of the Monmouthshire which had handed over its £3,000 in March 1794, the building work on the canal itself wasn't started until early 1797. Dadford junior was appointed engineer, a post he took up full-time when the Monmouthshire was completed at the end of 1798. The first stretch to be built was the embankment and aqueduct over the River Clydach at Gilwern. The length between Gilwern and Llangynidr was opened in November 1797.

Additional funds were raised in April 1799 by making extra calls on the existing shares and this enabled the completion of the section between Gilwern and Brecon. The main water supply for the line was then available from the River Usk via a weir just above Brecon. Alan Stevens notes that the water from the weir goes through a culvert that passes right under the town to the head of the canal.

By the time Brecon was reached in December 1800, the company was once again running short of funds. The cost so far was £120,000, including the tramroads. But some income was forthcoming from the shipment of coal to Brecon along the opened route and the company paid a dividend of £1 17s. 6d. in 1802. However, more cash was needed if the line was to be extended to Pontymoile. A further Act was obtained in 1804 providing powers to raise another £80,000 and by the beginning of 1805 the canal, now with Thomas Cartwright as engineer, was extended to Govilon Wharf.

Once again construction work stopped through lack of cash. It was estimated that £50,000 was needed to complete the line and it wasn't until

This peaceful scene at Gilwern, thought to have been taken in the early 1900s, perhaps suggests that the once busy wharf had already declined

The Boat Museum Archive

1809 that this was forthcoming, mostly from one individual, Richard Crawshay, who granted a loan of £30,000. With this the third engineer on the project, William Crossley, resurveyed the line from Govilon to Pontymoile. The work started at Pontymoile in about 1810 working north and was completed on 7 February 1812. At the official opening, the company committee boarded a boat and passed from the Monmouthshire to the B&A amid 'the aclamation [sic] of a very numerous body of the inhabitants'. The total cost of the canal was about £200,000.

There were problems with scouring of the foundations of the aqueduct at Pontymoile and it had to be demolished and rebuilt before the canal was completely opened, but by 1813 the line was fully operational. Toll receipts were now totalling nearly £9,000 p.a. and trade was increasing, albeit rather slowly. Although dividends had reached £3 per share in 1808–09, none were paid between 1811 and 1817. This was done to allow the company to clear its debts. Even at this stage concerns were being expressed over low profitability, and as part of an ambitious programme to encourage traffic, a whole succession of tramroads were built to link with the canal. In other schemes to stimulate business, agreements were forged with ironworks to carry ore free if finished goods was then carried on the canal, drawbacks were given to encourage long-distance traffic, and other deals were made in which tolls were reduced in exchange for guaranteed trade. All this had the effect of increasing revenue to £11,021 in 1818–19 and enabling the company to pay a £2 dividend. But by the early 1820s trade was again becoming hard to find, so that even though the company was able to pay a £4 dividend in 1821–2, by 1823 revenue was down to £10,221. In that year the company carried 86,944 tons of cargo on the canal and 27,024 tons on the tramroads. Most of this traffic was coal, coke and iron, with lesser amounts of lime and limestone. The later 1820s saw a marginal improvement in the situation, with the dividend peaking at £9 per £150 share in 1825–6, albeit at the cost of cutting staff wages. But these were the peak years and from then on trade declined, a situation not helped by more aggressive competition for the Nantyglo iron trade from the Monmouthshire.

By 1833 trade was definitely on the decline and dividends were down to £4 per share. The B&A shareholders resolved to amalgamate with the Monmouthshire Company, a situation helped by the fact that the influential ironworks owner Joseph Bailey was on the board of both. Despite the continuing threat of unnecessary competition and the fact that the companies' representatives had established the basis of a deal, the B&A shareholders could not agree to the terms offered, in which the Monmouthshire was to guarantee dividends. Thus the proposed merger was dropped. There was, however, a better understanding between the two companies and both prospered for a brief period after the talks. During most of the 1830s the B&A was able to pay an annual dividend of £5 to £5 10s.

By the 1840s the influence of the newly-built railways was beginning to be felt. At one stage the committee considered selling the line to the Welsh Midland Railway. However, the WMR was dissolved without the deal going through. Instead, a survey was made to assess the potential for putting a railway line along the canal bank between Brecon and Pontypool. This also came to nothing. Various other plans and potential sales followed in the course of the 1850s, during which the section of the Monmouthshire Canal from Pontymoile to Pontnewynydd was closed and that company changed its name to the Monmouthshire Railway & Canal Company (MR&CCo). Many of the tramroads built by the Monmouthshire were converted into railways. Other railways such as the Brecon & Merthyr and the Newport, Abergavenny & Hereford were also built, taking much of the traffic from the canals. Dividends for B&A shareholders fell from £6 per share in 1855–6 to 10s. per share in 1861–2. A further reason for the steady decline in traffic along the line was the closure of many of the old ironworks connected to the canals.

The B&A responded in the only way open to it, and that was to reduce tolls. On 1 February 1863 the rates for coal were reduced from 2d. per ton per mile to just ¹/₂d. But with things not improving, the committee offered to sell the canal to the Monmouthshire Company for £61,200. This deal was finalized on 29 September 1865. Although there were advantages for the two lines joining forces against the competition, the Monmouthshire was also keen to gain access to the water that flowed from the River Usk into the B&A. The Monmouthshire not only needed this supply to keep its remaining line open but found an additional income in selling supplies to industry along the canal.

The independence of the new company did not last long. In 1880 the Monmouthshire (and hence the B&A) was taken over by the Great Western Railway Company, which was plainly more interested in railways than canals. By the turn of the twentieth century traffic was reduced to about a boat a week. The last commercial traffic to pay a toll on the B&A at Llangynidr was in 1933.

The Monmouthshire and Brecon Canal was taken over by the British Transport Commission in 1948 and, like many other canals, rescue was slow in coming. The BTC Act of 1949 closed the Crumlin branch of the Monmouthshire Canal and most of this is now derelict. A further Act of 1954 closed another portion near Cwmbran. In the Report of the Board of Survey for the Commission published in 1955, the two lines, under the title of the Monmouthshire & Brecon, were officially classified as 'Remainder'. This led to the abandonment of the line in 1962. However, this action drew the attention of some local enthusiasts to the potential of the waterway. After being taken over by British Waterways (BW) in 1963, restoration work was started. The 1968 Transport Act still considered the B&A as a 'remainder' canal and as such the line has no legal protection. Despite this, the line has gradually been restored to health. In 1969 there was agreement between the Monmouthshire and Breconshire county councils and BW on the restoration of the canal as an amenity. In 1970 a low bridge at Talybont was replaced by a steel drawbridge and the lock at Brynich was rebuilt and fitted with hydraulic paddle gear. Now the whole line from Pontymoile to Brecon is open for navigation and is a highly popular route through the very beautiful Brecon Beacons National Park.

The Walk

Start:	Talybont-on-Usk (OS ref: SO 113227)
Finish:	Brecon (OS ref: SO 045285)
Distance:	7 miles/11¹/₄ km
Map:	OS Outdoor Leisure Map 11 (Brecon Beacons National Park Central Area)

Outward: Brecon to Talybont-on-Usk via Red & White Buses no. 21 (enquiries: 0633-265100)
Car park: Brecon: large and well-signposted
Public transport: BR goes to Abergavenny and from there the Red & White Buses no. 21 goes to both Brecon and Talybont.

I joined the canal at the Traveller's Rest pub on the B4558, south of the village of Talybont towards Cwm Crawnon. A quick ascent of the steps from the beer garden brings you to the canal. Before starting the walk enthusiastic towpathers may wish to turn left along the left-hand bank, for about 1/3 mile to see the Ashford Tunnel. This, the only one on the B&A, is 375 yd long and is masonry lined. Boats were 'legged' through while the horses took the road that runs alongside. I'm told that halfway through the roof of the tunnel dips precariously and passing boaters have to duck rather rapidly or receive a hefty clout about the head. To start the walk proper, return to the Traveller's Rest and walk along the right-hand bank. Within a short distance you go under a road bridge (no. 142). This was formerly a stone arch bridge but it collapsed and has been replaced by this steel girder bridge.

Almost immediately the canal reaches Talybont Wharf (on the opposite bank). This was once a busy spot! What appears to be a road just above the wharf is in fact the end of the former Bryn Oer tramroad. The wharf was used to tranship both limestone from the Trevil limestone quarry and coal from the Bryn Oer collicry near Rhymney, some 12 miles distant. The coal was taken north along the canal to Brecon and, via the Hay tramway, to

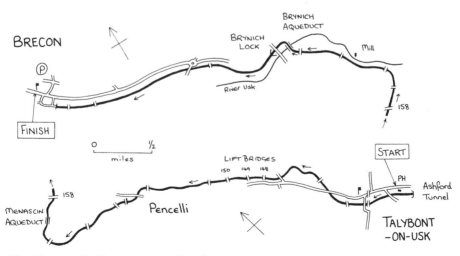

The Brecon & Abergavenny Canal

Kington and Hay. The limestone, meanwhile, was burnt in the limekilns which were positioned just beyond the tramway. The tramroad, which first opened in 1815, can be followed from bridge 143 up to the Talybont reservoir, a walk that is described in John Norris' canal guide (see below). After passing a pipe bridge, which carries water from Talybont reservoir to Newport, go under the former Brecon & Merthyr Railway. The White Hart pub is to the right. Bridge 143 (White Hart Bridge) follows shortly thereafter, its sagging arch now neatly supported by a series of wrought-iron rails. It is this bridge that should be crossed to follow the Bryn Oer tramroad. Those on their way to Brecon, however, should continue by following the canal sharp right and over a narrow aqueduct which crosses Afon (i.e. River) Caerfanell. The Star Inn on the right is recommended by CAMRA.

Talybont Lift Bridge (no. 144) is comparatively recent. Originally on this site was a wooden lift bridge. However, during the period when the canal was unnavigable, it was replaced by a fixed bridge which had clearance of just 2 ft. The present bridge is the second to be built here since 1970. The bridge is electrically operated. A motor winds a wire rope onto a drum to raise the deck. Mischievous souls will note that it only functions with the aid of a British Waterways key.

The route continues under Benaiah Bridge and then Chilson Bridge. This has a neat pile of stop planks positioned underneath. Along this stretch there are some simply wonderful curves and bends, as if the canal really wants to be a meandering river when it grows up. The whole ambience of peace and solitude is ably supported by the lush green Usk Valley to the right. To maintain a level course through such undulating country was a splendid feat but it was achieved at the expense of this winding route, beautiful for us but probably quite tiresome for a boatman with cargo to deliver. After bridge 147 (Cross Oak Bridge) you come to the first manually-operated drawbridge. Bridges 148–50 (Cross Oak, Gethinog and Penawr lift bridges) are all drawbridges of a similar style. These aren't the originals but were built to Dadford's design. The bridges are counterbalanced by means of weights on the end, mounted between the two upper beams, and are operated by a chain fixed to the end of one of the beams. This chain is worked by the lockgear-like system that can be seen at leg-level on one side.

After the third lift bridge, the remains of a drain plug and windlass can be seen on the near right. To the uninitiated this looks like two bits of old railway line stuck into the ground. There are several more between here and Brecon and only one, near the Brynich Aqueduct, still has most of its mechanism in place. The canal now goes under a stone bridge (Penawr Bridge) and on to a stop point, where the narrowness of the channel allows stop planks to be used to isolate sections of the waterway. Bridge 152 (Castle Bridge) is typical of all the fine B&A accommodation bridges; simple but with great character. Of interest here are the still extant metal

The Brynich Aqueduct carries the canal over the River Usk just south of Brecon

notices, erected by the Great Western Railway to warn potential users that the bridge has insufficient strength to carry a 'heavy motor car'.

The outskirts of Pencelli are announced by a large building to the left which sits near the remains of Pencelli Castle. This was formerly the seat of a medieval lord. Bridge 153 (Cross Keys Bridge) is newer than 152, can definitely take the weight of heavy motor cars and is none the better for it. Pass the canalside beer garden of the Royal Oak and go under Pencelli Bridge to reach a winding hole, a slipway and Pencelli Wharf. Just beyond the slipway is the final drawbridge (no. 155). Now go under two more bridges (Low Bridge and Llan-brynean Bridge) to reach Menascin Aqueduct. This entire stretch was extensively renovated during the winter of 1992, with the channel of the canal being relined to stop seepage. The towpath has also been upgraded. The modern method for laying canal channels in areas where leakage may occur is to lay a bed of concrete which is sealed with a plastic membrane and then protected by a thin layer of concrete on top.

Storehouse Bridge (no. 158) is a rather utilitarian structure dating from 1958 but it gives you your first view (left) of the Brecon Beacons and the start of some of the most scenic waterway in the country. Immediately beyond bridge 159 is the Cambrian Cruisers marina at Ty Newydd. The canal then swings left under bridge 160 and on to offer superb views to the right, firstly of Llanhamlach church and then of the River Usk which is now literally a stone's throw away. At one point the river passes over a weir to produce a series of white-water rapids. Bridge 161 (Bell Ear Bridge) has a sagging arch, thought to be due to subsidence of the abutments, and is neatly braced in the same way as White Hart Bridge was earlier.

Cross the canal at the next bridge (Turn Bridge: no. 162) to continue along the left-hand bank and on to Brynich Aqueduct. Just before it there is an overflow weir and, for those who have been looking out for one, an almost intact windlass. The aqueduct itself is a fine four-arch structure which was opened in 1800. Originally the channel had a masonry lining and was puddled with clay in the standard way. Only with renovations in the 1970s was the channel lined with concrete which can be seen clearly stretching out from the canal banks. The massiveness of the aqueduct and its superb position over the broad River Usk can only impress. It does, however, appear to be rather old-fashioned in style. It is, Alan Stevens points out, much more like James Brindley's Sow Aqueduct on the Staffordshire & Worcestershire than Thomas Telford's Pont Cysyllte on the Llangollen, even though it is contemporary with the latter. From the aqueduct there are good views of the river, and of the Brynich (road) Bridge to the left. It is possible to clamber down from the right side of the canal to the river's edge to see the aqueduct itself and, through its arches, Brynich Bridge.

After the aqueduct the canal makes a rather abrupt left turn and moves on to Cefn Brynich Bridge and Brynich Lock, which ends the $7^1/2$ mile long pound that has stretched all the way from Llangynidr. Like the other locks on the B&A, Brynich is about $9^1/2$ ft wide and 65 ft long. It has a rise/fall of 10 ft and some of the prettiest winding gear of any lock anywhere. After passing a small picnic area, the canal moves on with some magnificent views of the Brecon Beacons to the left. The highest peak in the range is the rather flat-topped mountain to the right of the others. This is Pen-y-fan and is 2,906 ft above sea level. From here on the towpath will be plagued by road noise as the canal bends right to meet the A40. Go under Brynich Bridge (no. 164) and then the bypass bridge which is so long that it could almost be classified as a tunnel. It is now a mile to the next bridge and 300 yd before you reach it is a long row of limekilns in the bank behind the towpath. These are invisible from the canal but if you walk on to reach a driveway you can turn left to walk back to see them. Meanwhile, to go on to Brecon continue along the canal past the rugby pitches and on to Watton Bridge (no. 165). This has a curious extra arch on the right-hand side which was built to accommodate the Hay tramway which followed the course of the canal into Brecon. On the other side of the bridge is the former Watton Wharf, now the site of some industrial premises.

Enter Brecon by walking past some terraced houses and going under the final bridge: no. 166, Gasworks Bridge. There is now just a short stretch of canal remaining and this has been used as a temporary mooring site. It is also home to Dragonfly Cruises. The canal comes to an end a little further on. Here a culverted feeder arrives from the River Usk, having been piped under the streets of Brecon from a weir half a mile upriver. The canal originally continued along the line of the road ahead for about 200 yd, ending in a side basin to the right. You should also go that way. Turn right at the end and follow Danygear Road round to pass a four-storey warehouse, dated 1892, which is one of the few surviving canal wharf buildings. At the end of the road, turn left to reach the High Street.

Further Explorations

The entire 33¹/₂ miles of the B&A is open for walking. Those lucky enough to hit a sunny weekend could well manage the entire length from Pontymoile to Brecon in two days, assuming that transport at both ends can be arranged. A halfway point would be somewhere between Abergavenny and Crickhowell. There are good bus connections all the way along and British Rail operates between Abergavenny and Pontypool, and on to Newport.

For those with less time there is a fine walk of just over 9 miles between Llangattock and Llanfoist which makes the most of the Red & White Buses bus no. 21, this time between Abergavenny and Crickhowell (enquiries: 0633-265100). The whole walk can be found on OS Outdoor Leisure Map 13: Brecon Beacons Eastern Area.

From the Crickhowell bus-stop walk past the Beaufort Arms and turn left. Continue past the pink-washed Dragon Hotel then turn right along Bridge Street. This passes the Bridge End Inn. Go across the River Usk via the fine bridge. At the far end go straight ahead and follow a signposted footpath into a field. This comes out past the Church of St Catwg to a road. Bear right and take the road which is signposted to Ffawyddog. Go uphill to reach the canal at Ffawyddog Bridge (no. 116) and turn left along the left-hand bank. From here there is a fine view back across Crickhowell to the Black Mountains.

The canal soon bends right and then left to go over a small aqueduct which crosses the stream from Nant Onnau. Shortly thereafter is Llangattock Wharf with its substantial limekilns. The wharf was built at the end of a tramroad from limestone quarries above it. The kilns were then supplied with coal from the canal. From here the B&A winds through some fine country and passes a series of bridges (nos 113–105) with great views over the valley to the left. The start of Gilwern (where the construction work on the canal began in 1797) is marked by bridge 104 (Auckland Bridge) and then Gilwern Wharf, home to Castle Narrowboats, on the far bank. At the far end of the Clydach dock is an arch through which the mile-long feeder canal arrives from the River Clydach. The dock here was an important canal/railway interchange basin and was served by a branch of the Glyngywyney to Brynmawr tramway. Coal from Gelli-felen, pig iron from Ebbw Vale and limestone from Black Rock in the Clydach Gorge were all transhipped here. Just beyond the basin, the line continues over Clydach Gorge on Thomas Dadford's massive embankment. The River Clydach goes underneath through a 100 yd long tunnel. The embankment also goes over the route of the old tramroad. If you take some steps to the left just after the basin, you can descend to see the tunnel. In fact you can go further still and John Norris (see below) describes a walk which starts at the tunnel. On the south (far) side of the embankment was the Clydach Iron Company's wharf, where the tramroad ran from the Llanmarch collieries.

At the end of the embankment the canal bends left and crosses a spill weir with a magnificent tumbling waterfall shoot which carries water down to the Clydach below. The next bridge, Gilwern Bridge, is actually more of a tunnel and towpathers are forced to walk up to and then across the road. The canal now leaves Gilwern and goes under the Heads of the Valleys Road (A465). This whole section between Gilwern and Govilon has seen some extensive renovation during the winter of 1992, with a concrete channel being laid. At Govilon the canal passes over two small aqueducts in quick succession. The first goes over a minor road and the second over the stream that comes down Cwm Llanwenarth. The towpath changes sides at Govilon Bridge and goes on to pass under the old skewed railway bridge, which once carried the Heads of the Valleys line, to reach Govilon Wharf. From 1805 to 1812, when the canal company was desperately short of cash, Govilon was the terminus of the line. It was also the terminus of the Llanvihangel tramroad which connected the canal with Hereford. This was built in 1819 to carry coal. The wharf was supplied by Bailey's tramroad which delivered coal and iron ore from Nantyglo and Brynmawr. All that remains of this feverish activity today is an old warehouse building and the boats belonging to members of the Govilon Boat Club.

The towpath changes sides at Govilon Quarry Bridge to return to the left-hand bank and the start of a stunning section of waterway. You're perched high above the valley with the town of Abergavenny now visible through the trees to the left and the tree-covered hillside continuing ever higher to the right. This lofty position is not without problems for the long-suffering British Waterways. The frequency of the stop gates and stop plank grooves along this stretch only suggests the problems. This whole length is very prone to breaches. The section just beyond the White House Turn (where

The formerly busy Llanfoist Wharf, near Abergavenny, is now a tranquil rural retreat

the canal appears to head straight into the hill, only to suddenly turn left) was breached in 1975. It took many of the downhill trees with it and blocked the line for some time. Following that breach the canal was lined with concrete, a move which will hopefully prevent such incidents in the future.

It is now just a short stroll to Llanfoist Wharf, one of the canal system's photographic hot spots. The warehouse and wharf-manager's house are still standing in this delightful place. Llanfoist was once the terminal for Hill's tramroad which brought iron ore up from Newport to Hill's Ironworks at Blaenavon and carried finished iron products back down. The tramroad contained no fewer than four inclined planes. The wood-decked bridge to the left is Tod's Bridge (no. 95). The tramroad crossed the canal via the bridge and passed down a further incline to join the Llanvihangel tramroad near Llanfoist village.

To return to Abergavenny don't go under Tod's Bridge but take the steps left which descend to Llanfoist village. At the road turn right and follow it around under the Merthyr road and over the Usk Bridge into Abergavenny.

Further Information

The Brecon & Abergavenny is a justifiably popular canal that is ably served by its own supporters trust:

The Monmouthshire, Brecon & Abergavenny Canals Trust,
6 Glamorgan Close,
Llantwit Major,
South Glamorgan CF6 9GG.

The trust produces a newsletter which updates members on canal matters, including the progress of the planned restoration of the line to Newport.

Two books which describe the delights of the B&A are:

Norris, John, *The Brecon & Abergavenny Canal.* J.R. Norris, The Birches off Wickham Hill, Hurstpierpoint, West Sussex BN6 9NP, 1991.

Stevens, R.A., *Brecknock & Abergavenny and Monmouthshire Canals. Towpath Guide No. 2.* Goose & Son, Cambridge, 1974.

There is also the more learned:

Gladwin, D.D. & J.M., *Canals of the Welsh Valleys and their Tramroads.* Oakwood Press, 1990.

3
THE BRIDGWATER & TAUNTON CANAL
Taunton to Bridgwater

Introduction

The short length of canal that connects the two West Country towns of Taunton and Bridgwater was originally built as just one part of an altogether grander scheme. In 1810 the Grand Western Canal was being built to join Exeter with Taunton, and the Kennet & Avon was being finished to give a continuous waterway between London and Bristol. The promoters of the Bristol & Taunton Canal saw the potential for a fully navigable route all the way from London to Exeter. Sadly this was a vision that with the dawning of an altogether different view on the wisdom of canal investment was never to materialize.

At its southern end, the B&T starts at Firepool Lock, not far from Taunton railway station. Here there is access to the River Tone and formerly a junction with the Grand Western Canal. The line starts by heading east under the M5 to Creech St Michael. Gradually the canal bends north through North Newton and Fordgate before winding around Huntworth and the westerly edge of Bridgwater to Bridgwater Docks. Beyond the docks, boats were able to enter the River Parrett and hence travel on to the Bristol Channel.

The countryside is flat and gentle between the two towns and this is a flat and gentle stroll. Keep it for a warm summer day and take a picnic.

History

In the days of canal mania, the people of the south-west seemed almost desperate to build a canal and schemes abounded. There were two prominent projects. The first aimed to connect the north and south coasts of the Devon/Cornwall peninsula in order to avoid the hazards of Land's End. The second proposed a continuous waterway between London and Exeter. As part of this latter scheme, at least two proposals were made to link Taunton with Bristol or the Bristol Channel. The first, the Bristol & Western Canal, was proposed in December 1792. The route was to be from the River Avon at Morgan's Pill, near Bristol, to Taunton. The rival Taunton & Uphill Canal would exit to the Bristol Channel at Uphill near Weston-super-Mare. In both cases, the opportunity was seen for a direct navigation between Bristol and Exeter, continuing from Taunton to the River Exe via the Grand Western Canal. Both groups surveyed their routes and both were defeated by vigorous opposition from landowners who feared that the schemes would affect land drainage or irrigation or both.

By 1810 the situation had changed. The K&A, which joined London and Bristol via the Thames and Avon, was now open and the GWC was finally under construction. The prospects for a continuous waterway between London and Exeter again looked good. John Rennie, who had been involved with both the K&A and the GWC, reviewed the proposed Bristol & Western Union Canal, later to be called the Bristol & Taunton Canal. 'No line of country', he claimed, 'can be more favourable for a navigable canal.' He proposed a level ship canal from a lock on the Avon at Morgan's Pill to run north of Nailsea through a 600 yd tunnel at Clevedon and one of 1,050 yd at Bandwell, to reach the south side of the River Parrett at Bridgwater. There a short branch would link canal and river. The line would then ascend two locks to cross the Parrett near Huntworth and run via Creech St Michael to join the GWC at Taunton.

As usual there were objections. A meeting of landowners in January 1811 concluded that it would harm their interests and hazard drainage. There were also objections from the conservators of the River Tone. The Tone navigation joined the River Parrett at Burrow Bridge to form a continuous waterway between Taunton and Bridgwater and thus to the Bristol Channel. This route was originally made navigable in 1717. The Tone had a some-what tarnished image as the navigation was in debt to its conservators who enjoyed a sizable repayment of interest. Despite this, the line was of great benefit to the people of Taunton who were able to import coal from South Wales and other goods from Bristol at greatly reduced cost. When the new canal was proposed, the conservators saw that their trade would be seriously affected and they decided on opposition.

The Bristol & Taunton Canal Act was passed in 1811. Rennie suggested a total of £410,896 and, in the rush of enthusiasm, the offer was fully sub-scribed. To placate objectors the company was required to not do any work at Clevedon or on the Parrett until it had finished the lengths on either side. On top of this the whole canal had to be completed within four years of the

Act being passed. As for the Tone, the new company had within three months of the Act to acquire the outstanding debt. This it eventually did but by 1822 construction work on the canal had still not started. This meant that the four year proviso had lapsed and with it went the desire to build the entire section from Bridgwater to Bristol.

By March 1822 a shareholder's meeting agreed that as the Tone regularly became unnavigable due to drought in summer or floods in winter, there was still potential for building a canal from Huntworth to Taunton as a useful alternative. Following an injunction against building the canal by some local farmers, the company obtained a second Act of Parliament in 1824. This confirmed the building of the new canal, allowed the company to lock down into the Parrett at Huntworth and to build a basin there. The new company promptly asked the GWC to extend its line to Taunton or allow it to build the intervening section itself. This might then provide the long-desired line from the Bristol Channel to Exeter. So, by 1824, the Bridgwater & Taunton was under construction. For a length of 13½ miles the engineer

The sale of hay could be an important additional source of income for many canals, particularly in the latter days when barge traffic may have declined. The hay collecting here was photographed near Huntworth

Kodak Museum, The Boat Museum Archive

James Hollinworth estimated the cost at £34,135. Locks were built at the slightly odd 54 ft x 13 ft (with a draft of 3 ft) which meant a normal craft load of 22 tons. The line was speedily cut with no significant problems. It was opened on 3 January 1827 when a barge, flags flying, arrived at a chilly Taunton to be welcomed by a cheering crowd. The canal had cost about £97,000 to build, with £57,000 having been raised from shareholders. The remaining funds were borrowed and weren't finally paid off until 1837.

The opening of the canal was not greeted with enthusiasm by the Tone conservators and there now followed a period of unseemly squabbling. The Tone conservators were intent on hindering the canal's water supply and at one point the B&T was reduced to breaking down the river bank at Taunton to forcibly connect the canal with the river. The conservators knew that the B&T was the more direct and least problematical line to Bridgwater and in order to maintain their share of business they reduced their tolls, even though the navigation still had unpaid debt (ironically now owed to the canal company). The B&T objected but lost the battle in court. In response it served notice on 28 August 1827 that it was going to take over the navigation under powers provided in the 1811 Act which were, it contested, renewed in the 1824 Act. It offered the payment required of it by the Act but was rejected. Things then got serious. In November 1827 the canal company forcibly took charge of the river, returned the tolls to their former rate and stopped all routine maintenance. The courts ordered that the B&T's action had been outside the time permitted by the 1811 Act. The conservators repossessed the river in July 1830 and once again reduced the tolls and kept the canal as short of water as they possible could. To cement this action, literally, a dam was built to separate canal and river at Firepool Lock below Taunton bridge. Although forced to take the dam down, the conservators threatened legal action on any boats or vessels entering the Tone from the canal.

The canal company made a final offer of purchase in November 1831. It was again rejected. The matter was then resolved in the House of Commons. An authorizing Act, passed in July 1832, enabled the canal company to purchase the Tone for just £2,000. The B&T also settled all the outstanding small debts, rebuilt two bridges, built a direct canal between the Tone and the Grand Western Canal, carried out various other odd jobs and, of course, cancelled the remaining debt which it now owed to itself. The conservators still retained the right to an annual inspection of the navigation and if it was not 'properly maintained' had the right to repossess the river. The B&T was now left with an open-ended requirement to keep the Tone maintained. At the time the river was carrying 39,516 tons p.a. and collecting tolls of £2,194. About 13,000 tons of this trade, mostly in coal, was unloaded at Ham Mills for towns as far south as Ilminster. However, when the Chard Canal was opened from Creech St Michael to Ilminster in 1841,

Bridgwater Docks, here shown in the early 1900s, were opened following the extension of the canal from its original terminus near Huntworth in 1841. Although largely unused during the course of the twentieth century, the area is now being redeveloped

M. Smith, The National Monuments Record

almost all of this traffic moved to the new route. Given that the B&T was preferred to the Tone for through traffic, almost the only business left on the river was that coming loaded or unloaded on the Parrett navigation.

While the traffic on the Tone was declining, that on the canal was increasing. By 1842 118,216 tons of goods were moved with a toll income of £8,239. But such riches were short-lived. In May 1836 an Act was passed for the Bristol & Exeter Railway, with branches to Dunball Wharf and Bridgwater (a branch to Taunton was approved in 1845). In response, the canal company came out fighting. In 1837 it obtained an Act which allowed it to extend its line from Huntworth to its present junction with the Parrett at Bridgwater. In doing this, the Huntworth lock and basin were closed, an extra mile of canal was built round the town and the ship dock was excavated. The full line as it is today was opened on 25 March 1841 with the peeling of bells, the firing of cannon, the playing of the national anthem, and the consumption of much 'roast beef and plum pudding'.

The new extension was a very risky venture given the likely traffic levels and the potential of the railway. With a cost of £100,000, heavy mortgages were incurred and as the profit in the first year was just £1,396, there was little hope of repayment. The situation was made worse when the railway between Bridgwater and Taunton was opened on 1 July 1842. This forced the B&T to lower its tolls and the company even agreed to pay the GWC a fee to persuade it to use the canal rather than the railway as the source of

imported coal. As the railway network, and the facilities it required, expanded, so the future of the canal looked ever more bleak. In 1845 the company
concluded that as it couldn't beat them it would join them and proposed the
Bridgwater & Taunton Canal Railway, in which the bed of the canal would
be converted into a railway track. Various other canal companies joined in;
for example, one plan in association with the Kennet & Avon proposed the
London, Devizes & Bridgwater Direct Line. The whole scheme was to be
called the West of England Central & Channels Junction Railway; able to
carry passengers from London to Penzance. None of these schemes came to
fruition, however, and the company appointed a receiver.

As a consequence of this action, debts to the GWC as part of the fee deal
worked out in 1842, weren't paid and in 1848 the latter started to use the
railway instead. This further reduced trade along the B&T. By the 1850s the
traffic had switched dramatically to the railways. In 1851 the company listed
its debts as £118,130. The company's receiver sought and obtained an
agreement with the Bristol & Exeter Railway Company which eventually led
to the sale of the B&T in 1866, the railway finally gaining possession on 8
April 1867 for a price of £64,000. This sum paid off the mortgagees and,
after the various debts had been sorted out, provided a small amount for the
shareholders. The railway company started its period of ownership with a
flourish. It built a new landing stage at Bridgwater dock where a steam crane
transferred coals to a horse tramway (later converted to a locomotive line)
which connected with its own main line. But in 1876 the Bristol & Exeter
Railway amalgamated with the Great Western Railway and the canal went
into decline. The opening of the Severn railway tunnel in 1886 meant that
the sea trade to the Bridgwater dock decreased, as coal now came from
South Wales by train. By 1890 the canal tonnage stood at 13,809 (compared with 59,806 in 1852) and was restricted to coal and timber. The final
coup de grâce occurred between 1896 and 1901 when the canal ran short of
water and many loads were forced to go by train. In 1905, when water levels
were restored, customers were loath to transfer back. By 1907, tonnage was
officially recorded as zero.

Thereafter the canal remained virtually unused. During the Second
World War the War Office turned it into a line of defence, building numerous pillboxes and replacing twelve swing bridges with fixed bridges strong
enough to bear the weight of army vehicles. The Tone meanwhile remained
navigable if unused. Both lines were nationalized in 1948 and allowed to
decline further, although the waterway was maintained as a water supply
channel for Bridgwater. In the review by the Board of Survey in 1955 both
the canal and the Tone were placed into Group 3, 'Waterways having insufficient commercial prospects to justify their retention', a valid comment
given the criteria applied. Although the Bowes Committee of 1958 thought
that the canal may be suitable for redevelopment, no significant improvements

resulted. Similarly, the condition of the canal did not improve with the transfer of ownership to BW in 1963. Although the Transport Act of 1968 still considered the B&T outside of those worthy of maintenance for commercial or cruising purposes, by that time the B&T Canal Restoration Group (later the Somerset Inland Waterways Society) had been formed and gradually attitudes and enthusiasms were changing. Today, BW is working with Somerset County Council, Taunton Deane Borough Council and Sedgemoor District Council to restore the line and to make it into the public amenity that it should be.

The Walk

Start:	Taunton BR station (OS ref: ST 228254)
Finish:	Bridgwater Docks (OS ref: ST 298376)
Distance:	14¼ miles/23 km
Maps:	OS Landranger 193 (Taunton & Lyme Regis) and 182 (Weston-super-Mare & Bridgwater)
Return:	BR Bridgwater to Taunton (enquiries: 0934-621131) – trains run on Sundays; or Southern National bus no. 21/21A – runs ½ hourly but not on Sundays (enquiries: 0823-272033)
Car park:	Taunton or Bridgwater BR stations
Public transport:	BR serves both towns.

From Taunton station turn left under two rail bridges. After these turn left past the Royal Mail pub. Turn left again along Canal Road with the cattle market to the right. After a further ¼ mile, the road reaches the canal at Firepool Lock, the site of the junction with the River Tone. Both the river and the lock are to the right of the bridge that crosses the canal at this point. Ahead the B&T canal heads towards Bridgwater. To the left of the bridge the course of the canal appears to make an abrupt turn into Firepool Lock. In fact, when operational this point (the line of a low brick wall) was a stop lock and the junction between the B&T and the Grand Western Canal.

The GWC was one part of the scheme to link the Bristol and English channels. In its original plan it was to run from the B&T to the River Exe. However, the canal faced considerable financial problems so that by 1838 the line that opened ran from Firepool Lock to Tiverton. It never reached either the Exe or the English Channel. The canal as built, however, was an extraordinary venture. It had no fewer than seven vertical lifts and an incline plane. One of those lifts, the Taunton Lift, lay just beyond the junction on the site of one of the railway lines behind you, in fact following the edge of

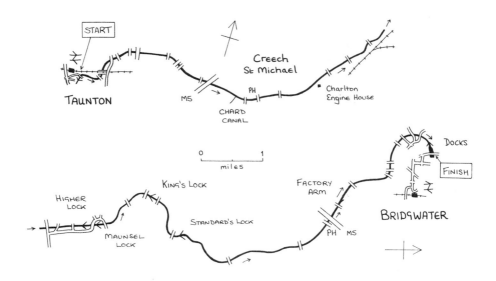

The Bridgwater & Taunton Canal

the present goods yard. This raised boats 19 ft from the B&T up to that of the GWC. The GWC was never a financial success and in 1853 was leased to the Great Western Railway. In 1864 the GWR bought the canal, only to close it three years later.

Start the walk along the right-hand bank. The tall GWR water tower to the left stands on some old limekilns that were supplied with both coal and lime from the canal. The water tank was also supplied from the canal. The B&T starts for Bridgwater by passing through an area of industrial works. At Obridge it goes under a pipe and then a railway bridge before swinging right to pass under a new road bridge and then Priorswood Bridge. This latter concrete structure is one of the last remaining obstacles to navigation between Taunton and Bridgwater. Fortunately, the replacement of the bridge is agreed in principle and there is even money allocated to the task. By 1993 the blockage should be lifted. At Bathpool the canal goes under the A38 and passes the St Quintin Hotel. You are now leaving Taunton behind and heading into the country. The last obstacle, it seems, is the M5 which you now go under. From here it is just a short stroll to the pleasant village of Creech St Michael. As the housing starts on the left side of the canal, keep an eye open on this side for an ivy-covered pillbox. Shortly after this is the site of the former junction with the Chard Canal.

The Chard Canal was opened in 1842 and closed just twenty-six years later. The 13½ mile long line was engineered by the relatively unknown Sydney Hall and comprised four incline planes, two aqueducts, three tunnels and two locks. At the time, Chard was an important wool and lace town and the canal carried 20–30,000 tons of coal p.a. to it. But 1842 was not a time to start building canals. Just four years after the opening of its new waterway, the company was considering converting it into a railway. This didn't occur and it was left to the opening of the Bristol & Exeter Railway's branch to Chard in 1867 to finally put pay to the Chard Canal. The line was bought by the Bristol & Exeter and closed in 1868. Not much remains now but here on the right are the buttressed walls of an embankment which carried the canal towards an aqueduct over the River Tone. At the junction there was formerly a stop lock, a towpath bridge and a lock-keeper's cottage, all of which have disappeared.

The B&T now heads through Creech St Michael and under Creech Bridge. The next two bridges, North End and Foxhole, were at one time swing bridges. Indeed, before two world wars there was a large number of swing bridges along the B&T. Most of these were destroyed during the war when the War Office, fearing a German invasion, converted the canal into a defensive fortification. It was this same move that produced all the fortified pillboxes and peculiar pyrimidal concrete obstacles that litter the entire length of the waterway, as indeed they do on the Kennet & Avon.

The canal now enters more open countryside. In 1/4 mile the towpath passes the large and partially derelict Charlton Engine House. This building, which dates from 1827, formerly housed steam engines which were used to

One of the curious things about the locks on the B&T is the fact that the ground paddle mechanism is fitted with a substantial counter-weight, the chain of which runs over the pulley to the paddle

pump water from the River Tone into the canal. From here the peace and quiet becomes even more intense and is only periodically destroyed by trains thundering past the Cogload Junction and Viaduct to the right. The canal meanwhile does a quick shimmy before going under the A361 (turn right here for the Railway Hotel pub) and past Black Hut Bridge (the black hut was formerly used by canal lengthsmen). After crossing an embankment the line reaches Higher Lock, Maunsel Lock and then King's Lock. The rise for each is about 5 ft 8 in. At King's Lock there is a rack of stop planks which are used to block off a section of canal in the event of an embankment burst or maintenance work. The groove for the planks is just above the upper gates. The canal has now circled round North Newton and returned to a northerly course to go through Standards Lock. One of the curious things about the locks on the B&T is the fact that the ground paddle mechanism is fitted with a substantial counterweight, the chain of which runs over the pulley to the paddle. This, with the reduction gear, makes operation of the paddle relatively simple. The next bridge is Fordgate Swing Bridge, a typical example of the type of bridge that predominated along the canal before the war. From here the line runs on to Huntworth Road Bridge and then onto Mead's Swing Bridge where the Boat & Anchor Inn sits in the shadow of the M5, which again spans the canal. Cross the canal at the next, Crossways, swing bridge. Before the completion of the Bridgwater Docks in 1841 this was the terminus of the canal. The original basin has been filled in

The tunnel-like Albert Street cutting in Bridgwater has, in the past, been subject to subsidence. The problem has now been stabilized but the walls still bear these massive supporting timbers

but was apparently close to a pillbox on the far side.

The canal is now approaching Bridgwater and the walk continues through Hamp, with a former brick pit (now a pond) to the left and the River Parrett to the right. The canal soon swings left, away from the river to go under three bridges, before bending back right to enter the tunnel-like Albert Street cutting. At one point it seems that only the huge wooden cross-buttresses stop the sheer cliff-like walls of the cutting from crumbling into the canal. From here it is a short stroll right through a cutting to Bowerings Animal Feed Mill and the much-redeveloped Bridgwater Docks. The docks were opened in 1841 and consist of two basins, the larger main one in front of you and a smaller outer one in the distance. Water from the tidal River Parrett was used to maintain water in the dock, which was separated from the river by the two locks on the far side. Thus the docks form a kind of floating harbour. The basin is being developed as a marina and, in theory, boats enter or leave it to the canal via the restored Newtown Lock.

Walk along the left-hand side of the basin and out to the road. Here it is possible to see the small outer (tidal) basin and the tidal gates and barge lock to the River Parrett. The tidal gates, which are now closed off by a concrete dam, could be used by coasters up to 180 ft by 42 ft but only those not exceeding 32 ft beam could then pass into the inner dock area. The barge lock measures 54 ft by 13 ft. Turn right cross the bridge, of a type known as a bascule bridge which is raised and lowered using a counterbalancing weight, and go round to Ware's Warehouse, one of the few original buildings still extant on the docks. To the right here is an original hand-operated crane. This is the end of the walk. To reach your return transport continue along the road to some traffic lights. Turn left over the River Parrett and then right to reach the bus station. For the railway station, walk past the bus station and through a car park. Then go along New Road and turn left along Eastover. Continue straight across the busy A38 and along St John Street. When this road bends right go straight on to reach the station.

Further Explorations

Having walked the entire length of the B&T, you may be fooled into thinking that there's nothing left to be done. There are, however, a surprisingly large number of canals in the south-west and the West Country branch of the Inland Waterways Association publishes a series of leaflets that describe walks along (among others) the Stover and Hackney canals, the Cann Quarry Canal, the Tavistock Canal, the Grand Western Canal, the Chard Canal and the Dorset & Somerset Canal. Perhaps the most historic of the canals in south-west England, however, is the Exeter.

The Exeter Canal dates from the time of Edward I, when Isabella de Fortibus, Countess of Devon, had a weir built across the River Exe, an act which forced boats to unload at Topsham where tolls were payable to – you guessed it – the Countess of Devon. Navigation was restored in 1290 but between 1317 and 1327 the river was filled in and further weirs built. It wasn't until 1539 that Exeter Corporation obtained an Act to render the river navigable. This wasn't very successful and in 1563 John Trew of Glamorgan was engaged to build a canal, the first to be built in Britain since Roman times. The Exeter Canal left the Exe near the city walls and ran to Matford Brook just below Countess Wear, a length of 3,110 yd. From there the river was improved to Topsham. It was opened in 1566 and cost £5,000. The canal had three pound-locks, the first on any British waterway, with vertically-rising guillotine gates. There was also a single pair of gates at the sea end. The canal wasn't easy to use: the approach was awkward, the line could only be entered at high tide and the river above Trew's Weir tended to silt up. In fact, by the end of the Civil War it was in a pretty bad state. As a result, in the 1670s the seaward end was extended 1/2 mile towards Topsham, a new entrance was built to take 60 ton craft and the rest dredged. Further improvements in the early eighteenth century meant that the canal could take coasting vessels and small deep-sea craft up to 150 tons. The three old locks were removed and the Double Locks built instead: a single lock of large size that served also as a passing place. There was still only a single pair of gates at the entrance, called Lower Lock, the sill of which was 4 ft below that on the Double Locks. There was also a pair of floodgates, called King's Arms sluice, where the canal entered the Exe.

The eighteenth century was one of great prosperity for the Exeter. Some 500 boats regularly used the line, shipping coal, slate, timber, woollens, cider, groceries and goods from southern Europe. In the middle of the century receipts averaged £747 p.a. and, by the end, £2,335 p.a. By the 1820s this profitability stimulated moves for further improvement. Under the control of James Green, the line was straightened and the rest dredged. Green also engineered the extension to Turf, 2 miles further down the estuary, where a proper entrance lock was built. At the same time the banks of the canal were raised so that it could take craft drawing 14 ft and carrying 400 tons. The cost of these improvements was £113,355 but the subsequent receipts increased considerably, reaching £8,550 in 1842–3. Coal imports increased and coal ships now entered the canal instead of unloading into lighters. Two ships a week arrived from London and there was both a coastal and a foreign trade. Goods could reach the capital in a week, although in bad weather it sometimes took a month.

The opening of the Bristol & Exeter Railway in 1844 had an adverse affect on trade and in 1846 the dues were reduced by a third. The Exeter

Basin was connected to the South Devon Railway from 1867 but traffic never recovered. By the turn of the twentieth century 275 vessels were still using the canal but receipts were down to £1,624 p.a. This level of traffic was sustained into the 1960s, 55,431 tons (mainly timber and oil) being handled in 1960. The last vessel to use the basin was the *Esso Jersey* which delivered oil to its terminal in 1972.

A walk of approximately 9½ miles starts at the Exeter Quay (OS ref: SX 923919 on Landranger 192). There is a car park near the National Maritime Museum. Alternatively, the quay can be reached from the centre of Exeter using the City Nipper bus G. The museum is contained within the warehouses on the quay on either side of the canal basin. The basin is 900 ft long, 17 ft deep and widens from 90 ft to 120 ft and was (and is) the Exeter terminus for the canal. Before starting it's well worth turning left to investigate the quay. Cross the River Exe by the blue suspension bridge (Cricklepot Bridge) and then turn right to cross the wooden Mallison Bridge. Here is a range of small shops and cafés as well as the headquarters of the Exeter Canal and Quay Trust in the wharfinger's house.

To start the walk, set off along the left-hand side of the Exe past the public conveniences and on past the Port Royal pub. Ahead on the river is Trew's Weir, built by John Trew to feed his canal, and across the river is the canal entrance with the single gates of King's Arms sluice, which gave boats access to the quay before the canal basin was built. Further on the path moves away from the river and passes between houses. Here turn sharp right to go over another suspension footbridge. This leads to a second bridge and then on to reach the canal. Turn left and walk along the towpath (note that there is towpath on both sides), past a swing bridge and on for about a mile to Double Locks. This massive structure (despite its name it is a single lock which replaced three separate locks in the early eighteenth century) is 312 ft long and 27 ft wide at the gates and is broad enough for two ships to pass. Charles Hadfield suggests that it is the largest manually operated lock in the country. Nowadays it is the only lock along the canal. Walk on past the Double Locks Inn, which dates from 1701. At Countess Wear the towpath crosses the busy A379 by passing, in short order, a power-operated lift bridge and a swing bridge. Shortly thereafter is Exeter sewage works and wharf. You may find moored here the vessel *Countess Wear*, a 265 ton sludge carrier that was built in Poole in 1963. This is the last commercial ship on the canal and it still makes regular trips to the sea where it dumps sewage sludge. After passing under the massive structure that carries the M5, the line moves on to the Topsham side lock. This was built to enable lighters and other small boats to move from the canal into the Exe to reach the town of Topsham. The lock has been derelict since 1976. Before the improvements made by James Green in the 1820s, the seaward terminus of the canal was above the position of the side lock, although it is not possible to determine precisely where. The final stretch of just over a mile leads on to the Turf Lock, the entrance to the canal, and the Turf Hotel. The lock was built on piles driven down through both the clay and underlying bog to the rock beneath. It cost £25,000 and was opened in 1830. The lock is 131 ft by 30 ft 3 in. Just above it there is a basin where larger vessels could tranship into lighters. It also acted as a temporary harbour where vessels could hold out for more favourable tide or wind.

To return to the quay, you can either cross to the opposite side of the canal and walk back along the other towpath, or you can continue along the sea wall into Star Cross where trains go back into town. If walking back, stay on the left bank of the canal to pass Trew's Weir and the Welcome Inn to reach the King's Arms sluice and the Maritime Museum.

Further Information

The Bridgwater & Taunton Canal is supported by the Somerset Inland Waterways Society which can be contacted at:
18 Lonsdale Road,
Cannington,
Bridgwater,
Somerset TA5 2JS.
The society has the objective of advocating the use, maintenance and development of all the inland waterways of Somerset.

The history of both the B&T and the Exeter Canal can be traced in:
Hadfield, C., *The Canals of South-West England*. David & Charles, 1967.

4
THE BUDE CANAL
Bude to Marhamchurch

Introduction

The Bude Canal is one of the oddest of the South's waterways. Situated in one of the least populated and least industrialized parts of the country, it appears to have no obvious chances of success. Yet here it was built and here it enjoyed a modicum of prosperity. The Bude was a curious mixture of conventional waterway and nineteenth-century inventiveness. The canal rode the hilly hinterland in a bold and uncompromising fashion. There are no fewer than six railed inclined planes on which the specially built tub-boats were hauled out of the canal and up the hills; an ingenious, if not necessarily reliable, solution to an otherwise intractable problem. And yet the prime cargo was nothing more extravagant than sand.

The canal begins, naturally enough, at Bude, a small town on the northern coast of Cornwall about 25 miles north of Bodmin. From there the line runs south to Helebridge where the first incline took the tub-boats up 120 ft to Marhamchurch. The canal then went east to Hobbacott, where the second incline raised the line by 225 ft. At Red Post the canal divided. On the northern arm, it crossed the River Tamar by the Burmsden Aqueduct and then took the Vealand incline up 58 ft to the summit level. Near Burmsden, the line divided again. One branch went up to the Tamar Lakes, the other continued to Blagdonmoor Wharf just north of Holsworthy. The southern arm from Red Post followed the course of the River Tamar south in the general direction of Launceston. The line is primarily downhill and thus takes in the Merrifield incline (a fall of 60 ft), the Tamerton incline (59 ft) and the Werrington incline (51 ft). The canal ends at Crossgate, just a couple of miles to the north-east of Launceston.

This wild, sometimes bleak, country is a real contrast to that which surrounds most canals. It is recommended to those who regularly walk the often wilder and bleaker inner-city stretches of the waterway network.

History

As every gardener knows, the acidity and physical make-up of soil greatly affects its fertility. Poor-draining, acidic earth won't produce good crops, let alone support the good life. The moorlands of Devon and Cornwall are therefore not promising places for farmers. Those who, in the eighteenth century, sought to establish farms on the plateaus of the region had a pretty lean time. The exceptions were the farmers who lived around the small, north coast town of Bude. For centuries the people who lived in the area dressed their soils with sand from the beach with miraculous results. The sand, with its high shell content and hence high calcium carbonate level, both neutralized the soil and improved its structure. The result was seen in better yields and richer farmers. At the appropriate time of year in the early 1700s, the beach at Bude was packed with people loading the precious 'manure' onto carts and wagons or into packs on mules and asses, which then took the tortuous and lengthy route home into the hills. The difficulties of doing this, and the wear and tear on the local roads, prompted thoughts of a more efficient system and attention focused on the prospects for a waterway.

It was John Edyvean who first suggested a Bude Canal. Edyvean, who had been building the St Columb Canal (near Newquay) for a similar purpose, proposed the line to a meeting in 1774. The original plan, surveyed by Edmund Leach and John Box, proposed to link the Bristol and English channels by rising up from Bude and passing over the moorland to join the River Tamar at Calstock, about 6 miles north of Plymouth. Although only 28 miles as the crow flies, the proposed canal ran for 90 miles, meandering around the contours. In this instance, a circuitous route was seen as advantageous in being able to reach the largest number of customers spread throughout the hinterland. Even this early proposal saw inclined planes as the most efficient way of raising loads up and down the many hills along the way. There were to be five, each consisting of trucks pulled along rails using specially designed engines. Goods would be transferred from the proposed small wooden tub-boats to the trucks and back again at every incline. The total building cost was put at £40,000.

The idea of the canal was welcomed by the local population and an Act to enable the construction was passed in 1774. However, by August 1775 enthusiasm seems to have cooled and nothing was started. In 1778 John Smeaton, who had been called in to pass comment, said that the hills and valleys of Cornwall weren't exactly an ideal place to build canals and estimated the cost of building Edyvean's line as nearer £119,000. To overcome this problem he drew up his own plans for a line between Bude and the

Tamar that was much shorter than the original proposal: 9$^1/_2$ miles to the Tamar with 15$^1/_2$ miles of river navigation to Greston Bridge, where the river would continue to Calstock. This was estimated as costing £46,109.

Further plans and estimates followed. In 1785 Edmund Leach resurveyed the original line and estimated a cost of £88,740. He also put some thought into the operation of the proposed inclined planes and pointed out that transferring loads from boat to truck and back again at each plane was cumbersome, time-consuming and costly. He proposed a boarded plane on which two cradles carrying the boats would run on dry rollers. The cradles were to be hauled up using a water wheel or men in a treadmill.

In April 1793 a local dignitary and inventor, Lord Stanhope, and some other local notables held a meeting to discuss a canal from Bude to Holsworthy and Hatherleigh. John and George Nuttall were asked to survey the route and reported on 25 October on a 75 mile long canal to carry 2 ton tub-boats. In this plan the Nuttalls included a proposal from Stanhope on the operation of the inclines. The small boats would be put onto wheels and pulled up the hill on rails using a horse. This design was later amended following a suggestion from Robert Fulton. The power to lift the boats, he suggested, could be obtained by adding water to a descending bucket. He

The American engineer and inventor Robert Fulton devised a water wheel-powered canal incline which he described in *A Treatise on the Improvement of Canal Navigation* (1796). A method derived from this was used on the Marhamchurch incline

British Waterways

also suggested that the boats themselves could be wheeled so that they could run up a short slope to drop into a huge tank or caisson which would then be pulled up the incline using the bucket-in-a-well system.

Like many projects proposed during the course of the Napoleonic War, the Bude Canal was put on ice for over twenty years. It wasn't until 1817 that James Green arrives in the history books. With the help of Thomas Shearn and the support of the new Lord Stanhope, he again surveyed a possible route. Green planned a canal on which tub-boats carrying 5 tons could be drawn along in groups of four by just one horse. The hillsides would be scaled using inclined planes which he reported as a third of the cost of locks, using a third of the amount of water and being five times quicker to use. The first 2 miles to Marhamchurch would take larger vessels. There the sand would be transferred to tub-boats to rise up an incline. One line would then go to Holsworthy and Blagdonmoor and perhaps even on to Okehampton. Another would turn south to Tamerton Bridge. A third line would run to Alfardisworthy, where a feeder reservoir would be built. The cost would be £128,341. Green suggested an annual toll revenue of £15,083, about 80 per cent of which would be derived from sand. A further £25,000, Green reported, would be needed to connect Tamerton Bridge with Launceston. Green also suggested that an extension of the Okehampton line to Crediton was feasible. With renewed enthusiasm an Act was passed in 1819 to give the newly formed Bude Harbour & Canal

Tub-boats in the canal basin at Bude. These simple vessels held 5 tons of sand and were moved up and down the inclined planes by means of wheels fitted to their hulls
British Waterways

Company powers to raise £95,000, with a further £20,000 if needed, to build a 46 mile line from Bude to Red Post, from where one route would go via Holsworthy to Thornbury (with branches to Alfardisworthy reservoir and Virworthy) and the other via Tamerton Bridge to Druxton. On 23 July 1819, with the ringing of bells and the gathering of 12,000 onlookers, Earl Stanhope laid the first stone of the breakwater and dug the first clod from the canal. This was followed by the consumption of 'ten hogsheads of cider and many thousand cakes'. Despite the occasional difficulty, by 8 July 1823 the harbour and canal as far as Tamerton Bridge were open.

The first weeks were encouraging. By May 1824 a hundred boats were reported to be on the canal and trade was brisk. Work on the line between Tamerton Bridge and Druxton was also underway. To finance this the

The beach and sea lock at Bude Harbour

company borrowed £16,000 from the Exchequer Bill Loan Commissioners (a kind of early job creation scheme). Two years later, however, it had to borrow a further £4,000. The final line was eventually finished in 1825, although the Blagdonmoor to Thornbury section was never actually built. The completed canal was just 35½ miles in length and the final cost was £118,000. There were six inclined planes: Marhamchurch, Hobbacott Down, Vealand, Merrifield, Tamerton and Werrington. Each had a double line of rails along which the tub-boats ran by means of small wheels fitted to their undersurfaces. Boats were drawn up the inclines by a chain which passed round a drum at the top. The power came from water: a bucket-in-a-well system at Hobbacott Down and water wheels everywhere else. When working, the planes were a great success. Unfortunately, virtually from the start they were prone to breakdown. Hobbacott's bucket system suffered numerous stoppages in its first year of operation due to the bucket chains breaking. Marhamchurch suffered from chain and gear wheel breakage. All the inclines suffered from broken tramrails. The canal generally had problems with leaks and slipping banks. By 1827, the repair and maintenance costs were still exceeding income from tolls. By 1830 the money available to John Honey, the resident engineer, was becoming tighter and he was often unable to do repairs or pay the labourers or taxmen. The farmers and traders were also complaining that the tolls were too high for them to ship sand at a profit. Indeed, many farmers had returned to collecting by cart. In 1831 tolls were lowered and trade recovered. By 1838 the canal was carrying nearly 60,000 tons p.a., 90 per cent of which was sand, and the remainder coal, culm, limestone, slate and building materials. The total revenue from tolls, harbour and basin dues and rents was £4,300, with expenses at £2,600. No dividend had yet been paid and the company still owed the commissioners £21,037 for principal and interest, and £3,200 of other debts. Despite the problems the beneficial effects of the canal were being felt. Sand at Launceston was now about 75 per cent cheaper than it was before the canal was built and land values surrounding the line were increased.

In 1838 a severe storm damaged the breakwater at Bude and it was rebuilt by the company. The funds to do this were raised by agreeing with the commissioners to divert trading surplus from the repayment of the loan to the repairs. Thus the debt to the commissioners increased further, but with trade remaining consistent it was reduced to £3,000 by 1864 and was repaid by 1870. Receipts peaked at £4,716 in 1841 and profits at £1,493 in 1859. But maintenance costs consistently made the operation of the canal uneconomic and forced the company to keep the tolls at rates which made the shipping of sand barely profitable. It was during the 1850s that the improvement in the roads and the introduction of artificial fertilizers began to impose themselves on the company's activities. And then the railways arrived. In 1865 the Launceston & South Devon Railway reached Launceston from Tavistock. In 1879 the South Western reached Holsworthy and in 1898 it was extended to Bude. The gradual incursion of competition from artificial fertilizers, imported along the railway, forced the canal company to reduce its tolls. By 1876 canal tolls were down to £1,897 (plus £152 in rent). But as maintenance costs were now low and the debt paid off, the first dividend was paid to shareholders at 10s. per share. It was to be the first of just eight paydays. By 1877 the success of 'artificial' lime meant that there was only one major customer left on the canal, Vivian & Sons. From 1878 sand-carrying went into terminal decline.

In 1884 Vivian & Sons decided to give up the trade and the canal company proposed to follow suit; an abandonment bill was proposed to Parliament. Although this was withdrawn and Vivians continued to use the canal for a time, the interest of the company increasingly focused on selling water from the Tamar Lake reservoir at Alfardisworthy to the towns of Stratton and Bude. In 1891 the Bude Harbour & Canal (Further Powers) Act enabled the company to close the canal from Marhamchurch to Brendon Moor, including the Druxton and Holsworthy branches, on 14 November. The plan was to maintain the length to the reservoir as far as the Hobbacott plane, from where water would be piped down to the towns. However, by 1894 the towns had been unable to raise any money and the scheme was dropped. By this time, large areas of the canal had been sold to local farmers. In 1895 the company tried, unsuccessfully, to sell itself to the London & South Western Railway. By 1898 the railway to Bude was open and canal traffic had virtually ceased. The end of the company came in February 1899, when agreement was reached with the Stratton & Bude Urban District Council to sell the remaining canal, including the Tamar Lake reservoir, for £8,000. This move was authorized by the Stratton & Bude Improvement Act of 1901 and the company formally changed hands on 1 January 1902. The line was finally abandoned as a waterway in 1912.

Most of the Bude Canal is now in private ownership, dry and undetectable. The two intact areas are the first 2 miles from Bude to Marhamchurch where, for some of the length, pleasure boating is allowed and the towpath open. This section is owned by the North Cornwall District Council. The second is the length from Venn to the Tamar Lakes reservoir. This latter section is not open to boats and is used solely to conduct water to Stratton and Bude. Virtually all of the rest has been bought by farmers or landowners and incorporated into fields. There is little chance of the Bude being fully restored, although the first 2 miles from the sea should remain open to the public. Other areas may, however, become more accessible, particularly after the recent formation of the Bude Canal Society.

The Walk

Start and finish:	Bude (OS ref: SS 206064)
Distance:	5^1/$_2$ miles/8^1/$_2$ km
Map:	OS Landranger 190 (Bude & Clovelly)
Car park:	At Bude Wharf near Bude Castle and the library
Public transport:	From BR at Barnstaple, take the Red Bus Company's no. 1 or 2 to Bideford and then no. 85 to Bude (enquiries: 0288-45444)

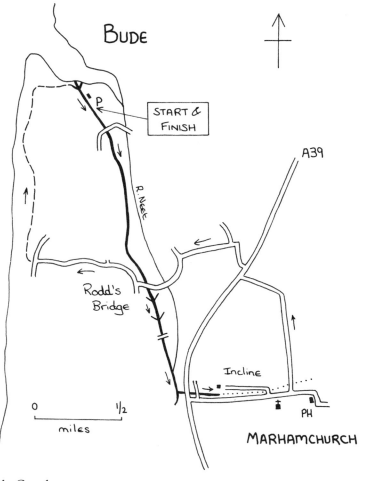

The Bude Canal

The walk starts at the Lower Wharf where there is a local museum in the Old Forge. The museum is open in the summer and contains various bits on the old canal, including a model of an inclined plane. The forge itself is just one of a group of old canal buildings that includes the harbourmaster's office, a one-storeyed stone building which has been used as a salt store, a tannery, a lime warehouse and other warehouse buildings.

To start the walk turn right towards the sea. Surprisingly there are still many remnants of the canal's heyday extant along the quay. On both sides there are mooring rings and on this side it is still possible to make out the rails of a 2 ft gauge tramway that leaves the wharf and goes down a slope to the beach. The railway was used to move sea sand up from the area around the nearby River Neet. The trucks were pulled by horses and then tipped into the tub-boats and barges moored at the quay. Originally the line had a 4 ft gauge but this was replaced in 1910. On occasions the shifting sands expose a turntable at the foot of the bridge. The tramway wasn't finally abandoned until August 1942. At the end near the lock there is a good view of the beach and the breakwater. The latter, built to protect both the harbour and the entrance to the canal, stretches north of Compass Point to Chapel Rock. It was built in 1838 to replace the original which had been washed away in a storm. The lock is a substantial structure: 116 ft long and 29 ft 6 in wide and able to take ships of 300 tons. Green's original plan was that barges 50 ft x 13 ft x 3 ft 6 in and able to carry 50 tons would be taken through the sea lock and allowed to settle on the beach as the tide went down. There they would be filled with sand, brought in again on the rising tide and towed to Marhamchurch.

Cross the canal via the lock gates and turn left to walk along the right-hand bank of the canal across the Company Wharf to Falcon Bridge, near the Falcon Hotel. Originally on this site was a swivel bridge which allowed ships to move from the lower wharf to the upper one (Sir Thomas Acland's Wharf, on the opposite bank just after the bridge). The present bridge severely restricts boat access to the rest of the still-extant waterway. Cross the road and continue along the left-hand bank of the canal. On the opposite side are the old lifeboat house and a renovated canal warehouse. The lifeboat house was built in June 1865 in memory of Elizabeth Moore Garden. It's altogether a peculiar site for a lifeboat house and the launching of the boat tended to damage Falcon Bridge. The large three-storeyed building was one of the first to be built on Sir Thomas Acland's Wharf. Sadly, a number of the other canal buildings have been demolished but once there was a shipbuilding yard, a sawmill, a steam laundry and a number of limekilns.

Continue past the winding hole and into more open country with the River Neet now running close on the left. Normally the barges were horse-towed to Marhamchurch, although at one time a steam-tug was tested. So much damage was done to the banks that it was soon withdrawn. The walk continues past a cast-iron milepost. Shortly after this, cross the canal via Rodd's Bridge. Walk on along the right-hand bank after passing through a gate. Within a short distance are the remains of Rodd's Bridge Lock, and a little farther along the one at Whalesborough. Both locks could handle barges and measured 63 ft x 14 ft 7 in. They have both been stripped down and their upper gates replaced by a concrete weir. The upper ground paddles are, apparently, still used to regulate the level of water in the canal to Helebridge. The pound between the two locks was recently dredged and the Bude Canal Angling Association intends to stock it with fish.

Above here the canal becomes more overgrown. After passing a small accommodation bridge the River Neet and the canal merge for a short distance. On the opposite bank is the weir over which the River Neet flows on its way to the sea. Continue out to cross the A39 and follow the path onwards. Cross the canal bridge and turn right to descend to the left-hand bank. From here you get a good view of the original Helebridge with its numerous rope score marks as well as Helebridge Wharf, which was the canal barge terminal. Goods were unloaded here either for storage and sale or transhipped onto tub-boats for passage further along the line. On the south side of the wharf is a small building which at one time was used for barge repairs and sailmaking. There is also a warehouse and the former home of George Casebourne, the company's engineer from 1832 to 1876. The cottages on the left bank of the wharf were formerly stables for the barge horses.

The path follows the north side of the basin to the bottom of the Marhamchurch inclined plane. Near the foot of the incline are the remains of Box's iron foundry (at which the canal's ironwork was cast). Bill Young tells us that the lintel over the end doors of the foundry is part of an old incline plane railway line. The incline itself, of which there is little trace, had a lift of 120 ft over a length of 836 ft. The tub-boats used on the canal from here on measured 20 ft long by 5 ft 6 in beam and were unique in having wheels fitted to the bottom. This enabled them to be pulled up the incline along rails. The motive power for the Marhamchurch incline was supplied by a 50 ft diameter overshot water-wheel which sadly no longer exists and would be almost impossible to restore without substantial funds. The mechanism was able to raise a boat in just 5 minutes using some 32 tons of water.

To reach Marhamchurch walk up the dirt track which runs to the left of the incline. At the top the track crosses the course of the canal (again with little to be seen) to reach the road. Turn left into central Marhamchurch. The return route starts at the Bray Institute. However, those desirous of lunch should continue straight on for a couple of hundred yards past a small shop to the Buller's Arms. To return, go back to the institute and take the turning which goes past a highly ornate chapel. On the right are new buildings and Old Canal Close, so-named because it was built on the canal bed. Now follow the road out to the A39. Turn right and then left towards Bude. This road bends left. When it bends back right, go straight on along a lane. At a T-junction turn left to go past the abutments of an old railway bridge, over the River Neet and then over the canal at Rodd's Bridge. The quick route back is to turn right here and return along the canal. The more exciting route is to carry straight on through Upton to the coast road. Here join the coastal path by crossing the road and bearing right. This obvious and well signposted route will take you back to the canal sea lock at Bude.

Further Explorations

Problems with access plague the rest of the Bude Canal. Many parts of it can, however, be seen from, for example, various road bridges. If you are interested in seeking out the remnants then Bill Young advises on what is to be seen if you have obtained permission from the landowner.

One route which isn't a right of way but is open for walkers through the good offices of the North Cornwall Council and South West Water goes along the Bude (feeder) Aqueduct to the Lower Tamar Lake. Ideally, the walk (which is 3 miles each way) is best undertaken with like-minded fellows who park at the Lower Tamar Lake (SS 294108 and well signposted from Kilkhampton) while you park in the tiniest of parking spaces at SS 288084, where a lane between Moreton Mill and Broomhill crosses the feeder.

Start along the left-hand bank from the lane bridge (Morton Bridge). The overgrown waterway was primarily built to act as a feeder from the reservoir to the main line and was never used extensively for navigation. It is still in water as it was bought by the Stratton & Bude Urban District Council on 1 January 1902 for use as a water supply for the two towns. The walk is along a cleared towpath through what must be some of the quietest and most peaceful countryside in southern England. The first half passes four bridges: Dexbeer, Gadlock, Wooda and Aldercott. At one point a wrecked mainte-nance tub-boat lies mouldering in its watery grave, only its wooden bones still showing above the vegetation. The canal shortly goes around a sharp bend before reaching Virworthy Wharf. This was formerly the limit to navi-gation. The canal buildings, a warehouse, a wharfinger's cottage and some stabling, have been renovated and are now private houses. The towpath from here on has been made into a waymarked trail from the lake car park. As part of the improvements the towpath sports some pleasant bronzes, the first of which is a representation of a tub-boat passing along the waterway.

A bronze on the towpath walk near the Tamar Lakes shows a train of tub-boats being towed along the Bude Canal

Continue over a lane, to the bottom of Lower Tamar Lake. The lower lake was built to supply the Bude Canal by damming the River Tamar. The upper lake, the dam of which can be seen in the distance, was built in 1975. Lower lake covers some 28 ha (about 70 acres) and holds 885 million litres (195 million gallons). Both lakes make good spots for picnicking and there are additional marked trails that go all the way round them. To complete this walk, however, cross the footbridge which goes over the large overflow weir and then turn left to the car park where, if you're lucky, your companion car can whisk you back to Morton Bridge. Otherwise, enjoy your picnic and then stroll gently back along the line.

Further Information

The Bude Canal Society has been formed to promote interest in and access to the waterway. The society can be contacted c/o:
  Mrs Audrey Wheatley,
  Tregea,
  Lower Upton,
  Bude,
  Cornwall EX23 0LS.
The society holds regular meetings and publishes a newsletter.
  Those interested in the history of the line should obtain a copy of:
Harris, H. & Ellis, M., *The Bude Canal.* David & Charles, 1972.
  Driving tours of the rest of the Bude Canal can be found in the pages of:
Young, Bill, *Walking the Old Bude Canal,* published by Bill Young, Longmead, Diddies Road, Stratton, Bude, Cornwall EX23 9DW, 1991.

5
THE CHELMER & BLACKWATER NAVIGATION
Beeleigh to Heybridge Basin

Introduction

The route of the Chelmer & Blackwater Navigation from Chelmsford to the sea is the result of just over a century's worth of unseemly squabbling between two towns. The natural route would have passed along the River Chelmer through Maldon to the River Blackwater at Heybridge Creek and then simply down to the sea beyond Collier's Reach. But the most obvious routes aren't always those that result from man's complex political machinations.

The Chelmer & Blackwater Navigation starts in the centre of Chelmsford just ¹/₂ mile south-east of the railway station at a spot called Springfield. From here the line goes almost immediately through Springfield Lock to join the River Chelmer and then heads east via Barnes Mill and Lock into the flat Essex countryside. The navigation now bends north on a new cut past Sandford Mill, before returning to an easterly course to the north of Little Baddow. At Ulting the tiny All Saints Church sits on the north bank among the trees. From here the navigation undertakes a southern loop to Hoe Mill before continuing on to Beeleigh, where the Chelmer meets the Blackwater at a waterway crossroads. While the combined rivers overflow south through Maldon, the canal continues in a south-easterly then a north-easterly direction, bypassing Maldon by taking a course through Heybridge. After returning to a south-east slant, a straight cut of over a mile ends at Heybridge Basin and the sea lock to Collier's Reach and the North Sea.

The result of the Chelmsford–Maldon feud is a fascinatingly contrived line and the wonderfully bewildering waterway crossroads at Beeleigh is worth an hour of anyone's time.

A boat-borne funeral procession on its way to Heybridge Cemetery along the Chelmer & Blackwater, *c.* 1905–10

Essex Record Office

History

The desire to make the River Chelmer navigable was established long before the age of canal mania. Andrew Yarranton, as long ago as 1677, proposed and planned the canalization of the 14 miles of river up to Essex's county town of Chelmsford. His plan, which included a survey and the brief for a Parliamentary bill, was well received in Chelmsford but was bitterly opposed by the people of Maldon. The main thrust of their concern was the inevitable loss of income from tolls, duties and wharfage if traffic passed them by on a direct route to Chelmsford. This opposition found favour among the owners and tenants of the watermills who were dependent on the river for their power supply. As a result the scheme, although innovative and far-sighted for its time, was dropped.

Nearly half a century later, on 13 and 14 July 1733, John Hore surveyed the river. Hore was an experienced navigation engineer from Newbury who had already worked on the Kennet, Stroudwater and Bristol Avon navigations. Hore proposed two schemes, the first to simply make the river navigable for £9,355, and the second to cut a wholly new navigation for £12,870. Despite the difference in cost, Hore favoured the latter as a way of avoiding conflicts with the landowners and millers. He justified this with a profit forecast of £382 p.a. However good an engineer he was, Hore had a reputation for being somewhat lax with finance. It was therefore relatively straightforward for those opposing the scheme to develop their own forecast with equal positiveness which predicted an annual loss of £430 10s. Again, the objectors proved powerful and the scheme was abandoned.

In 1762 the idea of a navigation was revived and surveys were carried out independently by John Smeaton and Thomas Yeoman. Both came up with a similar cost estimate; Smeaton's was £16,697. As before, the proposals were supported by Chelmsford, opposed by Maldon and had the same end result. But Chelmsford persisted. In 1765 Yeoman produced another plan and this was the basis for an application for a Parliamentary bill. Yeoman recommended widening the river at the surface to 30 ft and at the bottom to 20 ft with a depth of 4 ft. Locks were to be 70 ft by 14 ft 1 in. The cost estimate was £13,000. On the basis of Yeoman's plan, an Act was passed on 6 June 1766 'for making the River Chelmer navigable from the Port of Maldon to the Town of Chelmsford'. The navigation was to be built within twelve years but nothing was to be started until 25 per cent of the money was in hand. Thereby lay the end of this attempt, for despite numerous efforts it was found impossible to raise the necessary funds.

In 1772 yet another scheme was proposed, this time by an émigré Dutchman, Peter Muilman, at a meeting attended by various town dignitaries at the Chelmsford Coffee House. This plan was later amended to one in which a new cut from Maldon to Chelmsford was to be built, again with a view to avoiding problems with millowners. Muilman even offered to pay for the survey from his own pocket. However, all that actually came from the renewed excitement was another failure.

By 1792 canal mania was rearing its head across the nation. The success of inland navigation in other parts of the country and the increasing economic expansion during the period made the inhabitants of Chelmsford feel that they were losing out. A new proposal was launched with an observable increase in confidence. The plan took the navigation along the Chelmer to Beeleigh, where a new cut went away from Maldon to join the original course of the Blackwater which entered the estuary at Heybridge Creek. As this exit was perilously close to Maldon, an additional cut was proposed to leave the Blackwater at Heybridge and run for about a mile to a place now known as Heybridge Basin. From there a sea lock would give access to the estuary. This new line was surveyed by Charles Wedge under the direction of John Rennie. A further survey was carried out, again under Rennie's direction, in 1793 by Matthew Hall. Both surveys showed that by bypassing

Maldon the length of the navigation would only be increased by 2 miles to 13¹/₂ miles.

The people of Maldon were livid! They quickly published a pamphlet in which they alleged that the proposals were unsupported by evidence of public utility. For once, this opposition came to nought. The Act received Royal Assent on 17 June 1793 and established 'The Company of the Proprietors of the Chelmer and Blackwater Navigation'. The proprietors were authorized to raise £40,000 in £100 shares and a further £20,000 either by the issue of new shares or by borrowing. Tolls were set on a mileage basis, ranging from ¹/₄d. per mile per quarter of oats, malt and other grains, to 2d. per mile per chaldron of coal and 2¹/₂d. per mile per ton for other goods. Stone for roadmaking, other than for turnpikes, was to be carried free. The Act came into effect on 15 July 1793 and work began shortly thereafter. The project was nominally under the direction of John Rennie but was controlled on a day-to-day level by his assistant, Richard Coates, who had previously worked on the Ipswich & Stowmarket Navigation.

There were two main cuts: the stretch of 2¹/₂ miles between Beeleigh and Heybridge Basin; and the ¹/₂ mile from the river to the terminal at Springfield, Chelmsford. Altogether thirteen locks were built: from the Chelmsford end these were Springfield, Barnes Mill, Sandford, Cuton, Stoneham's, Little Baddow, Paper Mill, Rushes or Weir, Hoe Mill, Ricketts, two at Beeleigh and the sea lock. The Beeleigh locks were needed because the navigation diverged from the course of the Chelmer through a short cut to join the Blackwater, along which the line of the navigation continued before entering the final cut to Heybridge Basin. The Blackwater was diverted at this point into the Chelmer at Beeleigh Falls. The two locks at Beeleigh are therefore sited on each side of the present course of the Blackwater, providing protection at times of flood or drought.

By April 1796 work was sufficiently advanced to allow the lower part of the navigation to be used. On Saturday 23 April 1796 the brig *Fortunes Increase*, loaded with 150 chaldrons of coal from Sunderland, entered the line. The cargo was unloaded at a coal yard at Boreham and moved on to Chelmsford by road. The first load to be moved in the opposite direction left on 26 April, when a barge loaded with 150 sacks of flour departed from Hoe Mill for London. The final stretch to be completed was the cut from the river to the terminal basin at Chelmsford. This took another year. It wasn't until 3 June 1797 that coal barges arrived at Springfield Mead. As is usual with these things, they arrived in a grand procession with colours flying.

By the time the navigation was finished the cost had risen to £50,000. What was worse was that virtually immediately problems arose. In December 1797 serious flooding led to shoals in the river which impeded the movement of the barges. This problem got worse with every flood until 30 March 1799, when a meeting of the navigation committee resolved to send Lord Petre, one of the most important local landowners, to see John Rennie. Rennie was to be told that the committee had a claim on him to resurvey the navigation and suggest a cure for the defects without further cost. Rennie accepted responsibility, although his response implied that it was Coates who was really responsible. He carried out a survey on 25 May 1799 and remedial work was undertaken. By 1805 further problems arose when the millers at Moulsham, Barnes, Sandford, Little Baddow, Paper and Hoe claimed damages for the loss of water when their local locks were used, adding that the general condition of the locks was resulting in considerable leakage. Rennie was again called in and, following a survey on 25 November 1805, he suggested improvements at a cost of £4,918. In spite of these efforts the millers still seemed to be complaining in 1807.

Collier's Reach and the entrance to Heybridge Basin

Although there were these minor hiccups, on the whole the navigation performed well. Traffic was vigorous, even though in February 1799 the navigation froze over. Goods carried along the line included coal, timber, bricks, stone and general cargo inwards and grain and flour out. Success was such that in 1842 the line carried 60,000 tons of cargo. The barges used were 60 ft by 16 ft and could carry 25 tons. They were a special flat-bottomed type with a draught of just 2 ft. This was necessary to ensure good access on what was one of the shallowest waterways in the country. Some of the barges, which had no cabin accommodation, were apparently still horse-drawn until the 1960s. The cheap transport of coal into the area led to the establishment of a gasworks in Chelmsford in 1819, only the second to be built in Essex and the first inland. The navigation was also successful financially. Dividends were paid into the twentieth century and reached a peak of 5 per cent in the 1820s. In 1846 the original £100 shares were selling for £66, although by 1918 they were only worth £17 10s. The fall in both share value and dividend payment after 1838 was due, perhaps predictably, to the opening of the Eastern Counties Railway through Chelmsford in March 1843. But there is no doubt that the foresight of the original promoters helped Chelmsford grow from a small market town to the important centre that it is today.

After the Second World War the only traffic on the C&B was the carriage of timber to the sawmills and timber-yard of Gilbert Brown & Son Ltd, here photographed in the early 1900s

Essex Record Office

The C&B was never taken over by a railway company and, as a consequence, was not nationalized in 1948. The line remains independent and is owned by the Company of the Proprietors of the Chelmer & Blackwater Navigation Ltd. The navigation was in use commercially until 1972, although after the Second World War the only traffic was the carriage of timber to the sawmills and timber-yard of Brown & Son Ltd at Chelmsford Basin. The barges made the return journey empty. Gradually the proprietors are becoming aware of the leisure potential of this attractive waterway and have themselves started operating a trip boat during the summer. Following the 197th AGM in 1991 the proprietors viewed the line on board the cruise boat *Victoria*. The year had seen dredging work, the fitting of new lock gates, the extensive planting of trees and repairs to a listed building. The year also saw the first commercial trade for twenty years, when Blackwater Boats of Sandford Lock started to hire boats. The prospects for the navigation may therefore be bright and the line may yet play a part in the leisure time boating trade.

The Walk

Start and finish:	Langford (OS ref: TL 838090)
Distance:	7^1/$_2$ miles/12 km
Map:	OS Landranger 168 (Colchester)
Car park:	On road near St Giles Church, Langford
Public transport:	Chelmsford has a main-line BR station with frequent trains to London. Eastern National buses run to Langford and Heybridge (enquiries:0245-353104)

Facing away from the church, turn right to go past the old mill building for 50 yd. Just after Mill Cottage, and just before the Essex Water Company works, turn left to follow a footpath sign along a metalled road. The road bears left. Here the 'new' course of the River Blackwater can be seen to the right. The original course of the river ran further to the north-east via the mill at Langford. This 'new' river is therefore the overflow from the weir which became the main course of the Blackwater when the Langford mill was closed. The water to the left among the trees is not a river at all but is in fact a canal known as the Langford Cut. This was the only branch canal on the C&B and was originally called Mr Nicholas Westcomb's Navigation because it ran across his land to the mill at Langford. As it was entirely private it did not require an Act to sanction its construction. The line was surveyed in 1792 and opened for traffic in July 1793, about three years

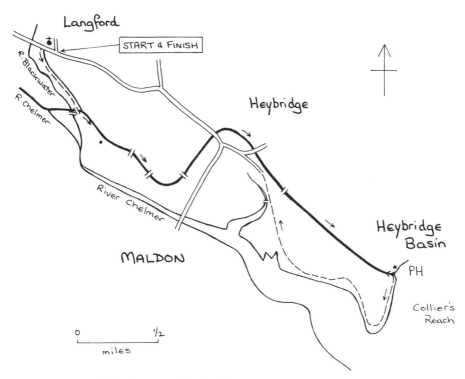

The Chelmer & Blackwater Navigation

before the C&B itself was complete. Originally the cut ran to join the Chelmer near Maldon, a distance of about a mile. It's now a little shorter as it joins the C&B at Beeleigh. The cut carried considerable traffic during the first half of the nineteenth century. By the second half traffic had all but gone and the last barge to reach the mill did so on 9 August 1881. The mill, which is actually not the original structure but a rebuild after a fire, closed in 1918 and was later purchased by the water company.

The route continues along the driveway to reach a bridge over a river. This is the Chelmer & Blackwater Navigation, a wholly new cut opened in 1796. To the left the Langford Cut peacefully joins the C&B. To the right is an extremely complex waterway junction that will need explanation. To get a good view, cross to the opposite bank and turn right. Walk up to the foot-bridge that passes over the weir and face upstream. Straight ahead is the new line of the River Blackwater that has flowed down from Witham, Braintree and beyond. To the left is the River Chelmer that has been rendered navigable (as part of the C&B) to Chelmsford. Behind you is the main downstream course of the Chelmer as it makes its way towards Collier's Reach and the North Sea. Flowing past you and on to the right is the new cut of the C&B on its way to Heybridge. This extraordinary complex is in fact a type of waterway crossroads. The main traffic lane went left–right, although water mostly still flows front–back. The flood weir that you are standing on is known as Long Weir and maintains the level of the canalized waterway. The original wooden footbridge that crossed the weir was built at the end of the eighteenth century, only to be washed away in 1947. If you continue for a short distance up the Chelmer you will see Beeleigh Lock, the last on the Chelmer portion of the navigation.

Return to the old brick accommodation bridge crossed earlier. Underneath it are the gates of Beeleigh flood lock. Walk on along a metalled lane with the C&B to your left. Opposite the entrance to Langford Cut and barely discernible on the right is the line taken originally by Mr Westcomb's pride and joy as it wended its way to join the River Blackwater. Those who purchase Pathfinder Map 1123 will see the course still marked.

To continue the walk, keep on the road until it bends right towards the Maldon Golf Club clubhouse. Bear left onto the grass to keep the navigation close left. At this point on the left-hand bank of the C&B it is possible to find a small stream entering the waterway. This was the original course of the Blackwater before the intervention of the mill at Langford. The C&B now follows the line of the river to Heybridge. The path eventually leaves the golf course behind some trees to reach a bridge. Cross here and turn right to follow the left bank. Go through a gate and to an old railway bridge. Here a new flood lock has been built. The path continues under the bridge as the navigation bends left. When I was here a complex rerouting of the right of way diverted the path left, over the railway bank and across a field on the opposite side. Whatever route you take, once around the corner, the waterway continues to a road bridge. Before reaching the bridge, note that a stream leaves the navigation from the right bank. This is the Blackwater wending its way out to sea via Heybridge Creek. From here on the C&B follows another new cut.

Bear left up to the road to Black Bridge. You are now in Heybridge. Cross the road and continue alongside the C&B, staying on the left-hand bank. The course bends right near Jeakins Crash Repairs and passes some old industrial units to reach another bridge (Wave Bridge). After crossing the road, continue down to the towpath on the left bank. The large factory building on the right-hand bank is William Bentall's agricultural engineering works or Heybridge Ironworks. Bentall built his factory here in 1804 to take advantage of the cheap land and the C&B which permitted the ready import of raw materials and the export of the finished goods around the world. In the nineteenth century this was one of the most important agricultural engineering firms in Essex. Among the implements exported from the works was the highly successful Bentall broadshare plough. The building on the canal bank was built in 1863 as a warehouse for the factory and is now a designated 'ancient monument'.

Continue into increasingly open country. At one point the route passes some entrance gates to the left. This is Beeleigh Cemetery, possibly the only one in the country to be served via its own canal wharf. Sadly the gates now look anything but pearly and clearly haven't been used for many a year. The canal, meanwhile, carries on along a straight course with numerous moored boats.

At the end of this section, about a mile from the road bridge, the navigation bends slightly left to reach Heybridge Basin. Opened in 1797, the basin has a tidal lock which gives access to the Blackwater estuary. It was built originally to hold sailing ships. Here the boats transhipped timber, coal and provisions to the broad, shallow-draught 25 ton barges which then travelled up the C&B to Chelmsford. A two-storey bow-windowed lockhouse flanks the entrance lock on the other side, a row of weatherboarded cottages and the Old Ship (formerly the Brig Inn and renamed in 1906) on this.

The sea lock was built to handle vessels up to 107 ft by 26 ft, allowing access for boats drawing 8 ft at neap tides and 12 ft at springs. The lock was lengthened after the Second World War to accommodate coasters that brought timber onto the C&B. If you continue to the far end of the lock, to see Collier's Reach and Northey Island on the far side, you may note that the lower gate of the tide lock is a type of caisson which slides sideways. It's electrically powered and chain-operated. Further round to the left is the Jolly Sailor pub. To continue the walk, cross the lower gate of the lock and go up to the sea wall embankment with the estuary to the left and a small lake (some old gravel workings) to the right. The town of Maldon can be seen to the right. Continue along the sea wall as it bends right with fine views of the Chelmer/Blackwater estuary and of an area of Maldon known as the Hythe to the left. The path eventually bends round, with Heybridge Creek (the original course of the Blackwater) to the left and some holiday huts to the right, to arrive at a minor road. Turn left and walk up to the main road. Turn right here past the post office and turn left just after Bentall's Mill to return back along the C&B to Langford. If in a hurry, when you reach Black Bridge it is possible to turn right to walk up the road to Langford (about 1¼ miles), although the original route is much more pleasant.

Further Explorations

The stretch of canal between Maldon and Chelmsford makes for a fine 12 mile walk. Eastern National bus no. 31 runs between the two towns at roughly half-hourly intervals (enquiries: 0245-353104).

From central Maldon, walk along Silver Street which is to the left of All Saints Church. This goes past The Blue Boar Hotel and Maldon Court School to Beeleigh Road. Continue into a private road and then bear left along the fenced pathway. Continue straight on and across Maldon bypass. After this, go straight on to reach a field. Keep to the left edge and go over a stile. The path now passes Beeleigh Abbey. Continue along the drive to a T-junction. Turn right and follow the lane into Essex Water Company property. After Beeleigh Falls House, the driveway turns right. Here go left along a path and over a footbridge across the River Chelmer. After crossing, bear right and follow the path around to the C&B with the waterway cross-roads to your right. Bear left to cross the C&B near Beeleigh Lock. Now turn left to walk along the right-hand bank.

The path winds through open country to reach Rickett's Bridge and Lock and then Hoe Mill Bridge, where the public right of way changes banks. Pass Hoe Mill Lock and then walk along the course of a new cut while the river takes a more southerly (and lower) route. The canal overlooks the river for a short stretch before the two rejoin near a substantial weir. From here the C&B turns north-west to Ulting where the delightful thirteenth-century Church of All Saints sits on the right bank with the churchyard descending to it. The navigation continues towards Chelmsford via Rushes Lock and under a footbridge. The next bridge is a road bridge and is followed by Paper Mill Lock, home of the C&B Navigation Company. Here you will

The diminutive navigation-side church at Ulting

also find Chelmer Cruises, which offers trips along the canal. At the next road bridge, Boreham Bridge, the footpath reverts to the right bank. After passing another small footbridge the C&B goes through Little Baddow Lock, where there is a fine mill pool and weir waterfalls. The line shortly turns abruptly left to Stoneham's Lock.

By this stage the tranquillity which has enveloped you since Maldon is disappearing rapidly. Turn left with the C&B to your left and the (relatively) new A12 on the right. After passing Cuton Lock, with its associated pillbox, and a small footbridge, go under the A12 and, step by step, peace is gradually restored. The towpath reaches Sandford Mill Bridge and Lock (from where Blackwater Boats operates) and then another small road bridge before bending right to Barnes Lock. The towpath continues over a small footbridge and into a field. Follow the course of the C&B round to the next road bridge which carries the 1932 vintage Chelmsford bypass.

The River Chelmer now heads south. You should follow the cut right to Springfield Basin. Just beyond the diversion is an old accommodation bridge and the last lock on the C&B (Springfield). At the end of the pathway the site of Chelmsford gasworks is still visible on the left. Brown's Timber Yard on the right of the basin is now occupied by Travis Perkins. To complete the walk, follow the path between the industrial sites to a road (Navigation Road) and turn left. This passes Travis Perkins and crosses Wharf Road to a T-junction. Turn left and walk on into central Chelmsford.

Further Information

The C&B is not operated by British Waterways but by its own company:
  The Chelmer & Blackwater Navigation Company Ltd,
  Paper Mill Lock,
  Little Baddow,
  Chelmsford,
  Essex CM3 4BF.

Although there is no C&B society as such, the navigation has its supporters in the Chelmsford branch of the Inland Waterways Association. Those interested in finding out more should firstly contact the head office of the IWA in London. The address is given in Appendix B.

For those interested in the history of the canal, the following can be recommended:
Jarvis, S., *The Rivers Chelmer and Blackwater*. Lavenham Press, 1990.

6
THE GRAND JUNCTION CANAL
The Tring Reservoirs

Introduction

The state of the nation's canals in the 1950s can be no better demonstrated than by pointing out that the Grand Junction, or the Grand Union as it became, is the only canal in this book which would have survived in its entirety, excluding the branches, if the British Transport Commission had had its way. All the rest were considered as having 'insufficient commercial prospects to justify their retention'. The canal earned this respect because of its strategic position as the main canal line between London and the north. But fear not. The GJC has no conceit. You can walk along it just like any other mere mortal waterway and find it every bit as charming.

The GJC starts at Braunston where it forms a junction with the Oxford Canal, a prime route north to Birmingham and the Potteries. The canal then takes an easterly course through the Braunston Tunnel to its second junction at Norton where it meets the Old Union, the Leicester line to the River Trent. From here the GJC gradually turns south to pass Weedon and the Northampton arm at Gayton to Blisworth. The canal now goes through the Blisworth Tunnel to Stoke Bruerne. After passing the course of the Buckingham branch the line leaves Northamptonshire to enter Buckinghamshire via the Ouse Aqueduct near Wolverton. The former Newport Pagnell branch is passed at Milton Keynes as the line heads south to Fenny Stratford and Leighton Buzzard. The Aylesbury arm meets the canal at Marsworth near the Tring reservoirs. From Berkhamsted the canal goes through Hemel Hempstead and Watford to Rickmansworth and Uxbridge. After passing the Slough arm, the canal bears east through London's suburbs to reach Bulls Bridge, from where the Paddington arm heads into the centre of the city. The main line, meanwhile, reaches its southerly end when it joins the River Thames at Brentford.

A stroll around the Tring reservoirs provides a fine varied walk with plenty of canal interest, bags of wildlife and a good pub. Need more be said?

History

One of the key objectives of canal developers in the late eighteenth century was to provide a navigable link between London and the Midlands. That initial connection had been made with the completion of the Thames & Severn Canal which, with the Stroudwater Navigation, joined the rivers Severn and Thames for a journey from Birmingham to London of 269^1/$_2$ miles. Just a year later, the opening of the Oxford Canal reduced that distance to 227 miles. However, neither line was exactly direct and both were subject to the vagaries of navigation on the upper Thames, which was often flooded in winter and drought-ridden in summer. As a consequence, even though London was no longer isolated from the industry and minerals of the North, a shorter, less-troublesome route was still a key objective.

A more direct waterway from Braunston on the Oxford Canal to London was first planned at the end of 1791. The proposed line to Brentford, known at that time as the Braunston Canal, would undercut the Oxford route by 60 miles and be both quicker and more reliable. Prospects for trade were good and had been recently bolstered by the completion of the Coventry Canal between Atherstone and Fazeley, thereby giving access to

Edward Powell's horse-drawn pair *Water Lilly* and *Forget Me Not* at Braunston Top Lock in 1913 with his and a friend's families posing for the camera

British Waterways

the Trent & Mersey Canal. Goods and minerals from Birmingham, Manchester, Liverpool and the Potteries could all now reach Braunston and, it was proposed, would wish to take the new line to London. By early January 1792 the plan was being so widely discussed that canals were being planned to form junctions with the proposed line. The Old Union was promoted in February 1792 to join the new canal. This would feed traffic to and from Leicester, the Trent and the Derbyshire coalfields. The Warwick Canals to Birmingham and Stratford were planned with the new London route in mind and provided the prospect of a much shorter route to Birmingham. The only people against the whole idea were, not unnaturally, the Oxford Canal Company. As soon as plans for the new canal became apparent, the Oxford decided to back an alternative route, the London & Western, or Hampton Gay Canal to run from Hampton Gay, 6 miles north of Oxford, to Isleworth via Thame, Wendover, Amersham and Uxbridge. For a while the Hampton Gay drew support but it couldn't fight the sheer inevitability of the much more direct route and the scheme soon disappeared into obscurity.

The first survey for what was becoming known as the Grand Junction Canal was made in 1792 by James Barnes who, ironically enough, had previously worked on the Oxford. This was followed by a second survey by

The New Mill on the Wendover arm near Tring was once wind-powered, as this 1900 photograph shows

British Waterways

William Jessop in the autumn. Although the two routes were broadly similar, the southern section which was originally to run from Watford via Harrow to Brentford was diverted via Uxbridge. The main engineering efforts were identified as being in the high ground at Braunston, Blisworth, Tring and Langley Bury near Watford, and in the provision of water to the summits at Braunston and Tring. The canal was to be a broad line of 90 miles, able to take barges capable of carrying 70 tons.

The GJC received its Royal Assent on 30 April 1793 with powers to raise £600,000. The Act, and a subsequent one in 1794, detailed branches to Daventry and Watford (although neither was actually built), Northampton (to meet the Old Union), Buckingham, Aylesbury and Wendover. Among the promoters of the new canal were the banker William Praed, the Marquess of Buckingham, the Duke of Grafton, the earls of Clarendon and Essex, the Earl Spencer and the Hon. Edward Bouverie. Later the company coopted numerous MPs; in 1812 five were on the company's board. These dignitaries gave the company a lot of political clout, a situation that didn't exactly enhance its popularity. With the passage of the Act, Praed was made chairman, Jessop was appointed chief engineer and James Barnes was made resident engineer.

Work began almost immediately at both ends of the line. By December 3,000 navvies were reported to be at work. The cutting of Braunston and Blisworth tunnels was a priority and they were started straight away. At the southern end building began at Brentford, where the line used the River Brent as far as Greenford from where it ran east to the River Colne, which it then followed to Watford. This southern end of the canal was open to Uxbridge on 3 November 1794. Meanwhile at Braunston problems arose when the line was discovered to run through quicksands, but by 21 June 1796 the tunnel was open and the canal finished from Braunston Junction to Weedon. In the south, the end of 1797 saw the line reach Hemel Hempstead. This included the construction of some highly ornamental sections through Cassiobury and Grove parks near Watford. The southern works reached Tring in 1799 and Fenny Stratford on 28 May 1800. By September the southern line was open to the bottom of Stoke Bruerne Locks and trade was possible over the whole canal except for the tunnel at Blisworth.

The completion of the work to this point was only possible with the provision of extra funds. The company obtained an Act in December 1795 which allowed it to raise a further £225,000. This still wasn't enough. Further Acts, in 1798, 1801 and 1803, raised an additional £150,000, £150,000 and £400,000 respectively. The cost of the canal was by now was approaching four times Jessop's original estimate. It was Blisworth Tunnel that was proving to be the GJC's Achilles heel as various sections collapsed. At one stage Jessop suggested building twenty-nine locks over the hill but

Barnes, supported by Robert Whitworth and John Rennie, called in as consultants, proposed a new tunnel on a slightly different line. A brief financial hitch meant that the work couldn't start until the autumn of 1802 so a temporary toll road was built instead. This was later replaced by a double-track tramroad which ran from Blisworth Wharf to the bottom of the Stoke Bruerne Locks. The new tunnel, built with underground drainage channels to prevent subsidence, was finally completed on 25 March 1805 and with this the whole canal was open for traffic. As opened, the GJC main line was 93½ miles long with 101 locks. The branches brought these figures to 136¾ miles and 137 locks. The cost to the end of 1811, i.e. without either the Northampton or Aylesbury branches which had still to be built, was £1,646,000.

The decision to go ahead with the branch to Northampton proved problematical as the GJC waited to see whether the Old Union Canal would go there. It didn't and the people of Northampton forced the GJC to build a line, initially a tramway but later a 5 mile canal branch from Gayton Junction, opened on 1 May 1815. The line to Aylesbury was only built after the Marquess of Buckingham had insisted. It was completed in May 1815. One through line that was built independently of the GJC was that from Napton to Birmingham. This line was composed of two separate canal companies, the Warwick & Birmingham and the Warwick & Napton. These were both opened in 1800 and provided the GJC with a more direct route into Birmingham. Meanwhile, at the London end another branch was added following an Act passed in April 1795. The new line went from the GJC at Bulls Bridge, Southall for 13½ miles to Paddington. This important extension was opened on 10 July 1801. At Paddington Basin warehouses and wharves were built and it soon became a busy terminus. A further extension along Regent's Canal followed in 1812.

As soon as the main line was operational the GJC held meetings with its neighbours in order to convince them to widen their canals and thus improve carrying efficiency. This far-sighted scheme was supported grudgingly by most but some, such as the Oxford, were positively hostile. The issue was complicated by the proposal in 1796 for an entirely new broad waterway, the Commercial Canal, to join the rivers Dee, Mersey, Trent and Thames. The project received wide support and the GJC was seemingly relieved when the scheme was defeated in 1797. However, the increasing financial stringency of the moment was also enough to shelve any further canal-widening schemes. Indeed, it was then seen that there were advantages in two narrow boats being able to pass each other in Braunston and Blisworth tunnels.

The traffic along the GJC in the early part of the nineteenth century was highly promising. The company was able to pay a 3 per cent dividend for the year after the canal opened, when toll receipts totalled £87,392. Even though coal taken into London faced a ban and then a tariff, as part of a protectionist scheme for the existing merchants, coal transport rapidly became the single biggest business. The Canal Act of 1805 allowed 50,000 tons p.a. to be shipped into Paddington on payment of duty. Later this limit was removed and gradually the tariff, 1s. 1d. a ton in the 1830s, was reduced so that by 1890 it was abolished completely. But in 1810, 109,844 tons of inland coal and 22,209 tons of sea coal was carried. By 1821 these figures had increased to 149,004 and 39,804 tons respectively. But it was the enormous range of different cargo which was so impressive. Pig iron, for example, was brought from Shropshire and Staffordshire into the city for manufacturing, together with castings and pipework. Some 24,364 tons were shipped in 1801 and this grew to 55,694 tons in 1840. There was a steady trade in bricks, timber and lime (77,797 tons in 1810). Agricultural produce was brought from all around the Midlands and there was a busy hay, straw, vegetable and cattle market at Paddington. On top of this, there was salt from Cheshire, glass from Stourbridge, pottery from Staffordshire, manufactured goods from Manchester and Birmingham, and stone from Derbyshire and Yorkshire. Exports from London included groceries, raw materials and products from overseas, together with ashes, cinders and manure. In total in 1810, some 343,560 tons of cargo were shipped in or out of London, a figure that does not include cargo shipped locally. One of the surprising things about the figure is the relative equality of the north–south trade: 191,696 tons going south; 151,864 tons going north. In 1813, receipts reached £168,390 and the company issued a 7 per cent dividend.

Despite a few minor ups and downs trade continued at this level throughout the 1820s. In 1827, for example, receipts reached £187,532 from tolls and a 13 per cent dividend was paid. But as early as 1824, the outlook changed when John Rennie was asked to carry out a survey for a London–Birmingham railway. The canal companies held a meeting in November, where it was decided to have a joint study on the future of the railways and to evaluate ways of upgrading the canal lines. Despite this action the mere threat of a railway caused GJC share values to fall from £350 in 1824 to £225 in 1831. Luckily any further action on the London & Birmingham was in abeyance because of a financial crisis. This gave the GJC a chance to rationalize its toll schemes and to investigate the potential for using steam to haul boats. It was also able to consider ways of fighting the new railways. However, its campaign did not receive the support of the other companies and the Railway Act was passed in May 1833.

In 1835 the canal carried 192,859 tons of trade to London and 631,815 tons of local traffic. Improvements were made, including the doubling of the flight of locks at Stoke Bruerne and the building of new reservoirs at Tring. Despite good business, the prospect of competition from the railways meant that tolls were reduced in January 1837. Thus marked the start of the steady decline in receipts from the peak of £198,086 in 1836. On 12 November 1838 the London & Birmingham Railway was opened to Euston and because of the need to reduce tolls just to compete, receipts on the GJC dropped instantly. By 1842, earnings were down 43 per cent on the 1836 level. This was despite a steady growth in the amount of cargo carried. More cuts were announced in 1857, following discussions with the London & North Western Railway and the Great Western Railway and an agreement designed to maintain differentials. This arrangement was such that it allowed a small rise in tolls in 1859 but it did not prevent virtually the entire coal-carrying business, the prime activity for the GJC, being lost to the railways.

In an attempt to rekindle business, the company decided in 1848 to start its own carrying company. This it did by raising £114,550 in preference shares. The company also started to use steam cargo-carrying boats, usually with a butty in tow. These were uniquely unpopular with the other canal companies who claimed that they were being sailed too quickly and too recklessly. Such recklessness included the carrying of explosives and one incident with a GJC steamboat at Macclesfield Bridge on the Regent's Canal in 1874 led to claims totalling £80,000 and eventual withdrawal from the carrying business in 1876. Steam engines were also installed in the Blisworth and Braunston tunnels to tow boats through. By 1871 steam tugs were employed in the tunnels, a service that continued until 1936.

In the mid-nineteenth century income was falling but the amount of cargo carried was not. Through tonnage, i.e. not including local traffic, stood at 294,141 tons in 1845. However, by 1870 the steady decline in trade was becoming apparent, with just 135,657 tons being shipped to and from the capital. It is perhaps surprising therefore to find the GJC spending £107,000 in 1883 to build a new 5 mile long branch to Slough, a town more than adequately served by the GWR (although this turned out to be a highly profitable venture). The GJC responded to the general downturn by trying to form an 'amalgamation' of all the interactive canals in the south and east Midlands. With the development of various jealousies and factions, this brave attempt failed.

The GJC was obviously not defeated by this failure for soon after the company inspected the Grand Union and Old Union canals (the Leicester line), and made a series of suggestions on ways of improving and increasing the traffic. The Leicester line companies replied by offering to sell, a deal that was eventually agreed on 12 July 1893 (completed on 29 September 1894) – the Grand Union for £10,500 and the Old Union for £6,500. With prompting from the canal carriers Fellow, Morton & Clayton Ltd, the GJC set about dredging its new possession and improving some of the bottle-necks, such as the Foxton and Watford locks. The purchase also stimulated the negotiation of agreements with the Leicester and Loughborough navigations and the Erewash Canal for lower through tolls.

Despite this expansion the dawn of the new century was not a promising one for the GJC. More and more traffic was being lost firstly to the railways and then to the roads. Unusually for canal companies, the response of the GJC and the Regent's Canal was to enter into a period of collaboration. In 1914 the two companies formed a joint committee of directors and, after the First World War, the companies decided to merge. To do this the Regent's bought the GJC to form the (new) Grand Union Canal Company. This was achieved on 1 January 1929 at a cost of £801,442. The new company then bought the Warwick Canals for £136,003. The grouping was further expanded in January 1932 when the Leicester and Loughborough navigations and the Erewash Canal were also purchased for a total of £75,423. An agreement to purchase the Oxford Canal, however, fell through for various technical reasons. For the first time the inland waterway line from London to Birmingham and London to the Trent was under one roof.

The new Grand Union was committed to expanding the use of the waterway. The key parts of its plan involved making the canal to Birmingham suitable for barges and providing an improved design of barge to work it. The improvement budget for 1931 amounted to one million pounds (with the help of a government guarantee). One key area for improvement were the Hatton Locks north of Warwick which were all broadened. The company also (re)started its own carrying company by buying Associated Canal Carriers Ltd, and renaming it the Grand Union Canal Carrying Co. Ltd.

The history of the company after this period of hopeful growth is one of disappointment. Receipts were falling rather than increasing and the hoped for traffic never arrived. No dividends were paid on ordinary shares between 1933 and 1945, and in 1948 the company was nationalized and put under the control of the British Transport Commission (later British Waterways). Because of its strategic importance as the prime route between London and Birmingham, the Grand Union was never in the same kind of danger as other canals. Indeed, in the Report of the Board of Survey in 1955, the line was listed as one of the few 'Waterways to be developed'. The line suffered a slight demotion in the 1968 Transport Act in not being listed as a commercial waterway. However, it is considered to be a cruising waterway and is thus as well protected as any of the nation's canals can be.

The Walk

Start and finish: Tring BR station (OS ref: SP 951122)
Distance: 7^1/$_2$ miles/12 km
Map: OS Landranger 165 (Aylesbury & Leighton Buzzard)

Car park: At Tring station or on road near canal bridge
Public transport: Tring BR from London Euston or Rugby

Tring station can be found off the A41 Berkhamsted to Aylesbury road to the north of the dual carriageway Tring bypass. It is well signposted from the main road. Access to the canal is from a road bridge that is 200 yd south (towards central Tring) of the station. Go down the steps to the right of the road and start the walk along the right-hand bank.

The GJC reaches its 3 mile long southern summit here at Tring, at some 400 ft above sea level. The builders reached Tring in 1799 and the resultant cutting is 1½ miles long and 30 ft deep. It took five years to dig! Remember that the navvies at the time had just spades and wheelbarrows with which to do the job. The cuttings on the Birmingham & Liverpool Junction Canal at Woodseaves and Grub Street are deeper but Telford designed and built the B&LJC in 1825–35, by which time engineering techniques had advanced considerably.

The towpath changes sides at Tring Cutting Bridge and then continues for a further ½ mile before the sides of the cutting subside. After passing through a short zigzag, you arrive at some mooring spaces, a winding hole

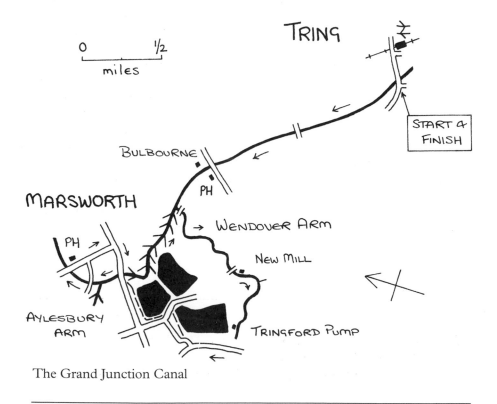

The Grand Junction Canal

and Upper Icknield Way Bridge. If craving refreshment, the appropriately named Grand Junction Arms pub can seen on the road. The walk, meanwhile, continues under the bridge and past British Waterways' (formerly the GJC's) Bulbourne maintenance yard, a collection of rather fine Victorian buildings, one of which has a rather ornate tower. Towpathers who have ventured north to see the remains of the Foxton inclined plane on the Leicester line near Market Harborough may be interested to know that the GJC built a small incline here in 1896 to test the idea before building the real thing.

The path now meanders past a picnic site to reach further mooring spaces, some cottages and Bulbourne Junction. Here is another BW maintenance yard, including a dry dock and a lengthsman's cottage. The GJC main line goes straight on through Marsworth Upper Lock. We, however, go under the roving bridge to turn left along the Wendover arm. The Wendover branch canal, originally 6³/₄ miles long, was authorized by an Act of 1794 and was finished in 1797. Its prime function was to act as a spring-fed feeder for the GJC main line, although the people of Wendover might have disagreed. The central section, which leaked badly, was largely unused by 1897 (when it was shown to be losing more water than it was adding) and abandoned in 1904. Today the arm from the GJC runs for just 1¹/₄ miles. At one time it looked as if the people of Wendover would be more adequately served by a main line: the London & Western or Hampton Gay Canal would have passed this way *en route* between the Oxford Canal at Thrupp and Uxbridge. The idea was well received and, at the height of canal mania in 1792, attracted considerable financial support. The scheme disintegrated in what appears to have been a wall of non-cooperation. In the end, the GJC guaranteed the Oxford annual receipts of £10,000 p.a. and the latter dropped the whole idea. In the event the Oxford never needed to ask the GJC to top up its earnings.

The Wendover branch feels both narrower and more remote than the GJC main line. The towpath changes sides at the first bridge (Gamnel Bridge) and continues alongside Heygates flour mill or New Mill. Although many of the buildings here are of relatively recent vintage at least two show their age with canalside loading bays. The line continues to pass the Tring feeder, one of the original small streams that fed into the Wendover arm and hence into the GJC. Within a short distance the arm reaches Tringford pumping station. The vast reservoirs that you will see shortly all feed to the station which then delivers the water into the Wendover arm and thus into the GJC. Observant towpathers will notice the feeder delivering the precious supply into the canal just opposite the main building. The Boulton & Watt engine started its work in August 1818. It was joined by a second engine, the York engine, in 1839. Since 1913 the task has been carried out by electric pumps originally powered by diesel generators.

The towpath continues to reach the remains of a lock where the current line comes to an end. Walk up to the bridge (Little Tring Bridge) to see the remains of the dry section and then turn right along the road. After ¼ mile take a right turn. Within a short distance you will pass a house on the right and the lane will bend left. Turn right over a stile and across a field to another stile. This path then leads left through some scrub and out to Tringford reservoir. As mentioned, the three reservoirs here provide the vital supplies needed to keep the Tring summit in water. They are interconnected by culverts to Tringford pumping station. The first reservoir on the Tring summit was built in 1802 on what was formerly the site of a marsh. That reservoir, Wilstone, is about ⅔ mile to the south-west (left) of here. It fed the canal via a pump which lifted water into the Wendover arm. This was supplemented in 1806 by the Marsworth reservoir, ahead to the right of centre beyond the road. In 1811 Wilstone had been expanded but this still wasn't enough. The reservoir in front of you, Tringford, was opened in

Tringford pumping station on the Wendover arm

1816, and in the following year Startopsend, the water to the centre-left, was added. The reservoirs were all fed from small streams, run-offs and even from Tring sewage works. Despite these efforts, by the 1830s the water shortage was becoming critical and the GJC was forced to impose severe restrictions on lock use: boats could only go through locks in pairs or face an extra toll. To avoid this, Wilstone was again expanded. Yet there were still problems during 1902 when water was so short that only 80–90 boats a week were allowed to pass to and from the summit, compared with 130 normally. This led to huge queues. Boats, crews and horses are reported to have littered the banks of the canal, each awaiting its turn. The reservoirs now form a National Nature Reserve for wintering wildfowl.

The footpath passes over a weir. Continue straight on and out to a minor road. Bear left down to the next reservoir, Startopsend. Bear left here and along the straight edge of the embankment. This turns 90 degrees right and on to a British Waterways car park near The Angler's Retreat pub. Take the steps that go down left through the car park and out to the road. Cross the road to The White Lion pub. Cross the bridge (the Lower Icknield Way Bridge) that goes over the GJC and turn left to go down to the right-hand bank of the canal. Within a short distance the canal reaches Marsworth Junction. Here on the left is the entrance to the Aylesbury arm.

This branch was opened in 1815, some ten years after the completion of the GJC. It descends by sixteen narrow locks for a total distance of 6¼ miles. The first lock seen from the GJC is, in fact, a staircase of two, i.e. it is two locks which share a centre gate. At one stage it was thought that the Aylesbury may be extended westwards to bypass the upper Thames. Two different schemes were proposed. The first, the Western Union Canal, ran from Cowley on the GJC to Maidenhead or Marlow and so to the Kennet & Avon. This disappeared under a weight of objections from the Thames Commissioners and various landowners. The second plan, the Western Junction Canal, would have taken the line via Thame to Abingdon on the Thames near the entrance to the Wilts & Berks Canal. This was authorized in 1793 but failed to interest the GJC which didn't believe that it would deliver any traffic. Only after the intervention of the Marquess of Buckingham was a line built at all and then only as far as Aylesbury. It never was extended to Abingdon. The walk from Marsworth to Aylesbury is described in a leaflet available from BW. The return trip is by bus (enquiries: 0296-23445).

Continue along the main line (a signpost here promises Braunston in 54¼ miles) and past a BW wharf and office on the left-hand bank. The towpath changes sides at the next bridge. It is then just a short distance to the following bridge where you leave the canal by turning right along the road. The Ship, on the canal just beyond the bridge, was once a pub but is now a small shop. The road goes into Marsworth village and on its way it passes The Red Lion pub (recommended in CAMRA's *Good Beer Guide*) and continues past a school and a small grocery store. At the T-junction, turn right and follow the road back to the bridge near The White Lion.

Cross over the canal and turn left to take up the towpath on the right bank. This leads to a lock. The double-arched bridge here seems an oddity but in fact dates back to the time when the locks along this stretch were all paired. Originally the locks were built wide to handle barges. However, they were commonly only worked by narrow boats and consequently used excessive amounts of water. To curb this waste, boats were forced to use the locks in pairs or pay extra, a move that was hardly popular with either the boat owners or their crews. In 1838–9 narrow locks were built alongside the broad locks and thus the amount of water wasted was reduced. Only later, as water supplies improved and engined boats began to tow a butty, were the narrow locks filled in.

After passing to the left of the third Tring reservoir, Marsworth, the canal twists and turns on its way up to Bulbourne Junction, picking up a series of locks as it goes. There are seven altogether between Startop's End Bridge and Bulbourne, a rise of 40 ft to the Tring summit. Again, as part of the measures introduced to conserve water, the Marsworth Locks were each built with side ponds where water from an emptying lock could be run into them and then back to partly fill the lock the next time it was filled. Along this stretch are also two fine lock-keeper's cottages. The towpath eventually returns to Marsworth Top Lock and the BW dry dock at Bulbourne Junction. Cross the Wendover arm by the bridge and walk on to the Bulbourne depot, then back to the starting point at or near Tring station.

Further Explorations

The towpath of the Grand Union Canal is due for wondrous things. British Waterways has plans to make it into a long-distance footpath between Birmingham and London (Gas Street to Paddington). This, the first National Waterway Walk, should open during 1993 to coincide with the 200th anniversary of the GJC Act. As part of the launch, an official guidebook will be published. However, you don't have to wait. Many riparian councils have published leaflets which describe towpath walks. Those interested should enquire of local authorities, tourist information offices, libraries or BW at Watford.

At the northern end of the GJC, there is a pleasant short stroll of 3 miles around the famous canal junction of Braunston (OS Landranger 152. OS ref: SP 537662). The village is just off the A45 between Daventry and

Rugby. Park in the High Street near the post office and start the walk by continuing along the road to a bus shelter. Bear right past a factory (National Starch & Chemical Ltd) and then turn right along a lane to reach The Admiral Nelson. Keep to the left of the pub and walk on to Anchor House. Bear right to reach the canal at Braunston Top Lock. Cross the small accommodation bridge and turn left past the lock cottage and on to reach the western portal of Braunston Tunnel.

They had quite a lot of problems building Braunston Tunnel. Quicksands led to subsidence and then they discovered that the line through the hill wasn't built straight so that there's a distinct kink in the middle. It was opened in 1796 and is 2,042 yd long. In the early days the tunnel was worked by 'leggers', men who lay on the deck of the boat and pushed it through by walking against the roof or the side of the tunnel. The horses, meanwhile, went over the hill via a country path. In 1870 the company tested a system in which moving wire ropes were used to pull boats through. The idea wasn't terribly successful and was replaced a year later by a steam tug.

Return to Top Lock and walk along the canal to pass three more locks and then The Admiral Nelson, built in 1730 and thus here long before the GJC. Two locks further on, you reach Bottom Lock where The Boat Shop sells provisions and other canal ware. Just beyond the shop is the old pump house. This formerly contained a steam engine which moved water back up the Braunston flight into the summit pound. A little further on, Braunston Marina to the left is situated on what remains of the old reservoirs.

Continue over the Horselcy Ironworks Footbridge and on past Butchers Bridge to reach a second Horseley Bridge. This second channel marks the former course of the Oxford Canal. Before the construction of the GJC, the Oxford Canal came through Braunston to reach this point, before turning left on its meandering way to Banbury. When the GJC was built the two formed a junction here. Those coming from London could turn left for Napton (where there is a further junction for Birmingham via the Warwick Canals) or go straight on for Hawkesbury (where the Oxford joined the Coventry Canal). The building ahead of you along the canal is the Stop House where tolls were collected and craft were registered before crossing the junction. When the Oxford was improved in the 1830s, the junction was moved and a highly convoluted length of canal bypassed. This reduced a 2³/4 mile long stretch to just ³/4 mile. You will see the new junction shortly. Most of the old waterway has disappeared into the fields but this short starting section now forms part of the boat-yard.

To see the new junction, continue past the Stop House, now a BW information centre, and then under the A45 road bridge. The new junction really is magnificent with not one but two Horseley iron bridges crossing a triangular canal entrance. This is the northernmost end of the GJC and the canal to the left and ahead is the Oxford. To the left is the route to Napton and ahead the line to Hawkesbury. The start of the southern Oxford is marked by another fine bridge, this time a brick one. This is a turnover bridge. Here the structure has its ramps built facing the same direction so that the barge horse can change canal sides without needing to disconnect the towrope.

To complete the walk, pass under the A45 for a second time before reaching the next bridge. This you have to climb up to in order to cross the canal. Continue over a stile and into a field. Walk up the hill, keeping the hedge close left to reach another stile opposite the church. Walk on to a road junction and turn left to return to the post office.

Purists will dislike the canal at Stoke Bruerne (OS Landranger 152. OS ref: SP 743499). The village must have more double-yellow lines than the rest of Northamptonshire put together and the canal has The Canal Museum, a shop, a café, a pub and swarms of people. Despite that, the museum is worth a visit and the canal itself has plenty of interest. A quick stroll north along the Tarmaced towpath takes you to Blisworth Tunnel, the

The canal at Stoke Bruerne

sixth-longest canal tunnel in England at 3,056 yd. The current tunnel was in fact the second attempt. The first, about 130 yd to the right of here, was abandoned before it was finished following a series of collapses. All in all it took twelve years to find a way through the hill (it was opened in 1805) and cost the company £90,000. As at Braunston, the tunnel was originally worked by leggers. Over the years the tunnel has continued to have problems with subsidence and, in the early 1980s, it was closed for four years for repair. Four million pounds was spent on relining the structure. A small display of the concrete blocks used is laid out near the tunnel entrance. The buildings near the portal were once a small stable and a maintenance workshop. Back at the museum, the Stoke Bruerne Locks still bear the marks of the time when the GJC attempted to speed passage by 'doubling up'. The old lock is presently home to an old boat-weighing machine. This peculiar mechanism was used on the Glamorganshire Canal when toll collectors wished to check the loading of a narrow boat as it passed a toll point. If you continue under the bridge, it is possible to stroll on to see the five remaining Stoke Locks and the pump house at the foot of the flight, which was used to return water back into the top pound.

Further Information

There is no Grand Junction or Grand Union Canal society as such but there is a Grand Junction Region within the Inland Waterways Association. It has branch secretaries at High Wycombe, Welwyn Garden City, Northampton and Milton Keynes. The address and telephone number of the IWA can be found in Appendix B.

The fullest historical account of the GJC is contained in:
Faulkner, A.H., *The Grand Junction Canal*. David & Charles, 1972.

7
THE KENNET & AVON CANAL
Bradford-on-Avon to Bath

Introduction

It is hard to believe that just a decade ago the Kennet & Avon was a derelict mess. Today the waterway is a superb amenity and a splendid sight. It is living proof that restoration of these so-called lost canals is not only possible but thoroughly worthwhile. The K&A is rich in wildlife, fine architecture and great scenery. Coupled with the fact that it's readily accessible and makes for good walking along the entire 86½ miles, it rivals any in the country.

The eastern end of the K&A starts at Reading, where it joins the River Thames. From this junction the former Kennet Navigation passes through the county of Berkshire to Newbury where the K&A 'cut' begins. After Hungerford the canal enters the rural charm of Wiltshire. Here are some of the line's greatest technical achievements: the oldest beam engines in operation at Crofton; the Bruce Tunnel at Savernake; and the magnificent flight of Caen Hill Locks at Devizes. From the headquarters and museum of the Canal Trust at Devizes, the path continues to Bradford-on-Avon. The way between Bradford and Bath passes through some great scenery and presents some more of the canal's wonderful history: two fine aqueducts and the fascinating water-powered Claverton pumping station. At Bath the canal joins the old Avon Navigation to pass through Saltford and Keynsham to reach the tidal Avon at Hanham Mills. From there the River Avon runs through the centre of Bristol and out to the sea.

This is as fine a route as you'll find in any walking book (let alone this one). It has regular public transport, loads of things to see, convenient and pleasant pubs and two fine towns at either end. It should be enough to tempt the staunchest of armchair towpathers out into the fresh air.

Claverton pumping station

History

By the time canal mania swept the nation, the idea of building a waterway to link London and Bristol was already nearly 200 years old. In 1626 an Oxford don, Henry Briggs, carried out a survey and suggested that if the River Avon was made navigable to Malmesbury and the Thames to Cricklade, then a short length of canal could link the two. How seriously the idea was considered is hard to assess but after the Civil War no fewer than five separate bills were introduced to Parliament, none successfully.

With the more visionary scheme dropped, it was left to two local developments to get the ball rolling. In 1699 Bath Corporation revived a plan of 1606 to make the Avon navigable from Bristol. Although there were objections, including claims that it would 'prejudice the Health of the Bath', the requisite powers were obtained on 22 May 1712. However, with further protest, the idea was shelved. Meanwhile, in 1708 a bill to make the River Kennet navigable from the Thames at Reading to Newbury was brought before Parliament. Supported by the towns of west Berkshire and Wiltshire, and fought vigorously by the people of Reading who feared loss of trade, the Act received Royal Assent on 21 September 1715. John Hore of Newbury engineered the line which followed the river via a series of artificial cuts. The 18$1/2$ mile long Kennet Navigation was opened in 1723 at a cost of £84,000.

By this time Bath was a major spa town and many of its fine buildings were under construction. The advantage of being able to import building materials along a new waterway became ever more apparent and objections to the 1699 scheme subsided. In 1724 the corporation assigned its interest in the river to the 'Proprietors of the Navigation between Bath and Hanham Mills' and employed John Hore as engineer. The work proceeded well and on 15 December 1727 the first barge reached Bath loaded with deal, pig lead and meal. The new line had cost £12,000. The Avon soon became an important waterway. Bath stone was exported around the country and coal came from Shropshire. There were also boats which for a shilling took passengers between Bristol and Bath in just four hours. At a meeting in Hungerford in March 1788, a line was discussed for a project then called the Western Canal which would join the Kennet and Avon Navigations. A second meeting, on 16 April appointed three engineers (Messrs Barns, Simcock and Weston) to carry out a survey. In the summer of 1789 they reported on a route that went from Newbury to Hungerford, Ramsbury, Marlborough, Calne, Chippenham, Lacock, Melksham, Bradford-on-Avon and Bath. After some discussion concerning the practicality of this line, John Rennie was asked to resurvey the route and advise. On 3 November 1790 he reported that the route was fine and the meeting resolved unanimously to proceed.

Raising funds proved harder than expected but with the arrival of canal mania prospects improved. In January 1793 Rennie was asked to undertake a detailed survey. He reported in July that a more southerly line through Hungerford, Great Bedwyn, Devizes, Trowbridge and Bradford was preferred as the water supply to the original route was suspect. The revised plan was approved at a meeting in Marlborough on 27 August 1793 with the cost estimated as £377,364 (including a branch to Marlborough that was never built). With this plan the committee, under the new name of the Kennet & Avon Canal, presented its bill to Parliament and received Royal

Boats at Mytchett on the day of the re-opening of the Basingstoke Canal, 10 May 1991

The eastern portal of Sapperton Tunnel near Coates on the Thames & Severn Canal

Cleveland House and Sydney Gardens on the Kennet and Avon Canal

British Waterways' Bulbourne depot on the Grand Junction Canal

Assent on 17 April 1794. The Act, for a 55 mile long canal, authorized the raising of £420,000 by 3,500 shares and the powers to raise a further £150,000 by mortgage should it be needed. Charles Dundas was appointed chairman and Rennie engineer. One of Rennie's first decisions was to build a wide (50 ton barge) canal rather than a narrow boat line. Some small alterations to the route were also made: Trowbridge was bypassed and the line at Crofton altered to reduce the length of the Savernake Tunnel. By October 1794 work had started at both Bradford-on-Avon and Newbury.

Generally the building of the K&A was harder than expected for business, geological and constructional reasons. The inexperience of many of the contractors, coupled with the largely unknown geology, inevitably meant that the work was more costly and more time-consuming than originally estimated. Furthermore, the Napoleonic Wars were having an adverse affect on the economy as a whole, causing inflation and reducing confidence. The company had problems raising funds and shareholders got behind with their payments. Despite this, in 1796 the K&A Company bought up virtually all of the Avon Navigation shares (one remained in private hands until 1885). This had the benefit of providing a cash flow.

By April 1797 the financial situation was such that the committee, following pressure from its bankers, instructed Rennie to reduce the works. Its position wasn't helped by the fact that the company treasurer, Francis Page, had secreted over £10,000 out of the accounts. To pay back the sum, Page and his brother offered to sell the Kennet Navigation to the K&A but were turned down, presumably because the company was short of ready cash. By 1799, with the canal open at the eastern end to Hungerford but with the western end inaccessible, the committee was forced to virtually halt the work. By December 1800 things looked grim. The line was still not ready and there was little hope of raising any funds to complete it. Only with the release of 4,000 new shares, after the passage of an Act in May 1801, was money forthcoming. This allowed the work to continue, if slowly, so that by 1803, with a grocer, John Thomas, acting as the resident engineer, the canal was open from Bath to Foxhanger, near Devizes (although there was no connection yet with the Avon) and from Great Bedwyn to Newbury. By 1804 Rennie had to report that although he had already spent over half a million pounds, it would still need an extra £141,724. A new Act of 1805 permitted a further injection of funds which were used to complete the central section and to start the Crofton pumping station and Wilton reservoir, both essential if the summit level was to be kept in water.

By 1809 virtually the entire line was ready, with only the flights of locks up Caen Hill to Devizes and down from Bath to the Avon still to be finished. With yet another Act passed on 3 June 1809 to raise a further £80,000, the final works were completed. On 28 December 1810 a bargeload of stone ascended the Caen Hill Locks and, without any ceremony or celebration, the line was open. It had, with the purchase of the Kennet Navigation in 1812, cost a total of £979,314.

Over the next thirty years the K&A was a success. It regularly returned profits and dividends and was constantly busy. The cost of shipping goods from London to Bristol was reduced by half, with a journey time for a horse-drawn boat of ten days (the 55 miles from Newbury to Bath took 3¹/₂ days). Later, with the use of high-speed fly-boats, the London–Bristol time was reduced to just five days. Over half the business at this time was coal coming from the Somerset Coal Canal, which joined the K&A near the Dundas Aqueduct at Monkton Combe. A large proportion of this went north along the Wilts & Berks Canal which left the K&A at Semington for Swindon and the northern Thames. Plans for similar linking canals were manifold and included a line to the Basingstoke Canal, more direct routes to Bristol and London, proposals for a new line to Salisbury and the Dorset & Somerset Canal which would have run from Bradford to Sturminster Newton. None of these projects were completed, the initial enthusiasm dying as the reality of canal-ownership dawned in a post-canal mania world.

The other reality that was to dawn was that of the railway. A line from London to Bristol was proposed as early as 1824. Luckily for the K&A, its arrival was delayed until 30 June 1841, when the Great Western Railway

Dundas Aqueduct at Combe Hay

linked the two cities. When the line finally opened, virtually all the through traffic was lost to the railway and the K&A's annual receipts fell from £51,173 to £39,936. To compete the company cut wages and made economies and was thereby able to reduce tolls: for instance, by 25 per cent in September 1841. It made no difference. Indeed, more railways were built, including, in 1848, a line that ran alongside the canal from Reading to Pewsey. Further toll reductions were made in order to sustain business. The lack of success of this move meant that by 1851 the company was unable to pay a dividend. In 1852 the end came. The GWR Act no. 1 allowed the railway company to take over the canal in its entirety. The interest of the GWR had started in 1845 when the canal committee had sponsored a review as to whether the canal could be converted into a railway. Immediately the GWR, fearing competition, made enquiries and was rebuffed. The K&A played hard to get for a while, even introducing a bill to Parliament for the London, Newbury & Bath Direct Railway in 1846. After the rejection of the bill, the company agreed to the sale on 18 March 1851 and this was enabled by an Act of 30 June 1852.

Despite the various legal constraints on the GWR forcing it to keep the navigation open, it paid no real attention to the canal, which gradually fell into disrepair. Trade declined from 360,610 tons in 1848 to 210,567 tons in 1868. The railway company's attitude to its canal subsidiary is amply reflected in the decision in 1864 to move the administration of the waterway to offices in Paddington, 34 miles from Reading. By 1877 the canal made a loss of £1,920 and profits were never seen again. Many of the canal-carrying companies filed claims against the GWR with the Board of Trade. C. Evans & Co., for example, complained about the lack of dredging along the line which rendered some parts difficult, if not impossible, to navigate. Despite the complaints and despite a Board of Trade report which confirmed that there was a problem, the GWR remained unconvinced by canal transport and allowed the line to continue its steady decline.

By 1905 annual traffic was down to 63,979 tons, a decline stimulated by the fact that tolls on the K&A were 50 per cent higher than on any comparable waterway in the country. By this time the canal was beginning to suffer from chronic water shortage. The closure of the Somerset Coal Canal and the Wilts & Berks meant that two important sources of water had been shut off. Improved land drainage also meant that there was limited surface water run-off. As a result many stretches, such as the 9 mile pound near Bath, were often difficult to navigate. The GWR's official view in 1905 was that railways were much more efficient and that there was little value in improving the canals. Following the First World War, increased levels of road traffic gradually took the rest of the business. In October 1926 the GWR announced that it proposed to close the entire line from Bath to Reading. This it would do by shutting all the locks (often by converting them into weirs) yet maintaining water levels. It was further suggested that local authorities should take control of that portion of the line that ran through their boroughs. The level of objections was such that the plan was withdrawn.

After further managerial re-organization in 1932, traffic declined to very low levels, with only the occasional pleasure launch ploughing through the weed. By the time that the railways were nationalized in 1948 the canal was only barely passable. From its poor post-war state, it fell into complete dereliction: locks fell apart, whole sections dried out. Eventually, in 1954 the BTC planned to close the canal (other than the Avon section). The response, at a time when the canal preservation movement was beginning to gain ground, was significant. The K&A branch of the Inland Waterways Association was one of the first to be formed and it soon led to the setting up of the K&A Canal Association to lead the fight against the planned closure. In 1955 the BTC Board of Survey put the K&A into category three of its report, namely 'Waterways having insufficient commercial prospects to justify their retention'. In the subsequent Act, the BTC was relieved of responsibility for maintaining the canal. The situation was rescued by the Bowes Committee in 1958. Although it saw no justification in raising the status of the K&A, it recommended that the line be considered for redevelopment. In 1962 the Inland Waterways Redevelopment Advisory Committee proposed that the canal be restored, albeit with little government cash, and that the canal association be allowed to play a role. With this the association became the K&A Canal Trust Ltd and the period of fund raising and restoration began.

It is the trust that has been the driving force in the revival of the K&A's fortunes over the last thirty years. From modest beginnings it has worked hand in hand with British Waterways and others to gradually bring life back to the canal. It is one of the great restoration achievements, which was officially recognized on 8 August 1990 when the Queen re-opened the canal. Once again it was possible to navigate from London to Bristol by way of the K&A canal.

The Walk

Start:	Bradford-on-Avon railway station (OS ref: ST 824606)
Finish:	Bath railway station (OS ref: ST 753643)
Distance:	9¹/₂ miles/15¹/₂ km
Maps:	OS Landranger 173 (Swindon & Devizes) and 172 (Bristol & Bath)

Return: Badgerline buses 264/265 from Bath stop at the Avon Bridge in Bradford and run regularly (enquiries: 0225-464446)

Car park: Well signposted at the railway station and in town

Public transport: Walk starts and ends near BR stations

From Bradford-on-Avon town centre, go along Frome Road towards Frome for 1/4 mile, past the Canal Tavern to reach a bridge that goes over the canal. On the left is Bradford Upper Wharf and the relatively deep (10 ft 3 in) Bradford Lock (no. 14). The first cuts of the K&A (as opposed to the river navigations) were made here in October 1794. Upper Wharf has a slipway and a dry dock which was formerly used as a gauging station where the carrying capacity of boats was assessed and tolls calculated. The stone wharfinger's house has been converted into a trust shop and the building also houses toilets. If visiting on a summer weekend, it may be possible to take a trip on the trust's narrow boat *Ladywood*.

To start the walk along the 'nine mile pound' (there being no locks between here and Bath) it is necessary to return to the Canal Tavern. Take

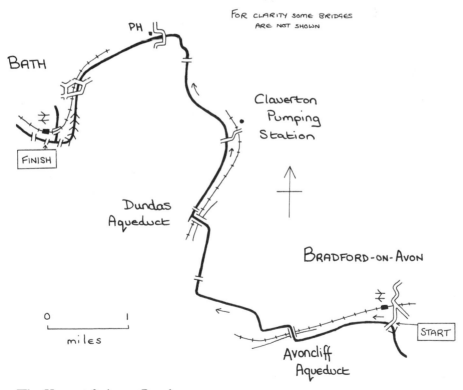

The Kennet & Avon Canal

the path that goes down the side of the pub to reach Bradford Lower Wharf and the canal. The towpath is on the right-hand bank. Within a short distance the canal bends left and passes a park, at the end of which is a fine fourteenth-century tithe barn. The course then curves away from the railway and the River Avon and passes Bradford-on-Avon Swing Bridge. The wooded hill to the left (Becky Addy Wood) was once the site of a quarry from which stone was brought down to the canal by means of a tramway. Within 1/2 mile the canal reaches Avoncliff where, after some buildings, the canal turns sharp right to Avoncliff Aqueduct. This is the first of two fine stone aqueducts along this section of the waterway. Built of local limestone, it was started in March 1796 and finished two years later. The poor quality of the stone caused the structure to crumble almost immediately and it is now much patched with brick and, in the 1980s, was substantially reinforced with concrete. This means that the sag in the centre of the span isn't as serious as it looks. The railway arrived over half a century later and led to the addition of the slightly less flamboyant extension on the northern side. In cold weather stalactite icicles reputedly hung from the ceiling of the aqueduct, making a significant hazard to passing trains. Railwaymen were employed to break them off every morning.

Before continuing to Bath, the towpath changes sides. Turn right to go down the slope towards the Canal Bookshop and the Cross Guns pub. After

Avoncliff Aqueduct in the 1950s. It was later restored

British Waterways

about 5–10 yd turn left to go under the aqueduct and then up the steps on the left. Here is Teazel's Coffee House and a towpath sign pointing to Dundas. Cross the aqueduct and turn left. The canal is now on the side of Winsley Hill and you can look down through the trees to the River Avon. This section is grounded on a type of oolite limestone which is highly liable to subsidence, making the canal bed prone to 'blowouts'. The problem was so great that the whole length was dry from 1954 until the early 1980s. The stretch was restored by firstly lining the bed with polythene and then covering it with concrete. The frequency of the landslips prompted the company to install a series of emergency stop gates (eighteen in 7¹/₂ miles) so that any given length could be isolated and repaired as quickly as possible. Over the years most of the gates have been replaced with stop planks (you'll see them in stacks all the way along the route) but stop gates are still in position on either side of the Dundas Aqueduct.

The canal straightens with some scrubby hedging to the left. Amid the undergrowth, down the slope towards the Avon, is Avoncliff Stone Wharf. These overgrown ruins are all that remain of the wharf for the Becky Addy Wood stone. In former times a railed way ran from the quarry across the aqueduct to end here, where the stone was transhipped onto barges.

Shortly, the canal turns right to Winsley Bridge. During the First World War convalescing soldiers stayed at the nearby Winsley Hospital. As part of the recuperation, they were given boat rides to Bradford and back using the Red Cross' narrow boat *Bittern*. After a gentle bend left, the canal passes through Murhill. Here, in the early 1800s, a tramway connected the wharf to Winsley quarries on the hill above. The path continues along a straight stretch for about ¹/₂ mile through woodland to Limpley Stoke Bridge. The view to the left through the trees goes across the valley to the River Avon, the Viaduct Hotel and the line of narrow boats that are moored in the Bath and Dundas Canal Company marina. After a bungalow and a derelict building, a lengthsman's cottage and barge horse stable respectively, the canal passes through a stop gate and turns sharp left. On the right bank is the former Conkwell Quarry Wharf but the attention is grabbed by the Dundas Aqueduct which takes the canal from the east side of the valley to the west before it swings west into Bath.

Charles Dundas was the first chairman of the canal company and memorial plaques to both him and John Thomas, a superintendent of works, were added to the aqueduct in 1838. Building work started here in August 1796 and, as at Avoncliff, stone from the company's own quarries was used. John Rennie preferred brick but was overruled by his committee. In the event Rennie was proved right. The Conkwell Quarry stone may have been handy but it was not of good quality and soon started to crumble. As a result, some rebuilding was needed virtually straight away and the aqueduct took nine years to finish. Despite this, it's a beautiful structure. When built, it had three arches and spanned 150 ft. A further arch was added at the western end when the railway opened. As at Avoncliff, extensive renovation was needed when the canal was restored and the rejuvenated aqueduct was re-opened in July 1984.

Cross the aqueduct to a footbridge. The narrow waterway on the left is the mooring site for the Bath and Dundas Canal Co. It was once the terminus of the Somerset Coal Canal. The SCC opened in 1805 to carry coal from the thirty or so Somerset collieries. The line ran from here to Midford where it split, one branch continuing to Timsbury via Combe Hay and Dunkerton, the other to Wellow. At Combe Hay the SCC carried out a series of trials with a boat lift which was capable of lifting a narrow boat 70 ft in just two minutes. The system relied on keeping a tank watertight, a challenge that proved beyond the technology available. It was eventually replaced with a flight of twenty-two locks. The SCC went into liquidation in 1893 and was converted into a railway. Ealing comedy fans will know the route as home to *The Titfield Thunderbolt*. Sadly, it closed in 1951.

All that remains of Dundas Wharf is a crane and two stone buildings, the left-hand one a tollhouse and the other a small warehouse. Go past the wharf crane and left of the old stone warehouse to Dundas Bridge, where there is another pair of stop gates. Cross the bridge and turn left to continue along the path on the right-hand bank. The path continues past Millbrook Swing Bridge and on to Claverton Road Bridge. Here you take a short diversion. Bear right down Ferry Lane to a railway crossing and then a small bridge. On the left is a millpond and the Claverton pumping station. This is a true wonder. The site was formerly home to a grist mill and was leased by the K&A when it was realized that water supplies to this end of the canal were insufficient to guarantee continuous navigation. A new building was put on the site and the pump, designed by John Rennie, was finished in 1813. It consists of a large waterwheel 18 ft in diameter and 25 ft wide, which is driven by the River Avon. The wheel powers two beams and pumps and lifts water 47 ft up to the canal at a rate of 100,000 gallons per hour. When the K&A was still a commercial concern, the pump worked twenty-four hours a day. Subsequently it was worked less and less, and in 1952 a log entered the works causing a jam which sheared off some of the wooden teeth from the pit wheel, so that it stopped working altogether. This splendid water-powered pump was then replaced by diesel power and left to gently rot. However, the K&A Canal Trust have, with the help of Bath University, restored the pump into full working order. If you're here on a summer weekend you may be able to see the pump operating (check with the Canal Trust for times). During the rest of the week, water is pumped using electricity.

Return to Claverton Road Bridge and turn right along the canal. Within a few yards the delivery leat from the Claverton pump pours water into the canal. The towpath continues under Harding's Bridge and on to Hampton

Wharf and Bathampton Swing Bridge. The canal does a double bend and passes Canal Farmhouse. The towpath now goes through a gate to reach Bathampton Primary School before going under Bathampton Bridge and out to a small green near the George Inn. The front door of the George originally opened where the canal now stands and was moved to its present position when the canal was built. The building itself dates from the fourteenth century and was originally part of Bathampton Priory.

The stretch from Bathampton to Darlington Wharf is a straight 1¼ miles and has two bridges, Candy's Bridge and Folly's Footbridge (with stop gates). At the end, the canal takes a sudden turn left and a wall borders the towpath side. Just over the wall is the railway and, beyond that, some abandoned housing. In 1839, the Great Western Railway obtained an Act to build its new line along this narrow ridge and to divert the canal south. If you look over the wall, you can see where the canal once went and how the railway has stolen the route. During the diversion the canal was closed for 383 hours, for which the GWR paid the K&A £7,660 in compensation.

The canal now passes Darlington Wharf. In the 1830s passengers were able to take the 'Scotch' boat, a wrought-iron vessel brought from Scotland, from here to Bradford-on-Avon. There were two trips a day with forty passengers. It took an hour and a half and those on board were serenaded by a 'string band'. A little further on, the canal passes into the short (85 yd) Sydney Gardens Tunnel which goes under Beckford Road (A36) and into Sydney Gardens. When the canal was built, Sydney Gardens was privately owned and the proprietors demanded both 2,000 guineas and some landscaping work for the right to take the canal through their property. The canal was therefore hidden at both ends in tunnels and the section in between has two ornate wrought-iron bridges from George Stothert's foundry in Coalbrookdale. The second tunnel is of particular interest because it goes underneath Cleveland House which became the headquarters of the K&A Canal Company. As you go through, keep an eye open for a small shaft in the ceiling which goes up into the building. It's said that messages were put through from the office to passing boats via the shaft.

At the end of the tunnel, climb the steps on the right to cross the canal behind Cleveland House. The path turns right and continues down a slope to reach the left bank. Within a few hundred yards the canal passes Sydney Wharf, where the Somerset coal was unloaded. Go up a cobblestone path to the side of Sydney Wharf Road Bridge and then to Bathwick Road. Cross to the right-hand side of the bridge and go down some steps. Turn right to continue along the towpath to Bath Top Lock with its accompanying Stothert wrought-iron footbridge. This is the first lock since Bradford and the downstream end of the 'nine mile pound'. From here the canal makes a steep descent to the River Avon which is under ½ mile away.

The canal now descends quickly to Second Lock, after which it broadens to form a reservoir pound. The path curves round right past a solitary chimney on the other side of a hedge, to reach Abbey View Lock and Horseshoe Walk Bridge. The loss of water into the Avon was always a concern to the company and two pumping stations were built to take water back up the line. One pumping station was built at the bottom of the Widcombe flight and the other was just here. The lower engine pumped water into the pound above Abbey View Lock and the upper engine pushed it above Top Lock. Trouble arose when the company pumped water from the Bottom Lock chamber with the gate open, effectively removing water from the River Avon. The millowners didn't approve and took legal action to have it stopped. As a result, in 1855 the pumps were removed. All that remains of the upper station is the ornamental chimney that can be seen on the city side of the canal near Abbey View Lock.

Cross Horseshoe Walk and continue along the path. Wash House Lock is followed by another wrought-iron footbridge. Just before the church, the canal bends right to Bath Deep Lock, an amalgamation of two old locks to form the deepest lock in the entire British canal system at 18 ft 8 in (5.69 m). To save water, a pump moves water back upstream after use. To avoid crossing the busy Pulteney Road Bridge, cross the canal at the downstream lock gate. Steps go down to a concrete path which goes under the road. The path continues under a second road bridge (Baptist Chapel Bridge) and out to Thimble Mill Basin, a 'lagoon' in front of the Bath Hotel. The path soon

A group of children pose near
Thimble Mill, Bath, in 1907
National Motor Museum

passes the final lock, Bottom or Widcombe Lock, which sits next to Thimble Mill, the lower of the two pump houses mentioned earlier. Go over Dolmeads Bridge to reach the western end of the 1810 K&A Canal and the River Avon, which appears massive in comparison.

From here both railway and bus stations can be reached by taking the higher of the two paths alongside the Avon. After 100 yd cross the green footbridge, Widcombe or Ha'penny Footbridge. To reach the station, walk on through a short tunnel. Turn left at the end for the BR station. The bus station is about 75 yd further on. If you prefer a prettier route, turn right at Dolmeads Bridge. The path soon reaches the famous Pulteney Bridge. Take the steps up to the road and turn left for the city centre.

Further Explorations

It's 92½ miles from the Thames at Reading to the centre of Bristol and, apart from the odd short stretch which can be readily got round, the whole length makes good, easy walking. It can be done either in one go as a week-long hike or as a series of day walks using public transport to get you back to the place where you started. The author's *Kennet & Avon Walk* (Cicerone Press) gives details of the route and the public transport available. The days can be organized as follows: Reading to Woolhampton; Woolhampton to Hungerford; Hungerford to Pewsey; Pewsey to Devizes; Devizes to Bradford-on-Avon; Bradford-on-Avon to Bath; and Bath to Bristol. It's a fine walk through great countryside and is recommended.

Those who believe a good walk is from their seat to the buffet car and back can also enjoy the eastern 40 miles of the K&A. BR's London to Westbury line runs parallel with the canal and affords views of such notable sights as the junction with the Thames, the Crofton pumping station, the western portal of the Savernake Tunnel, the locks at Monkey Marsh and Hungerford Marsh, and the Dun Aqueduct. True armchair canal-watching!

For those with more active feet, there are two walks which are clear 'musts'. The first is a walk of just over 3 miles up and down the Caen Hill flight of locks at Devizes. Towpathers should start at the Canal Trust Centre at Devizes Wharf (OS Landranger 173. OS ref: SU 004618). The wharf car park is well signposted from the through roads. The Trust Centre has a shop and museum, and there is a café nearby. From the centre, turn left and left again to reach Couch Lane and cross the canal at Cemetery Bridge. Turn left along the right-hand bank. This leads round to a lock and across Northgate Street. Pass three locks and the Black Horse pub before going through a small tunnel under Prison Bridge. Two locks further on is the Sir Hugh Stockwell Lock, otherwise known as Caen Hill Top Lock.

You are now at the summit of Caen Hill and have a fine view towards Melksham and Trowbridge. The hill is scaled by sixteen locks (Devizes has twenty-nine altogether in just 2¹/₄ miles) which lift the canal 140 ft. Each lock is provided with its own reservoir, or side-pound, to provide additional water to prevent the draining of the system when in use. Each side-pound is about 6 ft deep and measures 70 yd by 45 yd. By walking down the towpath alongside the locks it is possible to realize the scale of the construction as well as the effort required by boatmen to pass up the flight. The best record-ed time by a commercial boat was 2¹/₂ hours, although in a recent record attempt a crew scaled them in 2 hours 6 minutes. The locks were just about the last part of the K&A to be fully opened, both when the line was first built and when it was restored (in 1810 and 1990 respectively). During the course of the original construction, crews transhipped their goods onto a horse-drawn railway which ran beside the towpath.

Having reached the Lower Foxhangers Bridge at the bottom of the flight, it is possible to return by crossing at the bottom lock and walking on the opposite side of the side-pounds. You can't do this all the way and you will eventually come out at Queen's Lock, where there is a plinth commemorat-ing the re-opening of the waterway in August 1990. From here it is a simple matter of retracing your steps back to Devizes Wharf.

Crofton is close to Great Bedwyn (OS Landranger 174. OS ref: SU 264624). You can park close to the well-signposted pumping station. Contained within a dull exterior are two beam engines built to pump water from Wilton Water reservoir to the K&A summit pound. The Boulton & Watt engine was installed in 1812 and is the oldest working beam engine in the world that is still doing its original job in its original position. The other engine was built in 1846 by Harvey's of Hayle in Cornwall. After falling into disrepair, the Canal Trust has restored both. The Boulton & Watt was resteamed for the first time, in the presence of Sir John Betjeman, in April 1970. On summer weekends one or other is in steam and open to the public. It's worth a visit, though check first for opening times with the Canal Trust at Devizes. Each engine can pump 11–12 tons of water per minute into a narrow leat which disappears off from the back of the building.

For a pleasant walk of about 3¹/₂ miles, leave the pumping station to go down some steps that pass under the railway and out to a lock (Crofton Bottom Lock). Cross the canal here and turn right. Over to the left is Wilton Water, the reservoir used by the pumping station. The walk gently bends right and passes a total of nine locks in 1¹/₂ miles. At the top lock you will be able to see the incoming water from the pumping station leat. The canal enters a deep cutting which eventually reaches Bruce or Savernake Tunnel. At 502 yd, the tunnel is the only major one on the K&A. There's no tow-path and boats were pulled through by means of chains that hung from the walls. The horses, meanwhile, took a pleasant stroll up and over the hill.

For those with an OS map, it's possible to devise an alternative route back to Crofton. For those not in the mood for complex navigation, it's easier to turn round and gently amble back downhill to the pumping station.

Further Information

The canal you see today only exists because of the efforts of the K&A Canal Trust. It is well worth supporting.

Kennet & Avon Canal Trust Ltd,

Couch Lane,

Devizes,

Wiltshire SN10 1EB.

Tel: 0380-721279

This is also the number to call if you wish to know the opening times of the pumping stations.

The history of the K&A is related in:

Clew, K.R., *The Kennet & Avon Canal*, 3rd edition. David & Charles, 1985. I can heartily also recommend:

Quinlan, Ray, *The Kennet & Avon Walk. A walker's guide from London to Bristol*. Cicerone Press, 1991.

GEOProjects publishes a useful map of the entire canal. It is available locally or from: GEOProjects (UK) Ltd, Henley-on-Thames.

8
THE OXFORD CANAL
Tackley to Oxford

Introduction

The Oxford Canal is one of the oldest in the country. Surveyed by the doyen of canal engineers, James Brindley, the line has all the hallmarks of an early waterway as it meanders around the contours of the southern Midlands. Here is a canal that looks and feels much more like a river than a man-made waterway. It's rich in wildlife and justifiably popular. Perhaps its only drawback is the number of bodies that Inspector Morse seems to find in it!

The Oxford Canal starts at a junction with the Coventry Canal at Hawkesbury, 4 miles north of Coventry. From there the canal runs east via Ansty and Brinklow to the north of Rugby, before turning south to Braunston where the line meets the Grand Junction Canal. The main line of the Oxford now turns south-west to Napton where a second major junction, this time with the old Warwick & Napton Canal, offers a route to Birmingham and the West Midlands. The southern part of the Oxford then follows a convoluted line round Wormleighton, Claydon and Cropredy before reaching Banbury. The canal continues south by taking a more southerly line to Aynho, where the River Cherwell crosses the course of the canal in a kind of waterway crossroads. The line meanders past Upper and Lower Heyford and Enslow, where the River Cherwell and the canal merge for 1/2 mile, only to separate at Shipton Weir Lock. The canal continues its journey through Thrupp and the outskirts of Kidlington before reaching Wolvercote, where the Duke's Cut takes a course west to the upper Thames. The main line then enters Oxford, where boaters can continue on to the Thames via Isis Lock.

Oxford, of course, is worth visiting even in the absence of a canal and the navigation is certainly not the reason why most people go there. But it offers a fascinating alternative to the almost over-visited Thames and the northern stretches of this walk make for as fine a stroll as you can find anywhere.

History

On 29 January 1768 an Act enabling the construction of the Coventry Canal from that city to the Trent & Mersey Canal at Fradley was passed by Parliament. This first step towards forming a navigable waterway between the Midlands and London having been made, the second, to take that line from Coventry to the Thames at Oxford, was sure to follow. Sir Roger Newdigate, MP for Oxford University, initiated that second step. At a meeting in Banbury on 13 April 1768, Newdigate, the Mayor of Banbury and others decided to ask James Brindley to survey a line from Coventry to Oxford. At a meeting in Banbury on 25 October, the scheme was received with great enthusiasm and funds in excess of £50,000 were promised. The bill proposed to Parliament on 29 November was widely supported. It was said that the canal would half the price of coal in Oxford and that the improved availability of roadstone would do wonders for turnpike maintenance. The main opposers were those associated with the Newcastle coal trade and a clause was inserted into the bill which prohibited movement of Warwickshire coal from the canal onto the southern Thames. With this amendment, the Act received Royal Assent on 21 April 1769.

Skinner's boats unloading at the old coal wharf in Oxford Canal Basin

British Waterways

The Act authorized a canal from Coventry to Oxford via Braunston, Napton and Banbury. Powers were obtained to raise £150,000 in £100 shares, with permission to obtain a further £50,000 should it be needed. Most of the money came from Oxford, although James Brindley himself is reported to have invested some £2,000. A maximum toll rate was set at ¹/₂d. per ton per mile. James Brindley was asked to make a more detailed survey with Samuel Simcock as his assistant. Brindley's plan (for what was later decided to be a canal of 7 ft width) was agreed at a meeting in the Three Tuns on 3 August 1769 and the first sod was cut at the junction with the Coventry Canal at Longford. This rather peculiar junction, which required the Coventry and the Oxford to run parallel for about a mile, was the result of some not atypical non-cooperation between the two companies. The more obvious junction at Hawkesbury was rejected by the Oxford when the Coventry demanded compensation for the loss of 2 miles worth of toll earnings if the proposed junction was moved. The wrangling over the siting of the junction meant that one was not actually forged until 1777. The present, more sensible and direct, junction at Hawkesbury was made in 1802.

By March 1771 the first 10 miles were built and the company was receiving toll revenue. When Brindley died in September 1772, Simcock took over as chief engineer and it was he who saw the line to completion. By 1774 40 miles were open to Napton, from where a busy coal trade to the towns to the south was started. This stretch included the 412 ft long Newbold Tunnel and a twelve-arch aqueduct at Brinklow (now an embankment). The year 1774 also saw improvements in the water supply to the summit pound between Napton and Claydon, with the incorporation of Boddington reservoir, the deepening of the summit pound and the construction of Clattercote reservoir near Banbury. In 1787 this was supplemented by another reservoir at Wormleighton. Funds were, however, already running low and completion to Banbury was only possible with the passage of a second Act on 20 March 1775, which enabled the raising of an additional £70,000. This allowed work to continue south so that by May 1776 the line reached Fenny Compton, where Simcock built a 1,138 yd long tunnel. Cropredy Wharf was opened on 1 October 1777 and on 30 March 1778 200 tons of coal arrived at Banbury, an event celebrated with great festivity. The canal was now 63³/₄ miles long and had cost £205,148. But by 1780 money was so short that all work on the line had stopped, a situation mirrored by the failure of the Coventry Canal Company to complete its line north to Fradley. Thus the two canals, although connected to each other, were still separated from the rest of the canal network and from the Thames. The Oxford was at least earning money. In the year ending August 1780 income was £6,982, although of this half went to pay loan interest.

The future of both canals was resolved by a meeting at Coleshill on

20 June 1782. It was agreed that the Oxford would complete its line to Oxford and that a combination of the Trent & Mersey and Birmingham & Fazeley companies would complete the Coventry to Fradley Junction. Despite these grand plans, the money was still wanting and it wasn't until September 1785 that the Oxford was in a position to ask Samuel Weston to make a survey of the River Cherwell from Banbury to Oxford. Although shown to be feasible, it was the original plan for a new cut that was incorporated in an Act passed on 11 April 1786 which enabled the raising of a further £60,000. This Act also removed the restriction on the movement of canal-borne coal along the Thames south of Oxford. Robert Whitworth was asked to resurvey the Banbury to Oxford section and James Barnes was appointed as resident engineer for this last phase.

On 30 August 1787 the canal was open to Northbrook, just north of Tackley, and by January 1789 Samuel Simcock was preparing plans for the canal's terminus near Hythe Bridge in Oxford. The entire works reached a conclusion on 1 January 1790 when a fleet of boats entered Oxford and the band of the Oxford Militia played suitably celebratory numbers. The canal as built was 91 miles long and had cost approximately £307,000. In addition to the wharves in central Oxford, by 1796 boats were also able to navigate onto the Thames via Duke's Cut, a channel which left the canal near Wolvercote and entered the Thames near King's Weir. This had been built by the company following a suggestion from the Duke of Marlborough who

A picture taken by E. Temple Thurston during his trip along the Oxford Canal in 1910 or 1911. This picture was taken at Cropredy Lock and is thought to show his boatman, Eynsham Harry, on the right
Temple Thurston Archive, The Boat Museum Archive

owned the land. This route to the northern Thames was later supplemented by one to the southern, with the opening of Isis Lock in 1796.

With the completion of the Birmingham & Fazeley Canal and the Coventry as far as Fazeley, a line to Birmingham and the Potteries was now open and trade improved significantly. There were some initial problems with water supply but these were solved by the installation of steam engines to back pump water up to the summit pound. This shortage of water delayed the payment of any dividends during the early years. However, by 1795 a half-yearly dividend of 2 per cent was paid and it wasn't too long before the company was in a position to be much more rewarding to its investors. At this time coal was the major cargo to be carried along the canal. In the year ending 1792–3, 55,893 tons of coal were shipped from Coventry, of which 9,787 tons were brought into Oxford. In addition to this, coal from Wyken and Hawkesbury would also have been moved. In 1795 the Oxford company purchased its own coal wharf in Reading and later acquired wharves at Wallingford and Abingdon. The preponderance of coal as a cargo suggests that many boats must have made the return journey empty.

Although the Oxford Canal had significantly reduced the navigable distance from the Midlands (London to Birmingham via the Thames & Severn Canal was 269^1/$_2$ miles, compared with 227 via the Oxford), in 1791 the

Isis Lock and Bridge near central Oxford

route was still considered to be indirect and slow. In particular the Thames was difficult and often impassable. A new proposal for two Warwick Canals and the Grand Junction Canal in 1791 meant that the distance between London and Birmingham could soon be reduced to just 137 miles of modern, fast waterway. The Oxford's response to the threat was immediate. A proposal for the Hampton Gay, or London & Western Canal was published in direct competition to the GJC. The L&W was to run from Shipton-under-Cherwell via Thame, Wendover and Amersham to Isleworth. Although still 30 miles longer than the GJC, it offered a much more direct, Thames-free, line to London compared with the route along the southern Oxford. For a while it looked possible that both might be built. However, the political clout of the GJC meant that parliamentary interest concentrated on the GJC and the L&W eventually evaporated in return for the GJC guaranteeing the Oxford minimum receipts of £10,000 p.a., a guarantee which in fact was never called on. The two Warwick Canals, the Warwick & Birmingham and the Warwick & Napton, were open in 1800 and the GJC was opened to London in 1805. Traffic from Birmingham to London now passed along the Warwick Canals to Napton and then took the Oxford for 5 miles to Braunston, where it entered the GJC. The traffic along the 5 mile Napton to Braunston section was to become 'a nice little earner' for the Oxford company.

By 1801 the Oxford's receipts reached £37,996 and it was able to pay an 8 per cent dividend. Just eight years later these figures had risen to £79,438 and 25 per cent. About two-fifths of the traffic along the canal was coal shipped primarily from Wednesbury and Warwickshire. Other goods included limestone, roadstone, iron, salt and general merchandise. The canal was in all respects a success. By 1832 the use of the canal was such that yet more water was needed and the Boddington reservoir was enlarged. But perhaps the most important development was the modernization of the northern canal. The stimulus to update the line was the announcement in November 1827 of the London & Birmingham Junction Canal that would have run directly from the Stratford-upon-Avon Canal at Earlswood to the GJC at Braunston, thereby bypassing the Oxford completely. Although the L&BJ was never built, the Oxford went ahead with improvements which shortened it by 13⅝ miles, mostly by making direct cuts where Brindley had woven great loops. The work included a rebuilding of the junction at Braunston and a new, shorter Newbold Tunnel. Charles Vignole was employed as engineer. The work was enabled by an Act of 1829 and completed by May 1834 at a cost of £167,172.

The canal company entered the 1840s in a healthy position. Receipts in 1842 totalled £73,119 and a dividend of 30 per cent was issued. In that year 20,859 boats had passed through the Hillmorton Locks and 9,900 had passed through the summit pound at Claydon. By 1848 the company was able to pay off its debt. The canal had developed a fearful reputation for exploiting its position as the linking canal between the Midlands and the GJC, and for overcharging. Several meetings were held during the 1840s at which the other companies involved in the north–south route met with a view to increasing business. On each occasion the Oxford were accused of pursuing a policy which enabled the company to pay a 30 per cent dividend, whereas the other companies were only able to muster 7 per cent. In 1845 the position was such that the GJC and the Warwick Canals threatened to amalgamate and thus isolate the Oxford entirely. This finally sparked the Oxford into action. It promptly agreed to cooperate with its neighbours and to reduce tolls along its part of the line. But by that time, the effects of the railway were already being felt.

In 1826 a plan for a railway to run from London to Birmingham by way of Oxford and Banbury was published. This scheme came to nought but the London & Birmingham Railway, which almost directly followed the course of the GJC, opened in 1838. Even though this line was some considerable distance to the east, it immediately had an effect on trade along the Oxford. Toll receipts at Braunston were almost halved in just three years. Conversely, the opening of the Great Western Railway line from Didcot to Oxford in 1844 had a positive affect on the fortunes of the southern part of the Oxford, with a minor increase in toll receipts. It was not until 1850 that a GWR subsidiary continued the route north through Enslow, Aynho and Heyford to Banbury. The next phase, completed in 1853, took the line on to Warwick and thus to Birmingham. Also in 1850, the Buckinghamshire Railway was opened from Banbury to join the London & Birmingham. The same company then opened a branch via Bicester to Oxford. The competition for coal delivery was now intense and the Oxford was forced to reduce its toll rates to Banbury by 80 per cent. This maintained the tonnage carried along the canal but seriously reduced receipts. In 1848 420,000 tons of cargo were carried with receipts of £56,000. In 1858 the figures were 400,000 tons and £24,700. As a consequence the 1858 dividend was down to $8^{1}/_{4}$ per cent. By the 1860s, traffic along the southern line to Cropredy, Aynho and Oxford itself had fallen heavily.

Despite the difficulties, the Oxford remained fiercely independent and managed to maintain the amount of cargo carried to the end of the century. Indeed, at the beginning of the twentieth century, there were even suggestions that a new modernization programme be implemented as a way of once again competing with the railways. Needless to say, nothing came of any of these plans and the First World War saw the beginning of the end of the Oxford as a commercial waterway. However, in 1919 286,459 tons of cargo were still being carried and the company implemented a post-war maintenance and improvement programme. In the 1930s the new Grand Union Canal Company made an attempt to increase traffic on its London to

Birmingham route. As part of this it agreed to undertake work on the Napton to Braunston section of the Oxford. It improved the banking and rebuilt two bridges so that 12 ft 6 in wide barges could navigate between the old GJC and the old Warwick & Napton Canal. Included in this deal, the Grand Union guaranteed the Oxford minimum toll receipts for this section of its line. By the time of the Second World War the Oxford was still able to deliver an 8 per cent dividend. Coal still formed the bulk of the cargo carried. A wartime government report recognized the line's strategic importance and recommended that it be fully maintained just in case the GJC was blocked. But after the war the line declined considerably. The situation was not improved by nationalization in 1948. The canal now began to lose money and the Board of Survey classified the northern section of the canal as 'to be retained' and the southern section under 'waterways having insufficient commercial prospects to justify their retention'. Such was the concern that the Inland Waterways Association held its 1955 rally at Banbury with a view to promoting the Oxford as worthy of retention. The popular cause was justly supported and by 1968 the future of the southern section was recovered. The Transport Act listed the Oxford, all of it, as a cruiseway and thus gave British Waterways the remit to maintain the line to the required standard. Today the Oxford is one of the best-loved holiday waterways.

The Walk

Start:	Tackley BR station (OS ref: SP 484206)
Finish:	Oxford near BR station (OS ref: SP 508064)
Distance:	11¼ miles/18 km
Map:	OS Landranger 164 (Oxford)
Outward:	BR Oxford to Tackley (enquiries: 0734-595911)
Car park:	Oxford has numerous car parks but try one of the well-signposted park and rides
Public transport:	Trains from London and Didcot going north call at Oxford

This walk can be shortened to 2½ miles each way by parking near the canal at Enslow (OS ref: SP 479183), just off the Oxford to Banbury A423, and walking south to the Boat Inn at Thrupp (OS ref: SP 480157). This option would take in some of the most pleasant and interesting parts of the full walk. There is room for on-road parking at Thrupp for those lucky enough to have like-minded friends. Otherwise, lunch at the Boat should be succour enough for the return stroll.

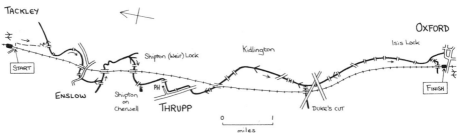

The Oxford Canal

From Tackley station, cross the railway via two pedestrian gates at the Oxford end of the platform. Follow the dirt road right and on for ¹/₂ mile to a T-junction. Turn left and after 200 yd bear right with the track. Go over a bridge which crosses one stream of the River Cherwell and on along a fenced path. This winds over a weir and round the perimeter of Flights Mill to reach the canal at Pigeons (or Pidgeons) Lock. Bear left down the side of the bridge and turn right to start the walk along the right-hand bank.

When I was here, the canal was blanketed with inch-thick sheets of ice. One intrepid boater had broken through to leave a series of mini-icebergs in his wake. The resultant aftermath produced a gentle Swiss cowbell-like tinkling sound which eerily reverberated around the freezing fog. There was a gorgeous, early morning tranquillity which was only disturbed by me stomping my feet to get warm. Unable to partake of the pub that formerly stood here at Pigeon's Lock, I put on my gloves and made off to go under bridge 214 and on to lifting bridge 215, Caravan Lift Bridge. The lifting bridge is one of those wonderfully simple inventions that make certain aspects of eighteenth-century engineering so enjoyable. The type found on the Oxford Canal (there were thirty-eight of them originally) was designed by James Brindley. The platform is counterbalanced by projecting beams which, when the bridge is open, lie horizontally and, when closed, rise accordingly. The water channel narrows to just 7 ft, i.e. narrow boat width, at the bridge to reduce the weight and cost. The whole thing, made mostly of wood, seems wholly in keeping with the pleasant pastoral nature of the canal; a highly functional object beautifully designed.

The canal swings round to join the River Cherwell to the right and you should continue along a narrow isthmus before going under the railway for the first time. The village to the left is Enslow, an important point on the canal as it is on the road which joins the villages of Islip and Woodstock. In the early days there was a flourishing coal trade from Enslow Bridge Wharf, a business that continued until after the Second World War. In the early twentieth century, the wharf also became an important canal/railway transfer point for cement brought along the canal from the works at Kirtlington, just north of Pigeons Lock. The narrowing of the waterway near the wharf suggests the position of a gauging point where tolls were applied.

The walk passes the old Enslow Bridge (from where if you cross you can reach the Rock of Gibraltar pub, formerly the wharfinger's house) and, after going over a stile, under the new road bridge. The inherently meandering nature of this, one of our earliest canals, can be seen as, once more, the line of the waterway swings east. Had the Oxford been built fifty years later by Thomas Telford we would perhaps be on a walk that would have been, say, 2 miles shorter. Brindley, however, did not have the technology to build the dramatic cuttings or embankments that embellish Telford's Birmingham & Liverpool Junction Canal. Brindley built his waterways to meander around the contours, following the easiest line. The Oxford is thus very much an example of an early canal which was later to be considered somewhat out of date and slow to use. Improvements in the nineteenth century meant that many of the meanders were bypassed with newer straight cuts. Some of the former can still be traced (see Further Explorations).

Heading south the canal comes, after Baker's Lock, to physically join the River Cherwell just south of Enslow. The Cherwell was made navigable from Oxford to Banbury, probably under the direction of Andrew Yarranton, in the late seventeenth century. It was still navigable here in 1777 and formed the natural route for this stretch of the canal. The towpath crosses the river by an iron bridge built in 1909. It is a splendid spot and remarkably remote. Kingfishers and herons abound and, even at midday, a tawny owl called from the small copse across the channel. A little further on along the combined waterway, there is a post with DIS written on it beside the towpath. You will see others along the way. As part of the company's by-laws boats within 200 yd of a lock with the water level in their favour had the right of way. The DIS post was positioned so that there was no dispute as to who was first to be within the required distance.

After going under the railway, the combined river/canal goes through a series of sharp turns through fine open countryside, only to separate again at Shipton (Weir) Lock. Here you turn right while the river goes straight on through weirs near the village of Hampton Gay. Shipton Lock has a fall of just 2¹/₂ ft and if a standard narrow lock had been built here it would not have passed sufficient water down the line to supply the next lock (Roundham) which was of more normal depth. Indeed, it may have acted more like a dam. So the lock chamber at Shipton was built in a hexagonal shape, which permits a much greater flow of water. A similar lock was built for similar reasons where the Cherwell and the canal cross at Aynho, just south of Banbury. According to a photograph in Hugh Compton's book, there was formerly a small lock-keeper's cottage to the right of the lock. Cross the canal via the bridge at the downstream end to take up the left-hand bank. The towpath passes a lift bridge to reach a rail bridge and the abutments of a former rail bridge. On Christmas Eve 1874, a train jumped the rails just here and crashed onto the frozen canal. Some thirty-four people died, many of whom are buried in nearby Hampton Gay churchyard.

After passing a single-gate stop lock and going under Shipton Bridge (no. 220), you arrive at Shipton-on-Cherwell with its loftily positioned church that stares down on passing boaters and towpathers. Anyone who has read Temple Thurston's *The Flower of Gloster* will recognize Shipton as the place where the traveller spent his first night on board boat. Here the author pondered the passage of time and professed an interest in having his bones laid to rest amid the green grass of the churchyard.

For a while the Cherwell runs alongside the canal but then the 'cut' broadens and turns abruptly right. This is Thrupp and the site of a BW maintenance yard. Cross the lift bridge (Aubrey's Lift Bridge) and continue past the houses and cottages along the right-hand bank. At the end of the village, the path passes The Boat, as fine a place to have lunch as any to be found. TV fans may recall that this is where Inspector Morse and Sergeant Lewis (sorry, I mean John Thaw and Kevin Whately) went to find breakfast after starting their investigation in Colin Dexter's *Riddle of the Third Mile*. The mutilated corpse had been found in the canal here at Thrupp by some passing holiday boaters. Nasty.

The canal soon bends left and passes another pub, the Jolly Boatman, before going under Sparrowgap Bridge and running alongside the busy A423 into Kidlington. Before entering town, the canal again bends sharply right to go under Langford Lane Bridge and into an industrial area of north Kidlington. Roundham Lock and Bridge are soon followed by a railway bridge. The quiet suburbs of Kidlington now dominate the left bank. Go under two more bridges (Bullers and Yarnton) and pass another lock (Kidlington Green Lock) before re-emerging into open country. Originally the canal wasn't coming this way at all but was going to run about a mile to the west through Begbroke and Yarnton. Samuel Simcock changed Brindley's planned route in January 1788.

The canal continues under a pipe bridge and the Woodstock Road (Kings) Bridge, while the ears are pummelled by the noise from the new A34 road way over to the left. The towpath passes a peculiarly positioned pillbox to reach Duke's Lock. Here there is a lengthsman's cottage and a roving bridge which passes over the entrance to Duke's Cut. This narrow channel, which runs to the Thames near King's Lock and Weir, was dug in 1789 at the request of the 4th Duke of Marlborough to link up with his Wolvercote paper mill. The mill is actually on a stream which runs parallel with the canal about 1/2 mile to the west. Until 1796 the cut was the only link between Oxford and the Thames. The cut enabled Warwickshire coal to be brought to the mill via the canal. The lock at the entrance is effectively a stop lock. The level of the canal is normally higher than the river but in flood the river may be 2 ft higher than the canal. Morse fans will of course recognize Duke's Cut as the site of yet more foul deeds. In *The Wench is Dead* Colin Dexter asks the inspector to investigate a murder which occurred in 1859. This, of course, he does with his usual seemingly somnolent style. Continue walking along the main canal to go under the substantial Oxford–Witney A40 road bridge that was built in 1933. A short distance further on is one of those peculiar juxtapositions that occur periodically along our waterways: an almost delicate and appealing eighteenth-century lift bridge and an elephantine and bullying twentieth-century dual carriageway road bridge.

You are now approaching the outskirts of north Oxford as you pass Perry's Lift Bridge and Wolvercote Lock. From here almost into central Oxford, the canalside is littered with a succession of houseboats. The next mile takes in a road bridge, Ball's Bridge, the last railway bridge and St Edward's Lift Bridge before reaching a factory which dominates the left-hand bank. Here bridge 239A is a lift bridge of a more modern kind, electrically powered and an important route from one part of the works to the other. There is now just over a mile into central Oxford and the factory gives way to suburban housing with gardens running down to the canal. Temple Thurston suggested that the building of these houses was like the growth 'of some unsightly fungus' nurtured by 'jerry-builders – men of execrable taste, whose only thought is to build for profit'. As he then goes on to say that they wouldn't last, he may well be surprised to see them still here and quite well thought of compared with what has come since.

After Frenchhay Lift Bridge the canal goes on to Aristotle Bridge, from where there is access to shops to the left. Bridge 242 is followed by a works bridge that runs from a metal castings works known as Lucy's Foundry, which has its own wharf and canalside loading area. The stroll through the suburb of Jericho is a popular spot for dog-walkers and joggers and leads to the fine cast-iron bridge that goes over the head of Isis Lock. The waterway to the right is the River Isis, a tributary of the Thames, the main channel of which is just a few hundred yards to the right. Boaters heading towards the river pass through the lock and turn right to go under the railway and out, along a channel called the Sheepwash, to the river. Although the canal was fully opened in 1790, the Isis Lock connection with the Thames wasn't built until 1796. It's construction allowed boats to avoid a convoluted section of the Thames west of Wolvercote.

Cross the bridge and walk on with the canal left and Isis right. The canal ends at Hythe Bridge, while the river continues south to enter the Thames. From 1839 there was a floating chapel moored on the river just before the bridge, positioned to enhance the moral and spiritual well-being of the boat people. In 1868 the vessel sank and a new on-shore chapel was built in Hythe Bridge Street. Originally the canal also continued under the road bridge and on to two wharves. If you go up to the road and peer over the wall, the site of Worcester Street Wharf can be seen. It is now a car park. The line of the canal bent left so that it continued across the open space and under Worcester Street to end at the New Road Wharf. The bridges here were built to barge width so that boats from the Thames could use the wharves as well as the narrow boats that had ventured along the canal. Interestingly, however, as a protectionist measure the Oxford company always steadfastly refused permission for the landing of coal from the Thames at either of its city centre wharves. Both wharves were infilled after the Second World War, although the land was actually sold to Lord Nuffield (William Morris) in 1936. Nuffield College now sits atop New Road Wharf.

Oxford town centre and the bus station can be reached by turning left along Hythe Bridge Street. The railway station is reached by turning right.

Further Explorations

Sadly, many sections of the Oxford's towpath are either in a very poor state or even non-existent, a problem that arose from the destructive wash of motorized vessels. This is a pity because it would undoubtedly make for fine walking all the way along. This should not deter the hardy towpather, who can savour the Oxford's delights with a series of short walks.

A walk of 7^1/$_2$ miles starts at Napton on the Hill (OS Landranger 151. OS ref: SP 464611). The village can be found to the south of the Leamington Spa to Daventry A425. Buses run twice daily from Leamington (G & G Travel enquiries: 0926-885555). For those with cars, park in the centre of the village near the Crown pub. To start the walk turn right from the pub past the small general stores and telephone box. The road bends right along New Street past the post office and Napton Christadelphian Meeting Room. It then turns left to form Thornton Lane and then Chapel Green. As the main road bends right again, go straight on along a much smaller lane which is signposted to Marston Doles. This pleasant country lane runs for just over a mile. Halfway along, those with keen eyes may spot a rundown building in a field to the right. This is the engine house that will be discussed later. At a T-junction turn right to reach a bridge over the canal. Cross and turn left through the gate to reach Napton Top Lock. Here turn left to take up the left-hand bank of the canal.

At Napton Top Lock the canal reaches its summit pound. The next lock south is at Claydon, 5 miles as the crow flies, 11 as the narrow boat meanders. The route, of course, goes north past a second lock before taking the first of the giant bends that mark Brindley's canal-building style. Within a mile is the Engine House arm, a narrow strip of water that leaves from the right-hand bank. This channel was built so that boats could take coal to a steam engine which pumped water up the canal from below Napton Bottom Lock into the summit pound. As you may have seen, the engine house is derelict and the job is apparently now carried out by electric motors.

The walk continues past six locks with fine views to Napton windmill before reaching Napton Bottom Lock. The canal now bends sharply left to go around Napton Hill with its massive quarry scar. Continue past a boat-yard and on to a main road bridge and the Napton Bridge Inn. Continue to the next two bridges and Napton Narrowboats yard. You will return to Napton by turning right at the second bridge. However, keen towpathers will want to walk briefly on past the boat-yard to see Napton Junction. Here the Warwick & Napton Canal (now the Grand Union) takes a line, left, to Warwick and then to central Birmingham. The Oxford, meanwhile, carries on to Braunston where it forms a junction with the Grand Junction Canal for London. Return to the bridge and turn left. Go over the crossroads and up the road which winds round some houses into central Napton and the Crown Inn.

For those who prefer their walks to be around a mile in length and to involve a pub, there are two offerings. Between Banbury and Southam, the A423 goes over the Oxford Canal near Fenny Compton. If you park near the road bridge at SP 436523 (Landranger 151), you can look down to the canal as it passes through a cutting. This part of the Oxford is known as the 'tunnel'. This is because when the canal was first opened here in 1778 a 1,138 yd long tunnel was built to take the line through the hill. The problem was that it was just 12 ft high and 9 ft wide and was responsible for endless traffic jams. In 1840 the land above the tunnel was purchased from Christ Church, Oxford and a central passing place opened up with two tunnels either side. In 1870 these were removed and hence the whole tunnel opened out into a cutting. On one side of the road bridge there is a fine cast-iron footbridge dating from the first 'opening out', which takes the towpath from the southern to the northern bank. A well-trodden path leaves the left-hand side of the road bridge on the other side of the road to go down to the canal. Once on the path turn right, cross the footbridge and walk along the left-hand bank, past Cowroast boat-yard and on to the George & Dragon pub at Fenny Compton Wharf, a total distance of just under a mile. The best option for returning is back along the canal.

In the 1820s the Oxford company was aware that the canal was showing its age. It was both slow to use and often highly tortuous as it relentlessly followed the contours around the countryside. As a consequence it set about a programme of improvements which took 14 miles off the 43 miles of canal between Hawkesbury and Napton. At Newbold-on-Avon (Landranger 140. OS ref: SP 487773) the former circuitous route of 1¼ miles was shortened

The entrance to the old canal tunnel and St Botolph's Church at Newbold-on-Avon

to just ¹/₂ mile by the straightening of the line and the construction of a new tunnel. From the B4112 at Newbold, go along a narrow lane at the back of the Boat Inn and past the Barley Mow to reach the towpath of the northern Oxford. If you turn left you can follow the curve of the canal round to the new Newbold Tunnel. It was built in 1834 as part of the improvement programme and is 350 yd long. It's a typical late canal tunnel: it has a towpath on either side and is wide enough for narrow boats to pass each other. If you continue onwards to the next road bridge, the original line of the canal can be seen on what appears to be a junction. The canal went left here and across the fields before swinging back into Newbold. To see the old tunnel, return to the Boat Inn and turn right to reach St Botolph's churchyard. Go through the gate and bear right past the church to a stile. Here is the old canal bed and, to the right, the red-brick entrance of the original 125 yd long Newbold Tunnel. It is considerably narrower than the new tunnel and does not have a towpath. The tunnel runs under the churchyard and the road to come out along the lane in front of the two pubs to join the present route. Surprisingly, the tunnel (although not the land above it) was still owned by British Waterways as recently as March 1991, when it was auctioned for £1,000 for Comic Relief.

Further Information

There is no Oxford Canal society as such. However, within the Inland Waterways Association, there is an Oxford branch. To find out about meetings first contact IWA head office, the telephone number and address of which can be found in Appendix B.

For more detail on the history of the canal, the best available source is:
Compton, H.J., *The Oxford Canal*. David & Charles, 1976.

Those who wish to read of the more seamy side of life on the Oxford Canal should head for:
Dexter, Colin, *The Riddle of the Third Mile*. Pan Books, 1983.
Dexter, Colin, *The Wench is Dead*. Pan Books, 1989.

9
THE REGENT'S CANAL
Camden Town to Paddington

Introduction

In the midst of the hustle and bustle of the nation's capital exists the most pleasant of watery retreats. The Regent's Canal has that peculiarly rural feel that it shares with the city's great parks. It's an urbanized pastoral haven, a rustic metropolis. Whatever it is, it's a lifeline. Not that it has always been fully appreciated. It was once viewed solely as a convenient repository, a place for everything that was no longer worth anything. Luckily, the councils, amenity groups and British Waterways have removed the junk that once encrusted the canal and the line is now a popular spot for joggers, lunching office workers, the inevitable fishermen and, of course, the ever-swarming towpathers.

The Regent's Canal leaves the old Grand Junction Canal at Little Venice, a lagoon a few hundred yards north of Paddington station. After 1/2 mile the line enters Maida Hill Tunnel which goes under the Edgware Road to re-appear near Lisson Grove and Lord's cricket ground. After a further 1/2 mile the canal reaches the northern perimeter of Regent's Park. Here the waterway curves around the outskirts of the park and through London Zoo to arrive at Cumberland Basin where it turns abruptly left. After passing Camden Lock and Hampstead Road, the line enters Kentish Town and St Pancras. It then continues under the Caledonian Road to enter Islington Tunnel and comes out near the Angel tube station and the City Road. The canal now enters the heart of London's East End, through Hackney and Bethnal Green to reach Victoria Park, where the Hertford Union Canal runs east to join the Lee Navigation. It is now just 1 1/2 miles through Mile End and Stepney to the Limehouse Basin and the River Thames.

Walking through cities isn't everybody's idea of a good day out but this short stroll has bags of interest and many surprises, not the least of which is the friendliness of the locals who, of those I met, seem to have a genuine interest in their canal. One to do instead of the Christmas shopping.

History

There was no doubt that by the end of the eighteenth century, London was still relatively isolated from the rapidly industrializing midland and northern towns. The need to improve the transport of coal and goods to and from the capital was becoming a major concern, remedied initially in 1790 when the Oxford Canal opened up the route from the Midlands via the Thames, and on 10 July 1801 when the Grand Junction Canal was opened to Paddington, a spot at one end of the 'New Road' which then formed the northern boundary of the city. From Paddington cargo was transhipped onto carts for haulage into the centre of town. It wasn't too long after the opening of Paddington Basin that it was suggested that a new canal be built to run parallel with the New Road from Paddington to the city and then the Thames near the expanding London Docks. The advantage of not having to tranship goods from canal onto cart was widely agreed. In July 1802 a line for the London Canal was surveyed from Paddington Basin to the New River (Islington), the Commercial Road and the Thames. Subscriptions of £400,000 were quickly raised by a committee headed by Sir Christopher Baynes, but with the protestations of local landowners and the refusal of the GJC to provide a water supply, the scheme was dropped.

By 1810 the idea was once again in circulation. The key protagonist at this time was a former committee member of the Coventry Canal and one-time canal boat owner, Thomas Homer, who suggested a line running parallel to the New Road from Paddington through Islington to the Thames at Limehouse where he planned to build a ship dock. John Rennie reported favourably on the scheme and Homer sought to enthuse the influential. At about the same time, the architect John Nash had been appointed to develop the Marylebone Park Estate, later to be renamed Regent's Park, for the Commissioners for Woods and Forests. Homer saw advantages in linking the canal with the estate and invited Nash to take responsibility for the canal. Nash was seemingly delighted with the idea of boats sailing gracefully through the park and accepted. Nash and his assistant, James Morgan, were joined by an engineer, James Tate, to examine Homer's proposed route and with minor adjustments agreed it. Within the park, the canal was to be combined with a broad ornamental lake which was to be a central feature. By August 1811 this link with the new Regent's Park was such that the canal was renamed the Prince Regent's Canal. In a prospectus the cost of the new navigation was estimated at £280,000.

By the time a bill was presented to parliament in the summer of 1812, Nash had changed his mind about having a commercial canal running through his new up-market gardens. As a result, the route was pushed out to the northern edge of the park, the banks were to be built up to nearly 25 ft to conceal the line and the towing path was moved to the northern side. With this change, and despite the protestations of the Paddington wharfingers who feared loss of trade, the bill received Royal Assent on 13 July 1812, granting powers to build the canal and to raise the necessary £400,000.

The canal proprietors met for the first time on 10 August 1812 at the Freemason's Tavern at St Giles-in-the-Fields. Charles Munro was made chairman, John Nash was appointed as a director and John Morgan as engineer, architect and land surveyor. Thomas Homer, who had started the whole scheme, was appointed superintendent. Morgan estimated that the line would cost £299,729 to build and the go-ahead to start the construction work was given. Curiously, none of these individuals had experience of canal building and the committee was forced to offer the design of much of the line, such as tunnels and locks, out to competition. This proved to be a fruitless exercise, with none of the entrants' designs being accepted by the committee. In the end, Morgan was instructed to draw up his own designs.

The construction work started on 14 October 1812, with James Tate as contractor. Interestingly, the Act demanded that the part of the canal that passed through Regent's Park should be finished within twelve months. This they just failed to do as it was completed by 11 November 1813. The stretch between Little Venice and Hampstead Road, with the exception of Maida Hill Tunnel, was finished six months later. The 2¹/₂ mile lock-free stretch between the GJC at Little Venice to Hampstead Road was officially opened on the Prince Regent's birthday, 12 August 1816.

On 4 April 1815 it was discovered that Thomas Homer had sequestered some of the company's funds, a charge he admitted before fleeing the country. He was later arrested and sentenced to transportation. The funds were never recovered and shareholders had to make up the missing amount. Financial problems also threatened the completion of the rest of the canal, with work being suspended for a while. Although £400,000 had been authorized, the company had only been able to raise £254,100. An Act in 1816, designed to produce a further £200,000, was passed but still the necessary monies were slow in coming. The problem was resolved through the receipt of funds from the Commissioners for the Issue of Exchequer Bills, a kind of early job creation scheme set up by the Poor Employment Act of 1817. Following a survey of the line by Thomas Telford, the commissioners agreed to loan the company £200,000 if it could raise £100,000 of its own. With the confidence of the investors thus restored, the extra funds were raised relatively easily. By the end of 1818 the Islington Tunnel was finished and it was firmly hoped that the line would be open by the end of 1819. In fact the end of 1819 saw the raising of another £105,000. This extra delay meant that it wasn't until 1 August 1820 that the Regent's Canal was officially opened from the Grand Junction Canal at Paddington through to the Thames at Limehouse. A flag-covered barge full of proprietors, including the Earl of Macclesfield, the company chairman, set off from Maiden Lane (now York Way, King's Cross) followed by two military bands playing 'enlivening airs'. The completed canal is 8½ miles long and drops 86 ft through twelve broad locks from Little Venice to Limehouse. There are thirty-six road bridges, nine footbridges, ten railway bridges and two tunnels. The new line had cost £710,000.

In its first year after opening, the company reported the carriage of nearly 195,000 tons of cargo and this steadily increased. In the year ending April 1829, it carried over 488,721 tons and in 1834–5 this had risen to 624,827, earning some £28,930 in tolls. The company was able to clear its debt to the commissioners of £235,000 in 1828 and the first dividend of 12s. 6d. per £100 share was paid in 1830. Contrary to expectations, nearly three-quarters of the traffic using the line came from the Thames Dock rather than from the inland GJC. The trade was primarily in coal, timber, road materials, bricks, lime and sand. The success of the line encouraged Sir George Duckett to build his Hertford Union Canal, which joined the Regent's with the Lee Navigation in 1830. However, the link line was never a great success, even after it was rendered toll-free for a period in 1831. In

The Great Northern Railway's interchange depot near St Pancras Lock on the Regent's Canal

British Waterways

1848 the line was blocked off to avoid water loss and in 1857 it was finally purchased by the Regent's Canal Company.

It was just seventeen years after the opening of the canal that the first railway from the Midlands reached London. Stephenson's London & Birmingham Railway of 1837 followed roughly the same route as the GJC and immediately took traffic from it. In 1852 the London & York railway was opened to a temporary terminus in Maiden Lane, or York Road as it became, and the Midland Railway reached London in 1868. The position of the Regent's, running from Paddington Basin across the lines of all the northern railway termini to the Thames, attracted considerable interest from those who saw an opportunity to convert it to a railway line. The first offer from a company so interested arrived in September 1845. In the same year the company's own engineer, William Radford, suggested that it make the conversion itself. The initial reluctance of the committee was overcome when an outside consortium offered to buy the line for a million pounds. However, the scheme for what was intended to be the Regent's Canal Railway floundered as the consortium failed to raise the necessary funds. Several other attempts to make the conversion were made but none of them was able to generate any financial enthusiasm, or, as in the case of the Central London Railway & Dock Co., were defeated in Parliament following opposition from, among others, the GJC. By this time the canal was a highly profitable concern. In 1876–7 trade along the line was some 1,427,047 tons p.a., with receipts of £92,877 and profits of £46,559. A large proportion of this income was derived from warehouse and port dues. It wasn't until 31 March 1883, when the canal was taken over by The Regent's Canal City Docks Railway Company for £1,170,585, that a conversion plan looked like becoming a reality. But even though the company changed its name in 1892 to the North Metropolitan Railway and Canal Co., the necessary capital of £2,500,000 wasn't forthcoming. By 1904 defeat was admitted when the name reverted to the Regent's Canal and Dock Company.

In 1905 the Regent's Canal carried 1,045,184 tons of cargo and had receipts of £92,000. The main business at the time was the shipment of coal to local gasworks. Despite this, more and more traffic was being lost to the railways and then to the roads. Unusually for canal companies, the response of the GJC and the Regent's was to enter into a period of collaboration. In 1914 the two companies formed a joint committee of directors. After the First World War the companies decided to merge. To do this the Regent's bought the GJC (which already owned the Old Union and Old Grand Union canals in Leicestershire and the three Warwick canals (the Warwick & Birmingham, the Warwick & Napton and the Birmingham & Warwick Junction to form the Grand Union Canal Company. This was achieved on 1 January 1929 at a cost of £801,442 for the GJC and £136,003 for the Warwick canals. The grouping was expanded in January 1932 when the

Grand Union purchased the Leicester and Loughborough Navigations and the Erewash Canal for a total of £75,423. For the first time, the inland waterway route from London to Birmingham and London to the Trent was under one roof.

The new Grand Union was committed to expanding the use of the line. The key parts of its plan involved making the canal to Birmingham suitable for barges and providing an improved design of barge to work it. The improvement budget for 1931 amounted to one million pounds (with the help of a government guarantee) and a new class of boat, the Royalty Class, of 12 ft 6 in beam, was in production. The company also started its own carrying company by buying Associated Canal Carriers Ltd, renaming it the Grand Union Canal Carrying Co. Ltd.

The history of the company after this period of hopeful expansion is one of disappointment. No dividends were paid on ordinary shares between 1933 and 1945, and in 1948 it was nationalized and put under the control of the British Transport Commission (later the British Waterways Board). The Regent's Canal is now used principally for pleasure craft, including privately-owned boats, trip boats and even a 'Waterbus'. The only working boats are those used by BW for maintenance. The canal is, however, widely recognized as a significant local amenity and since the 1960s it has undergone some marked renovation. The piles of assorted junk that once predominated have been removed and the towpath is now a popular spot for those attempting to escape temporarily from the noise and bustle of city life.

The Walk

Start:	Camden Town tube station (OS ref: TO 289840)
Finish:	Paddington BR station (OS ref: TO 265815)
Distance:	3 miles/4$\frac{1}{2}$ km
Map:	OS Landranger 176 (West London)
Return:	For Camden Town take the Metropolitan or Circle line from Paddington to King's Cross and change to the Northern Line going north
Car park:	There is no specific car park. Public transport is both frequent and convenient and may be preferred
Public transport:	Tube from Camden Town connects to King's Cross, Euston, Charing Cross, Waterloo and Mornington Crescent

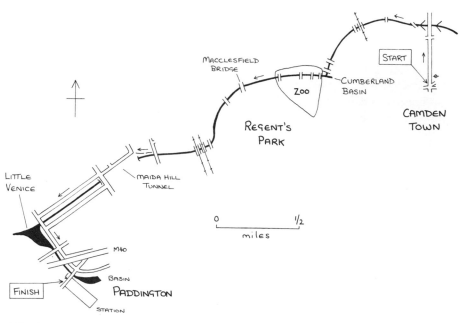

The Regent's Canal

Walkers on the Regent's should be aware that the towpath is open in winter between 10 am and 4 pm only.

Camden Town underground station has two exits. Take the right-hand one as you come up out of the darkness and turn right to walk up Camden High Street for about ¼ mile. At the canal bridge, cross the road and go through the gate on the left-hand side of the bridge.

From a busy shopping street, the scene immediately changes. Within earshot of the traffic in Camden High Street are the vestiges of an altogether different age. The first thing in view is Hampstead Road Top Lock. This is the highest of twelve that run east towards the Thames at Limehouse, some 86 ft below this point. Until 1973 all these locks were paired to speed traffic and save water. Today, only this lock is paired. In the others one side was converted into a weir to help avoid flooding. The small, castellated lock cottage on the left was built in 1815. At one time it was a BW information centre and may be again one day. At the far end of the lock the canal is spanned by a cast-iron footbridge. On the near side a restored winch challenges the mind as to what it must have been for until I reveal that it was moved here from the entrance to the Lee Navigation at Limehouse, where it stood between 1866 and 1968. It was originally used to open and close one of the lock gates there and was put here by the Greater London Industrial Archaeology Society.

Hampstead Road Top Lock

Cross the footbridge to take the northern, right-hand, bank. On the far side of the bridge it's worth noticing the deep grooves in the stone and iron-work caused by the constant abrasion of barge towropes. The cobblestones on the incline, known as scorchers, helped give the barge horses purchase on a potentially slippery surface. Just after the bridge a doorway leads off to the much-renovated Camden Lock Centre. Opened in 1973 on the site of an old timber wharf, the buildings were once stables and hay lofts. The trade in timber continued here well into the 1960s, with the last lighter unloading cargo about 1969. Today the buildings house a variety of shops and cafés, and on a weekend an open market.

The towpath continues over the canalside entrance to the large railway interchange warehouse on the right. This entrance was known as the Dead Dog Tunnel for reasons perhaps not worth dwelling on. Here goods were transferred from the railway to the canal and there was a huge area of underground storage used for wines and spirits, popularly known as the Camden Catacombs. After going under Southampton Bridge, a watersports youth club known as the Pirate Club occupies the southern bank. Just beyond is a horse ramp, built so that beasts which had fallen into the canal could make a ready exit. The canal company claimed that the railways were a particular cause of horses bolting, although why this should be a problem peculiar to the Regent's is a mystery. Meanwhile, the canal moves quietly through the busy backstreets of Camden Town and under two railway bridges that carry Robert Stephenson's London & Birmingham Railway to the nearby Euston station. Stephenson had problems with these bridges, as both had to be built without interrupting the canal traffic. There was also a relatively steep incline up to the canal in order to provide sufficient head-room. As early locomotives were incapable of tackling the gradient, the trains originally had to be hauled into Euston by a stationary engine that was connected to the carriages by cable.

After the Gloucester Avenue Road Bridge, the scene changes again and the canal enters a stretch where the gardens of some stately Victorian houses proffer their greenery to passing towpathers. Ahead are two more bridges (Grafton Bridge and Water Meeting Bridge) before the canal reaches the outskirts of Regent's Park. These 410 acres were originally part of Marylebone Park, one of Henry VIII's royal hunting grounds. It was John Nash who, between 1812 and 1827, restructured them into the park you see today. It was opened in 1838 and named after the Prince Regent (later King George IV). At first it was planned that the canal would pass through the centre of the park on its way to Paddington. John Nash certainly thought well of the idea originally but local objections and the realization that the canal would be populated by uncouth bargemen forced Nash to re-route the line along the park's northern perimeter. This decision explains the sudden right turn which exists here at the north-eastern corner. As the canal bends right, the Cumberland Basin is on the left. What is today a mooring area, including the rather fancy floating Feng Shang chinese restaurant, was once a 1/2 mile long arm that led off round the park to Cumberland Hay Market (although it also sold meat and vegetables) on the New Road near Euston station. The branch was filled in with rubble in 1948 and part of it sits under the zoo car park.

The canal now takes on the ambience of a royal park, with some fine trees and a pair of elegant cast-iron bridges. The first was made by Henry Grissell of London in 1864 and the second by R. Masefield & Co. of Manor

Ironworks, Chelsea. Between the two is another horse ramp. Beyond the bridges is the zoo landing stage of the canal Waterbus. The canal runs through the middle of London Zoo and, although there is no access, tow-pathers get a good view of some deer as well as Lord Snowdon's massive aviary. The 36 acre Zoological Gardens was designed by Decimus Burton and first opened to the public in 1828. Incidentally, the two upright stones by the side of the towpath mark the old parish boundaries of St Marylebone and St Pancras.

The canal continues around the park and under a footbridge to reach a fine-looking bridge with cast Doric columns. This is Macclesfield Bridge, so named after the first chairman of the Regent's Canal Company. It is also known as Blow-Up Bridge following an infamous accident in the early hours of Friday 2 October 1874. A steam tug, the *Ready*, was towing five barges from the City Road towards Paddington on its way to Nottingham. One of the barges, the *Tilbury*, was loaded with a range of goods, including several barrels of petroleum and 5 tons of gunpowder from the Royal Arsenal at Woolwich. Just before 5 a.m., a spark from the chimney of the *Ready* landed

The aftermath of the Regent's Park explosion, 1874. The remains of the columns, which were re-used when Macclesfield Bridge was rebuilt, can be seen in the bottom right-hand corner

Marylebone Public Library

on the deck of the *Tilbury*, igniting firstly the petroleum vapour and then the gunpowder. The explosion killed three crewmen, reduced the bridge to a pile of rubble and damaged houses in the vicinity. The bang was heard 12 miles away. The bridge's Doric columns were salvaged from the wreckage and were re-used in the new bridge. If you look carefully, you should see that the towrope marks appear, improbably, on both sides of the columns, showing that when re-erected they were turned round. The Grand Junction's carrying company eventually paid over £80,000 in compensation to more than 800 claimants, a considerable sum in those days.

The walk continues on under what appears to be a sturdy footbridge but is in fact an aqueduct for the River Tyburn, one of London's many underground rivers (it's even buried within the aqueduct). It was used by Nash as a feeder for his Regent's Park lake. Continue round the north-western fringes of the park, with many fine Regency houses adding an air of elegance to the route. A large house on the right is Grove House, also designed by Decimus Burton. As Park Road Bridge crosses the canal, the minaret of the Central Islamic Mosque can be seen over to the left. This is the border of the park and you should pass under two railway bridges, one part of the underground line to Watford and Amersham, and the other the BR line to Marylebone station. The Lisson Grove housing estate now dominates the left-hand bank as the canal bends gently right past a mooring site to St John's Wood Road. In normal circumstances it should be possible to continue through the short tunnel ahead. However, when I was here some repair work was being undertaken on the towpath. If this is still closed on your visit, take the pathway up right to the road. Cross and take the passageway to the right that passes alongside an iron-railing fence.

Whether on the towpath or on the footpath above it, the Maida Hill Tunnel is now directly ahead. Just before it is the site of the former wharf for the Marylebone (St John's Wood) power station. At one time this was a very busy spot as narrow boats were used to bring in the coal as well as to take away the ashes. The tunnel itself is 272 yd long and it is reputed that the spoil from its excavation was used as topsoil on the field of the nearby Lord's cricket ground. This was done because the canal passed straight through Thomas Lord's original site for the ground. The canal company also paid him £4,000 in compensation. There is no towpath through the tunnel so horses were uncoupled and brought up to road level to walk over to the other side. The men on the boat, meanwhile, would have laid on their backs on the deck and pushed against the sides of the tunnel with their feet to take the vessel through. An alternative method was a kind of inverted punting using a pole to push against the roof of the tunnel. Your route is somewhat easier. If on the towpath, go up the steps by the tunnel entrance to Aberdeen Place, which is where those who couldn't gain access have been for some time. Continue on along the course of the canal to Edgware Road

Junction House (the former tollhouse) at the western end of the Regent's Canal

where there are a small number of shops. Cross the road and bear right past a café into Blomfield Road where the canal re-emerges on your left.

A gate promises access to the towpath but doesn't deliver so continue along the pavement. The canal follows a straight course in-between two tree-lined avenues, each with some fine Regency houses. The line here is a popular mooring place and the gaily painted or plant-bedecked boats must make this one of the most attractive parts of town. The road culminates in a canalside cottage known as 'Junction House', formerly the Regent's Canal tollhouse, and the end of the Regent's Canal at the Warwick Avenue Bridge. The stop gates, once used to prevent water flow from the GJC to the Regent's, still sit in position under the bridge, although these days they are rarely closed unless for maintenance.

Cross the road to reach the Paddington Canal and Little Venice. This small lagoon is also known as Browning's Pool after the poet Robert Browning who lived in nearby Warwick Crescent. A good view of the pool and its island can be had either from the bridge or, if you cross the bridge, from Rembrandt Gardens. To reach Paddington station walk past the gardens and, following the signs, cross the road. The path passes under the massive complex of roadways and out to a small green where, by briefly turning right, there is a glimpse of the Paddington Canal as it sits underneath the

A40(M) overpass. To the right the line goes down to Little Venice, to the left on to the Paddington Basin. The basin was, of course, here some time before the trains and thus Paddington was an important inland port prior to the arrival of Isambard Kingdom Brunel. Many roads hereabouts still bear names like North Wharf Road and South Wharf Road in homage to a much more watery past. The basin is now fully enclosed and inaccessible to walkers, although there has been talk of redeveloping the area.

Continue on your original line to a huge roundabout on Harrow Road. Turn right into Bishops Bridge Road, cross this road and then turn left (again following the signs) into the rear of Paddington station. This is normally the taxi entrance but it affords a brief view left of the line of the Paddington Canal Basin. By following the path round, you will shortly arrive at the main line and tube stations to complete the walk.

Further Explorations

Virtually the whole of the Regent's Canal is open and towpathers can complete the entire line by walking from the Limehouse Basin to Paddington, a distance of about 9 miles. The towpath was much improved in the early 1980s under the aegis of the Canal Way Project and numerous information notices and plaques can be seen all along the canal.

Some of the most interesting features of the eastern end can be seen, however, in a shorter of walk of nearly 3 miles from the Angel tube station (Northern Line) to Camden Town. From the station, turn left along the City Road for about 1/3 mile to the offices of London Electricity, where you cross the base of the City Road (or Grand) Basin. This can be seen over to the left of an apparently pointless road bridge. The basin was opened in 1820 and was the most important of the canal's wharves, covering some 4 acres. At its peak the basin went beyond City Road but sections have been gradually filled in and the southern end is now occupied by office buildings. The basin was surrounded by factories and there were flour, timber and general trading wharves. By the turn of the twentieth century British Drug Houses (BDH) occupied most of the site. In recent years the old wharf buildings have been pulled down and new canalside houses built.

Continue along the City Road for a short distance and turn left along Wharf Road. After passing the rebuilt Pickfords Wharf, cross the canal itself and then turn left, just after an information notice, to reach the canal. Walk along to City Road Lock. From here walk on to the next road bridge, where you should cross to the left-hand towpath. There is now a good view of the 960 yd long Islington Tunnel. The tunnel was opened in 1818 and was originally a 'legging' tunnel. However, from 1826 the company provided a steam-powered tug which pulled boats through by working its way along a chain on the canal bed. This service survived well into the 1930s.

To reach the next stretch of the canal, you have to divert off the towpath and through the residential undergrowth. At the tunnel entrance walk up some steps to Colebrooke Row. Turn right and walk on to turn left along Charlton Place. Cross Camden Passage and then Upper Street. Go up Berners Road (with the Business Design Centre to the right), turning left and right into Bromfield Street to reach Liverpool Road. Cross to Ritchie Street and on, through a park, to Dewey Road. Continue over Barnsbury Road into Maygood Street. Walk on along an alleyway between houses to Muriel Street where the canal can be seen and a gap in the wall takes you back to the towpath.

Immediately left is the western entrance to Islington Tunnel. Further along, after some converted warehouses, is the 1¹/₂ acre Battlebridge (or Horsfall) Basin which abuts King's Cross station. St Pancras and King's Cross can be seen over to the left. You should now continue under three bridges and turn sharp right to reach St Pancras Lock, followed by St Pancras Basin, a coal wharf opened in 1870 by the Midland Railway. From there there is now just a short stroll on to two locks in close proximity, Kentish Town Lock and Hawley Lock. To reach Camden Town tube station (Northern Line), turn left at the next (iron-girder) bridge into Camden High Street.

Further Information

There is no Regent's Canal society as such but the Inland Waterways Association holds regular guided walks from Camden Town tube station. To find out the date of the next walk, contact the IWA head office (telephone number and address in Appendix B).

The most complete published description of the Regent's Canal is in:

Essex-Lopresti, M., *Exploring the Regent's Canal*, 3rd edition. K.A.F. Brewin Books, 1990.

However, there are sections on the canal in:

Hadfield, C., *Canals of the East Midlands*. David & Charles, 1981.

10
THE ROYAL MILITARY CANAL
Rye to Hamstreet

Introduction

The Royal Military Canal is a true oddity. Built at government expense in an impressively fast time, the canal, with its associated road and parapet, was designed as a military fortification against the rampaging French. Only after Bonaparte had decided to turn on Austria instead was the potential for commercial carrying considered. As such it never even looked like succeeding. It was, virtually from the day it was opened, a wonderful and notorious government folly.

The Royal Military Canal stretches like a necktie around the throat of Romney Marsh, separating that flat expanse of reclaimed land from the rest of Kent and Sussex. At its western end the canal reaches the sea at Cliff End, just 3–4 miles east of Hastings. From there it crosses the Pett Level to reach the orderly town of Winchelsea where, having joined the River Brede, the line continues to the Cinque Port of Rye. After winding around the base of the rock on which Rye sits, the line of the Royal Military then takes that of the River Rother to head north-east via Scots Float to Iden Lock. From here on the canal follows a new cut through Appledore and then goes more determinedly east past Hamstreet and Bilsington and on to the outskirts of Hythe. The final stretch of the canal is wedged between the town and the sea. It ends, in a rather inglorious way, on the sea front at Shorncliffe, just a few miles west of the centre of Folkestone.

This part of coastal Kent and Sussex has a peculiarly airy and remote feel, and in summer there's more sunshine than can be imagined. With a good pair of binoculars, a fine packed lunch and this book, you can easily lose yourself here (metaphorically speaking of course!).

History

In 1803 the brief truce between France and Britain (the Peace of Amiens) came to an end and the Napoleonic War resumed. With renewed hostilities came the fresh threat of invasion. British military experts saw the promontory at Romney Marsh as the most likely area for an assault, a figurative chin on the coast of Kent just waiting to be hit. The French punch was to come from Boulogne and was to be directed at Dungeness. The coastline around the town was almost ideal for beach landings and, despite the enthusiasm of the local people for fighting the French, it was a difficult area to defend. Between 120,000 and 180,000 French troops were reported to be awaiting the assault, supported by floating gunboats and other vessels that were to carry artillery, support supplies and some 6,000 horses.

By August 1804, the threat of invasion was at its height, and the new Prime Minister, William Pitt, was determined to improve the coastal defences. Although their efficacy was much contested, the first decision was to build eighty-six Martello towers as defensive forts along the south coast. A second, much vaunted, scheme was one in which the whole area of Romney Marsh would be flooded by the simple expedient of opening a

The line of the military canal around Rye follows the course of the River Brede to meet the River Tillingham on the southern side of the town, where this picture was taken probably about 1910

Hastings Museum

series of sluices. The suggestion for a canal as a military barrier against the potential invaders was made by Lt-Col John Brown, Assistant Quartermaster-General and Commandant of the Royal Staff Corps (a predecessor of the Royal Engineers). He had reviewed the flooding scheme and rejected it as impossible to implement. One report, for example, suggested that ten days warning would be needed, another that with a neap tide little if any of the land would be inundated. There was also the immense damage that would be done to the land; largely irrelevant if an attack actually occurred but severely embarrassing and expensive if the area was flooded as the result of a false alarm.

In a plan submitted to his chief, General Sir David Dundas, Commander of the Southern District, on 18 September 1804, Brown suggested that a vast ditch be built from Shorncliffe, a little under 3 miles west of Folkestone, to the River Rother north of Rye. The canal would not only be a line of defence but would provide the means of bringing up reinforcements to the threatened points, moving the troops either by barge or by the road which was to be made on the inland side of the canal. The road itself was to be screened from enemy fire by a parapet built of the soil dug from the canal works. The Commander-in-Chief, the Duke of York, approved of the scheme and agreed with Dundas that the line would not be without its commercial potential. The cost of construction was estimated at £80,000.

The speed of decision making was impressive. The scheme was forwarded to the Minister of War, Lord Camden, on 27 September. On the day before, however, the Duke of York had met with the Prime Minister to discuss the canal. Pitt was impressed by the scheme and ordered that the construction work should start immediately. The well-known canal builder John Rennie was appointed as consultant engineer and Brown was appointed military director. The following week, Rennie reported on his survey to the officer responsible for the direction and management of the works, the Quartermaster-General, Major General Brownrigg. In his report he praised Brown's original plan and suggested a sluice at Shorncliffe, culverts to carry flood water under the canal, and a navigational lock at Rye. Much of Rennie's efforts were expended on finding ways to remove excess water in winter and spring, and finding additional supplies for the summer and autumn. This was done by installing drainage pipes at the ends of the canal and by diverting some small streams into the line. It was also at this time that plans were circulated to extend the canal from the Rother through Rye and Winchelsea to Cliff End, about 3 miles east of Hastings. The ditch itself was to be 44 ft wide at the bottom, 62 ft wide at the surface and 9 ft deep. The extension beyond Rye was both narrower and shallower. A military road, running parallel with the canal some 52 ft to the north, was to be 30 ft wide and raised 3 ft above the level of the marsh. Rennie estimated the cost at £200,000.

To the east of Rye, the River Rother was incorporated into the military defences. This view shows the Union Channel junction, with the Cinque Port of Rye in the distance

William Pitt became personally involved in much of the organization and certainly assisted in convincing the local landowners to cooperate. At a meeting at Dymchurch on 24 October, he told them how the new canal would not only defend their country but would also act as a major drainage ditch for the winter months and a linear reservoir for the summer. This personal intervention won their support and agreement. The necessary land was handed over without recourse to the normal practice of valuation, haggling and legal document. Pitt's involvement in the project was such that locals soon started to call the canal 'Pitt's Ditch'.

The construction work, on what was then being called the Romney Canal, began on 30 October, just over a month after the original go-ahead and in the absence of any parliamentary authority. It was estimated that the canal would be ready by 1 June 1805 but virtually from the start it was clear that this was highly optimistic. Delays occurred which Rennie put squarely down to the incompetence of the contractors. However, difficult weather conditions and problems with flooding contributed to the delays. There were also misunderstandings as to who was actually responsible for what and as a consequence much of the work was inadequately supervised. By May 1805 the works were in severe disorder with just 6 miles completed, all activity stopped and the navvies dispersed. As a result, after a meeting between Pitt and the Duke of York on 6 June, the contractors were sacked and Rennie resigned. Lt-Col Brown was then given full control of operations. He took personal charge of the eastern end, with Lt-Col Nicolay in charge of the stretch from Winchelsea to Cliffe End. The project was henceforth treated as a field work under the Quartermaster-General's department. Work now proceeded much more smoothly, using military and civilian labour under the expertise of the Royal Staff Corps, who in 1805 were moved permanently to Hythe. The civilian labour on the project was paid 5s. 6d. a week and used mostly for excavation work. The soldiers built the ramparts and turfed the banks. For this they were paid just 10d. per day.

By 14 July 960 men were at work between Shorncliffe and Hurst: 700 cutting, 200 turfing the banks and sixty working hand pumps to keep the trench from flooding. By August this number had risen to 1,500. By the following August, 1806, the canal was open from Shorncliffe to the Rother. However, this stretch was partially rebuilt twice over the next couple of years. On 18 November 1808 a storm led to extensive flooding between West Hythe and Shorncliffe which washed away about 2 miles of the south bank of the canal. To prevent this recurring a strip of land between Twiss Bridge and Shorncliffe was purchased and the sea defences strengthened. On 9 January 1809 the regimental stores burnt down, considerably disrupting the work. The situation was made worse when, on 30 January, a storm coinciding with a spring tide, again saw a breach in the sea defences. This time the canal was filled with shingle and the Shorncliffe–Sandgate road was damaged. Despite these setbacks, the canal was virtually finished by spring 1809, although the defensive cannon, eighty Danish guns that had been captured at Copenhagen, weren't in position until July 1812.

The final cost of the Royal Military Canal was put at £234,310. This figure included £66,000 for land purchase and damages, which was still being sorted out in April 1807 when a meeting of landowners was held in Hythe. With accusations of profiteering being voiced, the government decided to put the whole matter on a more legal footing and passed an Act to establish the maintenance of the canal. The Royal Military Canal Act received the Royal Assent on 13 August 1807. The Act appointed commissioners to administer the canal. This illustrious group included: the Prime Minister, the Speaker of the House of Commons, the Chancellor of the Exchequer, the Principal Secretaries of State (including for War), the Commander-in-Chief of the Forces, the Master-General of the Ordnance and Quartermaster-General. One of the earliest actions (April 1809) of the commissioners was to reward John Brown with what must have been the princely sum of £3,000 in return for his good efforts. Brown was also made the director of the canal and provided with a house at Hythe.

By the time the RMC was ready, the threat of invasion had gone.

This print of Hythe from the canal bridge was first published in 1829. Apart from the bridge itself, it shows one of the passenger boats that operated between Hythe and Appledore, Iden and Rye at that time

P.A.L. Vine

Napoleon's plans had suffered a severe reversal at Trafalgar in 1805 and he was no longer in a position to assault Kent's chin. With the decline of the threat came an increase in the number of sceptics and cynics. Suddenly it was agreed that there never was a serious threat of invasion and that even if there was, that Napoleon's army would not have found the canal much of an obstacle. In 1807, almost in desperation at what was rapidly being seen as an appalling waste of public money, the government opened the canal to navigation in the hope of recouping some cash. The Royal Military Canal Act of 1807 enabled the commissioners to collect tolls. Only manure and produce of the owners or occupiers of the lands adjoining the canal was to be excused payment. The commissioners even owned a few barges of their own, sometimes towed by horses of the Royal Waggon Train (the predecessor of the Royal Army Service Corps). Over £84,000 was spent by the government between 1808 and 1809 finishing off the waterway, including covering the road and towpath with shingle, replacing twenty temporary wooden bridges and building a new bridge over the River Rother at Iden.

The canal was opened to the public on 18 April 1810. However, by then the line was already being used for a small amount of traffic. The wharf at Shorncliffe, for example, was busy as a landing place for timber and coal. The working canal was used primarily to carry shingle, sand, bricks, stone, timber and, perhaps almost uniquely for a British canal, a steady trade in hop poles. A convoy of army barges once did the trip from the Rother to

Hythe in 4 hours and at one time there was a regular barge service between Hythe and Rye run by a Mr Pilcher. The single fare in July 1818 was 7s. 6d. Obviously there couldn't have been much custom, for by 1819 it only went from Hythe as far as Iden and from 1822 only as far as Appledore. Even this truncated trip didn't last; it stopped altogether in 1833, seemingly because the RMC felt that too much damage was being done to the canal banks. However, a new boat may have continued operations during the 1840s. Tolls for use of the military road between Rye and Iden started on 9 July 1810, and between Rye and Winchelsea on 22 August 1812. Traffic in the first year after the opening of the canal for public use reached 15,000 tons, with receipts of £542. By 1812 canal receipts reached £576, and with road tolls and other income, primarily rents of canal land for sheep grazing, the total receipts were in excess of £1,000. Although the RMC's income reached a peak of £1,530 in 1836, traffic levels were never high and the government was never able to recoup its expenditure on the building and maintenance of the line.

The canal was thus never tested as a defensive structure. Certainly William Cobbett in *Rural Rides* thought that as Napoleon had succeeded in ransacking most of Europe he wouldn't have found too many problems with a 30 ft wide canal. Cobbett and others continued to grumble about the waste of money for many years and the canal soon became renowned as a kind of military folly. Its reputation wasn't wholly rescued until the late 1840s when both the Duke of Wellington and Lord Raglan commented on the strategic value of the waterway.

Although the planned military use of the canal was controversial, it had a major impact on the local villages. They now had a road as well as a waterway that allowed them to travel between Hythe and Rye, previously a tortuous trip at the best of times. The canal also provided drainage to the lands around Appledore, increasing the amount of rich arable land available and reducing the incidence of marsh fever.

In 1837 an Act transferred responsibility for the canal to the principal officers of the Ordnance (who controlled the Royal Artillery and the Royal Engineers). By this time the line was being looked after by an officer and sixty troops who were basically fulfilling roles more normally found on a conventional commercial waterway. There were twenty-one barges licensed to use the canal and able to carry between 21 and 38 tons. Annual income (£1,237) was sufficient to cover expenditure but a recent survey had shown that the canal needed extensive dredging if it was to remain navigable. The advantages of doing this can't have been very obvious to the officer in charge of the Ordnance, the Master General Sir Hussey Vivian. His view was that the canal was 'absurd as a means of defence . . . which ought never to have been dug'. This view clearly had an effect on the management of the RMC over the next few years, as a programme of what would now be called rationalization was implemented. The numbers of soldiers employed in maintenance work was drastically reduced and expenditure cut all round. Many functions, for example, were now filled by pensioners instead of sappers and at a much reduced cost. By May 1841 only seven Royal Sappers & Miners were engaged in work along the RMC and by October there were none.

The decline in traffic during the 1840s can be related to the opening of the South-Eastern Railway between Ashford and Rye, and, later, the Ashford to Hastings line. Coal traffic dropped sharply. From tonnage levels of 15,766 in 1847, cargo loads were down to 10,612 in 1849, with receipts halved. By now 60 per cent of the cargo was in shingle being taken inland to repair roads. The RMC increasingly relied on renting land to local farmers for sheep grazing. The average income from this source throughout the 1840s was £450–500 p.a. Two attempts were made to buy the RMC from the government during the course of the 1840s. The first, by the Hastings, Rye & Tenterden Railway Company, which wanted to convert it to a railway line, was defeated in Parliament. The second attempt was by the Lords of Romney Marsh and the commissioners of the adjacent levels. This proposal fell through when the Ordnance raised the issue of finance.

The outbreak of hostilities in the Crimea brought about another change of administration when the Secretary of State for War took direct responsibility for the RMC in May 1855. Among the first decisions to be made by the incumbent, Lord Panmore, was to reject another offer to buy the canal, this time from a Mr Smith. Although the present view of the War Department was that the line had little value as an effective defensive fortification, it would have at least the effect of being a modest hurdle for potential invaders. The key problem was that the government was now responsible for the drainage of the area and any relinquishment of ownership had to take that into consideration. In a survey in 1857, George Wrottesley proposed selling off the various assets, including, for example, the now fifty-year-old elm wood that grew along the banks of the RMC.

In the 1860s traffic along the canal continued at roughly 10,000 tons p.a. and income at about £1,200. This meant that there was a small but continuing loss on the venture of some £400 p.a. Thus, with no improvement in economics and a gradual disinclination to take the military purpose very seriously, the idea of divesting itself of the RMC grew more prominent in government circles. In August 1867 the Secretary of State was authorized to do so. There then followed a hectic series of exchanges. In February the *Kentish Gazette* announced that Seabrook Harbour & Dock Co. was about to buy the canal. In the end the War Department decided to give the canal away to anybody willing to take responsibility for the land drainage, and an Act to allow this was passed on 10 August 1872. The War Department was now able to agree to lease the stretch between Iden and Shorncliffe to the

Lords of the Level for a period of 999 years starting on 3 June 1873. The annual rent was to be a shilling. This was soon followed by Hythe Corporation obtaining an Act to purchase that stretch through the town for conversion into ornamental waters. In 1874 the War Department sold the canal west of Rye to four private owners. Meanwhile, the trade had all but gone. From 11,000 tons p.a. in 1867, traffic in 1887 was 158 tons. Despite a mini-revival in the closing years of the century, trade in 1903 had almost dried up completely. The last toll collected at Iden Lock was taken from the barge *Vulture* on 15 December 1909 for 27 tons of shingle.

The eastern canal is still owned by Hythe Council and used by local pleasure boats. The western end from Appledore to Iden Lock is owned by the National Rivers Authority. Interestingly, the parapet on the road side of the canal was not sold by the War Department until 1935 and even then was requisitioned during the Second World War, just in case the original purpose of the canal was to be tested after all. As part of the 1935 sale, the section between Appledore and Hamstreet was bought by Dorothy Johnston who then presented it to the National Trust. Today the canal is still in water and lengths are occasionally used by canoeists and other boats. The River Rother near Rye is used by water-skiers who tear around the tortuous curves near the sluices of the Union Channel. The rest is left to the birds and the towpathers.

The Walk

Start:	Rye (OS ref: TQ 919205)
Finish:	Hamstreet (OS ref: TR 001337)
Distance:	11 miles/18 km
Map:	OS Landranger 189 (Ashford & Romney Marsh Area)
Return:	Hourly trains between Hamstreet and Rye, including Sundays, on the Hastings to Ashford line (enquiries: 0732-770111)
Car park:	At Hamstreet station (free of charge at weekends)
Public transport:	The walk starts and finishes at BR stations

This walk can be shortened by using the BR station at Appledore for a walk from Rye of 8 miles. The drawback of this is that the last mile is along a relatively busy road with no pavement. Alternatively, park in the centre of Appledore and walk to Iden Lock or to Hamstreet, returning along the same route. The distances involved are then 7 miles and 9 miles respectively. Sadly there are no convenient pubs or shops at Iden Lock.

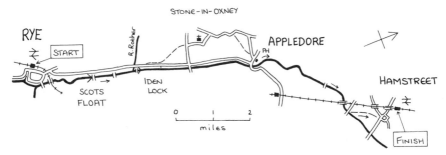

The Royal Military Canal

For the main walk the easiest approach from a traffic and car parking point of view is to leave your vehicle at Hamstreet and then take the train to Rye. Hamstreet station is on the A2070 Ashford Road just north of the intersection with the B2067. At Rye, leave the station to walk up a slight hill past some bus-stops to the post office on the right.

There are two minor diversions that you may consider worth taking before setting out. To see the River Tillingham and Rye Harbour, turn right past the post office. Walk down Wish Street, following signs for Tourist Information. Eventually this road leads to a bridge over the Tillingham. Here is the Tillingham sluice, a vertical gate that protects the land beyond from high tides. Just up from it is a smock windmill. A little further round on the road (with the harbour on your left) is a rather sorry-looking Martello tower. These towers were built at roughly the same time as the canal as part of the nation's defence against Napoleon. There were eighty-six of them altogether. This one was intended to guard the Tillingham and Brede sluices but is now dissolving into the surrounding scrap-yard. Turn left along the road opposite the tower and within 50 yd you will come to Brede sluice and lock, a large affair that protects the River Brede from the tides. The stretch which runs off right towards Winchelsea was, in 1809, the last part of the canal to be built. It was never used by commercial barges and, as it is much narrower than the rest of the canal, was thought by many to be not too serious as a defensive obstacle either. But of course by the time it was built the threat of invasion had subsided.

The second diversion is Rye itself. It's worth wandering through the unspoilt, if tourist-infested streets. There are a number of interesting places to visit and plenty of places to consume cups of tea. From the post office, go up Market Road to reach the town centre. Maps describing town walks are available in local shops or at the Tourist Information Office.

The walk proper starts at the post office. From the entrance, turn left to cross Station Road. Walk past a small Baptist church and the police station. Continue until you reach the massive town gate (Landgate). Turn right to go underneath and then bear left down a slope. Cross the road to pass to the right of the public conveniences and then bear left diagonally across the field to a road bridge across the River Rother. At the opposite end of the bridge, a sign for the Saxon Shore Way takes you left across the road and onto an embankment high above the river. The path continues to a railway bridge. This is the trickiest part of the walk as you have to go underneath it, a manoeuvre which could dampen the feet at high tide. Once on the other side, you realize that the walk has truly begun, with the Rother, acting as the Military Canal here, on the left and the vast, flat expanse of Romney Marsh to the right. The path now winds gently back and forth and passes over the sluices of the Union Channel, an important drainage ditch for the area of Walland Marsh to the east. After a further gentle bend, the massive Scots Float sluice and lock come into view. This is the tidal limit of the River Rother. The sluice that was here at the time of the building of the RMC was destroyed in a flood in 1812. Another was built here in 1831. The present structure dates from 1984. It's worth passing by the small brick hut to the bridge, where you can look at the sluice, formerly known as Star Lock, at close quarters. It consists of a pair of doors which open either way depending on the water levels. Boat owners wishing to pass through the gates have to give the sluice-keeper warning as there are occasions when there is simply insufficient water available to allow passage. This alone indicates that the key objective of the National Rivers Authority here is to prevent floods rather than permit navigation. In Edwardian times the sluice was the starting point for some early motor launch trips along the Rother to Bodiam Castle near Northiam.

To continue the walk, return to the right-hand bank, where a concrete road goes past the sluice-keeper's house and into a field. The Military Road, built at the same time as the canal to assist the rapid movement of men and artillery along the defence work, now runs parallel to the canal along the opposite bank. In just over 1/2 mile the path passes Boonshill Bridge to go over a stile and into a field. This then leads to the junction where the canal bears right while the River Rother bears left. The first bridge over the Rother here was completed in the spring of 1808. The present one dates from 1960. Iden Lock, to the right of the bridge at the start of the RMC proper, is the only lock on the cut section of the canal and was built entirely by the army, in fact the Royal Staff Corps under the command of one Captain Todd. It was 73 ft long and 16 ft wide with a pair of floodgates below the bottom gates to prevent excess water flowing back into the canal. Work on the lock itself wasn't begun until October 1807 and it was finished in September 1808 after severe problems with its foundations. Curiously, the quoin and coping stones were obtained from Dundee. The toll collector for the western end of the RMC was based here at Iden. Tolls were charged at 3d. per ton per mile for timber and 2d. per ton per mile for coal. Corn

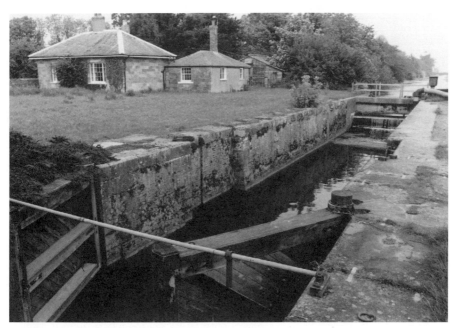

Iden Lock at the start of the canal 'proper'

was charged at $2^1/_2$d. The lock itself was used regularly until 1902 and was last used on 15 December 1909. In the early 1960s it was converted into a drainage sluice by the Kent River Catchment Board (absorbed by the National Rivers Authority). Next to the lock, on the northern bank, is an original army-built stone lock cottage and, over the Military Road, are the former toll collector's house and military barracks, now a private house.

A well-trod towpath proceeds along the right-hand bank of the canal all the way into Appledore. However, as this is not a right of way, I am unable to recommend it. It is worth noting, however, that the National Rivers Authority is currently investigating its options and that Kent County Council has a long-term strategy to provide a linear route along the length of the canal. For the moment, walkers may consider crossing the bridge here at Iden Lock and then turning right to go along Military Road for the 3 miles into Appledore. After about $^1/_2$ mile, the canal shifts a canal width left from its rigidly straight line. A series of these zigzags can be seen along this stretch of the canal. The system, devised ironically by a French military engineer called Vauban, was designed to allow troops to fire laterally along the canal from gun emplacements situated at each double bend. One advantage of walking along the road here is that a better view can be had of the protective defence works that border the landward side of the canal. The embankments themselves were made from the soil dug from the canal.

For those walkers on the road who may wish to get off, there is a small bridge about a hundred yards or so after the first zigzag. This takes a path away from the road to Appledore via the Saxon Shore Way. The route is quite interesting but not always adequately waymarked. The first part may be particularly confusing. After passing over the bridge, turn right and walk alongside the roadside ditch until you pass the boundary stone which sits on the side of the Military Road (erected in 1806 to mark the boundary between Kent and Sussex). Here turn left and then right alongside another ditch. Eagle-eyed walkers, aided by a map, should be able to follow the waymarks up Stone Cliff then along the road to Stone-in-Oxney. Turn right at a T-junction and then, after passing a pub, turn right again into a field with a brook to the right. This eventually reaches a road near some oast houses. Turn right. The route arrives in Appledore opposite the church in The Street. Turn right to reach the canal. Walkers who walked along the road, or indeed those naughty people who stayed on the towpath, pass some further zigzags, Stone Bridge and a few more zigzags before entering Appledore.

Appledore is a most pleasant place, with pubs and a village bakery. Before the arrival of the canal, the village was surrounded by marshland and prone to outbreaks of marsh fever. The canal drained the surrounding fields, supplied the first road link with Rye and the west and thereby enabled the village to thrive. In the early days water from the marsh was lifted to the canal using windmills and, latterly, steam pumps. Nowadays electric pumps at Shirley Moor, Appledore, Warehorne, Ruckinge and Bilsington drain the marsh into the canal; others drain it into the Rother. In return, the canal acts as a kind of linear reservoir for the drier summer months. For many years there was a regular barge service along the canal between Hythe and Rye. The building by Appledore Bridge now used by Hayne's garage, was originally owned by the proprietor, Nathan Bates, who used it to house the barge horses.

Just across the bridge and through a gate is the National Trust section of the canal, with full rights of way along this left-hand bank. This 3 mile stretch between Appledore and Warehorne was bought in 1935 by an Appledore villager, Miss Dorothy Johnston, who presented it to the National Trust. The military road has been covered with grass and it is a popular spot for afternoon strolls. Here, next to the canal, are the remaining stumps of the Huntingdon Elms that were planted by the War Office with a view to producing wood for use in rifles. The National Trust has taken great pains to replace the lost elms (which suffered in the 1970s during the first wave of Dutch elm disease), this time with oak and ash. But at the time I visited here, it was the profusion of May blossom that astounded the eye.

The fact that the canal was still considered to be a viable defensive position in 1939 is witnessed both at Appledore Bridge and along the National Trust section of the canal. At various points Second World War pillboxes were built to guard selected points. One pillbox manfully guards the Appledore Bridge, and another attempts to protect both the railway and the road bridges at Warehorne. At the Warehorne Bridge, the first after the railway crosses the canal, turn left and then immediately right over a stile and into a field. Take a line which passes diagonally left across the field, just to the right of the nearest pylon. This path eventually leads to the centre of Hamstreet. Turn left over the crossroads to the railway station, which is about 200 yd along the Ashford Road on the right.

Further Explorations

Sadly, not all of the Royal Military Canal is open to walkers, a position which is both illogical and frustrating. However, there are at least two other short stretches which can make a pleasant stroll.

There is a good walk of approximately 4 miles return from the Hythe terminus of the narrow-gauge railway, the Romney, Hythe and Dymchurch (OS ref: TR 153347). There are signposted car parks near the station. The walk goes to the small village of West Hythe, with a return along the embankment on the northern bank (there are rights of way on both sides). Take the narrow path to the right of the RH&D station near a café. The early stages of the walk afford good views of the trains to the left and there is an opportunity to see some of Vauban's zigzags. On reaching a road bridge, simply cross to the opposite side of the canal and return along the raised parapet. There are fine views over the marsh and a number of picnic spots.

At the opposite end of the canal, it is possible to walk from Cliff End (Landranger Map 199, OS ref: TQ 886134) to Winchelsea. From Cliff End, a rough lane firstly takes the left bank to a gate. The path then follows the course of the canal north, passing to the right the Pett Levels, famous for their rich birdlife. Follow the canal until you reach a road and then turn left into Winchelsea. Afternoon tea will no doubt restore tired walkers to a condition suitable for the return journey.

Further Information

The canal has no society to guard its interests and so the best bet is to join the Inland Waterways Association. The local group (the South-East Region: Kent & East Sussex branch) will then send you details of its meetings. The address of the IWA can be found at the end of this book.

For those who wish to learn more about the history of the canal, the standard work is:

Vine, P.A.L., *The Royal Military Canal.* David & Charles, 1972.

For a less learned but highly enjoyable read look no further than:

Godwin, F. & Ingrams, R., *Romney Marsh and the Royal Military Canal.* Wildwood House, 1980.

11
THE THAMES & SEVERN CANAL
Stroud to Sapperton

Introduction

The Thames & Severn Canal must have seemed like a good idea at the time. The busy wool town of Stroud already had the Stroudwater Navigation which linked it with the River Severn, and London was clearly in need of a waterway link with the burgeoning industry of the midlands. A line between Stroud and the Thames was thus built with lots of good intention and hope of a brisk trade. Then came the setbacks. Firstly there were constant problems with the waterway itself, and on top of that some bright spark built a bypass in the shape of the Oxford Canal.

The Stroudwater Navigation runs from the Severn at Framilode, not far from Gloucester, to Wallbridge on the outskirts of Stroud. Here it joins the Thames & Severn Canal which travels up the Golden Valley via Brimscombe, Chalford and Daneway to the Sapperton Tunnel. After emerging at Coates the line continues east to Siddington, just south of Cirencester, to where there was once a branch canal. The main line now turns south-east past South Cerney to Latton, formerly the site of the junction with the North Wilts Canal, a link route to the Wilts & Berks Canal at Swindon. The T&S meanwhile passes to the north of Cricklade and, in its final stretch, runs almost parallel with the Thames to Inglesham, just a mile south-west of Lechlade. Here it joins the Thames for all points south-east to London.

On a literally freezing January day, the towpath between Stroud and Sapperton was surprisingly busy with families, couples, joggers and, of course, towpath guidebook writers. In other words, it's highly recommended.

History

With the coming of the 1780s, it must have been more than evident that the largest potential market in the country, London, was physically isolated from the ever-expanding industry and canal network contained within the midlands and the north. At that time the capital still received the bulk of its goods by sea, a route that was always susceptible to inclement weather or passing pirates. The advantages of forging a waterway link cross-country must therefore have been a great talking point among canal entrepreneurs around the land and it is no surprise to find that the people of Stroud saw some potential in the extension of their newly constructed Stroudwater Navigation.

The Stroudwater, which joined the mill town of Stroud with the River Severn, had been a messy affair in incubation. There had been various schemes since Elizabethan times and during the eighteenth century a number of promoters had started and then failed to complete the works. Eventually, in July 1779, the Stroudwater was open from the Severn at Framilode, 8 miles south-west of Gloucester, to Wallbridge on the outskirts of Stroud. The new waterway was about 8 miles long and had twelve broad locks. Although the line had obvious potential in local trade, the opportunity to make the route part of a through line from Staffordshire to London was, it was widely agreed, just there for the taking. The Stroudwater Navigation Company, at a shareholder's meeting on 12 April 1781, ordered a survey to be made for a new line between Dudbridge, just west of Stroud, to Cricklade, John Priddey reported back in the following August. He advised that the canal could be built but that a better route was from Wallbridge to Lechlade, a village further down the Thames, about 12 miles east of Cirencester. He also informed the company that a plentiful supply of water could be obtained at Cirencester.

By 17 September, when a meeting was held at the King's Head in Cirencester, the proposed new line had attracted interest from the midlands. Among the promoters of the new line were individuals involved in the Staffs & Worcs, the Dudley and Stourbridge, and the Birmingham Canals. These companies saw opportunities for carriage of Staffordshire coal and goods manufactured in Birmingham. In addition, the people of Bristol viewed the line as an important route for their own products and, of course, Stroud businessmen saw potential for significant local development. Robert Whitworth, a student of Brindley, undertook the final survey and reported back on 22 December 1782. His recommendation was for a canal of $28^{1}/_{2}$ miles, with a rise of 241 ft from Stroud to the summit level at Sapperton by twenty-eight locks and a fall from Siddington near Cirencester to the Thames at Inglesham near Lechlade of 128 ft by fifteen locks. The estimated cost, which included a

1¹/₂ mile long branch to Cirencester, was £127,916. In drawing up his plan, Whitworth made the canal broad enough to take 12 ft wide Thames barges. Severn trows, which were 15 ft wide, would be able to navigate as far as Brimscombe, a few miles east of Stroud, where transhipment onto the Thames vessels would take place. The reason for this rather cumbersome action was that the Severn trows would be unable to navigate through the Thames locks and would have to tranship somewhere anyway. It was thus cheaper, in terms of construction costs, to get that transhipment as close to Stroud as possible.

At another meeting of the promoters on 17 January 1783, enthusiasm was such that almost 80 per cent of the required finance was promised within just three weeks. With this the Act received Royal Assent on 17 April. It authorized capital of £130,000, with powers to raise a further £60,000 if necessary. Josiah Clowes was appointed resident engineer, with James Perry acting as manager on behalf of the committee. Interestingly, the funds were raised from London-based speculators and those associated with the Staffordshire & Worcestershire Canal. This fact further emphasizes the enormous interest in the use of the line as a through route between the industrial cities of the North-west and London.

Among the most daunting of the engineering works planned for the line was a 2 mile long tunnel at Sapperton. The only tunnel of any significant length that had been built at this time was the Harecastle Tunnel on the Trent & Mersey Canal near Stoke. However, Harecastle had been built to Brindley's 7 ft narrow boat 'gauge' whereas Sapperton was to be a full 15 ft wide. It's no wonder, therefore, that many experts, including those of the Thames Commissioners, questioned the wisdom of building such a structure and recommended the adoption of narrow boat width. However, the fledgling company remained undaunted and confirmed its decision to have a broad canal in September 1783.

The construction of the T&S was, mostly, trouble-free. The contract for Sapperton Tunnel had been awarded to Charles Jones, who, perhaps optimistically, said that he could finish the work by the beginning of 1788. By January 1785 the line was open from Stroud to Chalford, and by the summer of 1786 the canal up the Golden Valley to the summit was in use. A wharf and coal-yard were built at Daneway Bridge and were doing a brisk trade. At Sapperton, meanwhile, building work was temporarily halted when Charles Jones became insolvent with only about a third of the tunnel finished. Other contractors stepped in and on 20 April 1789 the first boat passed through this most impressive structure. Beyond Sapperton, building work began in June 1785. The area near Coates consists of rocky ground, identified by Whitworth as difficult to keep watertight. To counter this the clay puddling was put down in two layers, each 2 ft thick.

The summit level, including the Cirencester branch, was complete by the end of 1787 and was in use as soon as the tunnel opened. The entire canal

The northern portal of Sapperton Tunnel *c.* 1902, with the lengthsman's cottage intact
Gloucester Record Office, The Boat Museum Archive

from Stroud to Inglesham was opened on 19 November 1789 at a total cost of approximately £220,000. The moment was celebrated in the inns of Lechlade and with a huge bonfire party. Traffic was able to travel from Brimscombe to Lechlade in three days and was able to reach London from Stroud in under two weeks. The line was not, however, without its teething troubles. Among other things, there was an acute shortage of water and the unusual problem that the summit level hadn't been built level.

Although the canal was opened with celebrations, the proprietors would have viewed developments to the north with trepidation. In 1790 one of the canal's prime purposes, as a route for coal from Staffordshire to London, had been removed with the opening of the shorter route along the Oxford Canal. London to Birmingham along the T&S was 269$^1/_2$ miles. London to Birmingham along the Oxford was 227. With the opening of the Warwick Canals and the Grand Junction Canal in 1805, that distance was down to 137 miles. The T&S responded to this development by concentrating its efforts on making its line a major link between Bristol and the upper Thames. The company sponsored a survey for a line from Bristol through the south Gloucestershire coalfield to join the T&S near Cirencester. In another scheme, the company considered a new cut from Inglesham to join the proposed London & Western Canal (which would have run from London to Oxford). With the Bridgwater & Taunton and the Grand Western Canals, a line was planned which would have provided a navigable route all the way from London to Exeter. However, with the more obvious line promised by the Kennet & Avon, this plan foundered.

Another problem for the T&S was the quality of the river navigations at either end of the line. The Severn and the Thames were often impassable through flood or drought. The Severn situation was to be remedied by the construction of the Gloucester & Berkeley Canal, which would provide an artificial channel to Gloucester and to the Severn at Sharpness. The Thames, however, was much more of a problem. The Thames Commissioners did much to improve the river during the course of the 1790s but the T&S saw its prime hope in the construction of bypass canals. The most promising of these was the Wilts & Berks Canal. Although this had been proposed as early as 1793, it wasn't until 1819 that the North Wiltshire line was opened to give the T&S a bypass as far as Abingdon. The final link that would have given an all-canal route to London, the Western Junction from Abingdon to the Grand Junction Canal, was never built.

Arguably the most crippling difficulty for the T&S was the lack of water at the summit pound. Weighty fingers were pointed at Robert Whitworth's final survey when it was realized that the feeders to the canal, most notably the River Churn, the Boxwell springs, the Coln and a supply at Thames Head, were simply inadequate. Some serious miscalculations had occurred which nobody had spotted until June 1790. A Boulton & Watt steam engine was installed at Thames Head in September 1792 to draw water from newly-dug wells but problems with leakage led to the canal being closed over the winter of 1791–2 for repairs at the summit level. These problems, of course, would have paled if business had been brisk. But it wasn't. Trade along the line at the turn of the nineteenth century was, at best, moderate. The traffic was mostly in coal imported at Framilode and distributed to wharves along the Stroud Valley, to Cirencester and to villages along the upper Thames. Even with this, toll receipts were low. In 1793, with a mortgage and bond debt of £97,200 and share interest owing of £28,600, the company's finances were in a poor state. The interest on the debt was serviced by raising more debt and the situation reached a head in 1809 when arrears on interest stood at £193,892. An Act in that year allowed the company to issue preference shares alongside its ordinary shares and the company freed itself of debt for the first time. In 1809–10, a maiden dividend of 1$1/2$ per cent was paid to owners of both types of share.

By 1810 both the Kennet & Avon and the Wilts & Berks canals were open, and competition developed for the Bristol to Thames traffic. It was inevitable that tolls would be reduced, as they were under various agreements in 1811 and 1815. In the first half of 1817 some 10,615 tons of goods were shipped eastwards along the line, of which 9,481 tons was coal, primarily from the Forest of Dean. The trade was peculiarly eastward-dominated: only 664 tons of goods, half of which came from London, were shipped westwards. Naturally, the logistics of this in terms of boats having to return empty or only partially laden, had a big effect on profitability. The new canals were, however, not all bad news. The opening of the W&B (connected to the T&S via the North Wilts Canal) meant that the troublesome portions of the upper Thames were avoided and, as a consequence, the traffic levels at the eastern end of the canal began to increase. There was even a scheduled fly-boat that operated between Gloucester and London. But increasing competition from road transport, as well as from the W&B itself, limited toll receipts. For the period 1814–20, average receipts were £4,477, while for the period 1831–7, average receipts were £6,375. This increase in trade was helped by extensive works on the upper Thames, including dredging and better worked flash locks.

In 1835 a proposed railway between Cheltenham and Swindon via Gloucester and Stroud forced the canal to take drastic measures. It sought, unsuccessfully, to sell itself to the Cheltenham & Great Western Union Railway (later bought out by the Great Western Railway). The new railway was fully opened, a matter which included the construction of a second Sapperton tunnel, in May 1845. By 1847 the number of boats entering and leaving the canal from the W&B was under half of that just ten years before. Railway competition was also being felt for other coal traffic and toll revenue dropped from £11,000 in 1841 to £2,874 in 1855.

In 1865 the T&S decided to turn itself into a railway between Stroud and Cirencester, quoting the Sapperton Tunnel as being a prime asset. However, a bill of July 1866 was defeated in the Commons and all thoughts of challenging the GWR in the Stroud Valley were lost. Edward Leader Williams surveyed the T&S on behalf of the Stroudwater Company in 1875 and reported that the canal was in a poor state of repair. Although the situation was denied by the canal company, a majority shareholder, Richard Potter, sought an Act in 1882 to close the canal and convert it into a railway. This action was fought by what became known as the Allied Navigations, a consortium of canal companies including the Stroudwater, the Sharpness, the Staffs & Worcs, the Wilts & Berks, the Birmingham Canal Navigations and the Severn Commission. On 27 February 1882 the alliance agreed to underwrite the operation of the T&S (cost at about £550 p.a.) and to manage it. With this intervention the bill to convert the line to a railway was lost. However, the valiant and unusually cooperative efforts of the canal companies came to nought when in 1883 Richard Potter secretly sold his shares to the GWR.

The GWR treated its new acquisition with scant respect, selling off property and allowing the canal to decline for want of maintenance. The situation reached a climax in December 1893 when the company gave two days notice for the closure of the line between Chalford and the Thames. This led to a hurried re-assembly of some of the Allied Navigations, together with other interested business folk. This new consortium successfully persuaded the GWR to give it the T&S free of charge in return for promising not to build a railway line on the canal bed. Under an Act of 1895, the county councils of Gloucestershire, Wiltshire and Berkshire and the urban district councils of Stroud and Cirencester formed a trust to operate the canal, with powers to raise £15,000 to render the line into a navigable state. The line was to be managed on a day-to-day basis by the Stroudwater. At the same time the Thames Conservancy spent some £20,000 on dredging its waters and rebuilding many of the locks and weirs.

Under new management, the T&S re-opened in March 1899, only to close again shortly thereafter through shortage of water. At this point Gloucestershire County Council took legal control of the line, still receiving subsidies from the members of the alliance and still managed by the Stroudwater. With great expenditure of local government funds (nearly £30,000 in all), the canal was re-opened, bit by bit, in January 1904. However, by now there was little interest in using the route and the phrase 'white elephant' was widely used. Eventually the reality of the situation dawned and the last commercial voyage on the T&S was undertaken in 1911. The line was kept in water until 1925 when the legal requirement for subsidy guarantees ran out. After some negotiation, it was agreed to abandon the canal from Chalford to the Thames in 1927. Although the Stroudwater company fought hard to maintain the rest of the T&S, this battle too was lost in 1933 and the canal has been closed to traffic ever since.

After abandonment, Gloucestershire County Council sold lengths to various local landowners. With this division of the spoils, the future of the canal looked bleak. However, in 1972 the Stroudwater Canal Society was formed (the Stroudwater had been abandoned in 1954) and this was expanded in 1975 to form the Stroudwater, Thames & Severn Canal Trust Limited. In 1990 this metamorphosed into the Cotswolds Canals Trust, with the objective of restoring both waterways for navigation. The trust has its work cut out as anyone who does the walk will discover. It will obviously have to surmount the original problems related to the leaky summit and an inadequate water supply, but it will also have to overcome the new problems of a derelict tunnel and large sections of over-built canal line. In 1976 the cost of rendering the canal navigable was put at eight million pounds. Today it will be even more than that and at the time of writing new engineering surveys are being undertaken. The clear objective, however, is inspiring the trust and it may someday once again be possible to navigate along this, the first inter-city route to London.

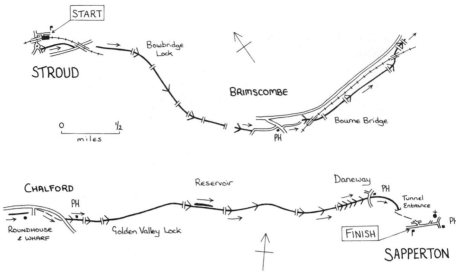

The Thames & Severn Canal

The Walk

Start:	Stroud (OS ref: SO 850051)
Finish:	Sapperton (OS ref: SO 947033)
Distance:	7¹/₂ miles/12 km
Maps:	OS Landranger 162 (Gloucester) and 163 (Cheltenham)
Return:	Stroud Valleys bus no. 64 from Sapperton to Stroud (enquiries: 0453-763421). No service on Sundays
Car park:	Stroud has well-signposted car parks and there is parking in Sapperton at the bus-stop near the school or near the church
Public transport:	Stroud has a main-line BR station

From Stroud railway station, take the road towards town and then turn left and left again to go under the railway. At the first small roundabout after the railway bridge, bear left to go past the Bell Hotel. Within a short distance a bridge goes over the canal. Sadly, the view to the right has been blocked by the new Bath Road. Formerly, apparently, it was possible to see the Stroudwater Navigation from here; the junction between the two waterways was just beyond a lock called Wallbridge Lower Lock. The navigation's basin and offices were to the left of the canal. Now all there is to see is a short, dismal length of canal, some scrubby concrete and the Bath Road.

Best therefore to get on with the walk by crossing to the left-hand side of the bridge and going down to the towpath which is to the right of the canal.

From the path the gates of Upper Lock can be seen. The decision as to which gauge should be used for the T&S was a tricky one. Severn trows were short and wide compared with the long and narrow Thames 'West Country' barges. Neither type would be able to travel the entire distance from Bristol to London without transhipment. The company could have decided on the use of a wholly new vessel able to undertake the entire trip but decided instead on building a special transhipment port at Brimscombe. Locks either side of Brimscombe are therefore of different sizes: all those to the west are built to take Severn trows, all those to the east built for barges. Upper Lock is thus a trow-sized lock 72 ft x 16 ft 6 in.

The area up from the lock is the site of Stroud Wharf, marked by the stone edging to the towpath and the old warehouse (built in 1828) on the opposite bank. From here the line winds around the outskirts of town. After going under the relatively recent Stroud bypass, the route seems to get lost as the canal disappears into a culvert. Follow the clear path that goes under the railway (the canal originally went beneath the steel span of the viaduct, though a restored line would go through one of the brick archways) and then go straight on with the road left and a wooden fence right. This leads back down to the left-hand bank of the canal, with the River Frome rushing headlong close to your left. Within a short distance, you reach the first of many mills to be seen between here and Sapperton: Arundel's Mill. This is marked by the presence of the sluice gates and an overgrown millpond. Like many of the other mills, when Arundel's was converted from water to steam power, it received its coal supplies directly from the T&S.

The path is now wedged between the canal and the River Frome. After going up and over a road bridge, you reach Bowbridge Lock. As at Wallbridge, the lock has been dammed to keep the length above it in water. To the left is a fine circular weir, and just beyond is the first of the many surviving T&S milestones which recorded the distance to both Wallbridge and Inglesham. The canal now bends round to reach Stanton's Bridge and Griffin's Lock. Above Griffin's Lock is Jubilee Bridge, an iron footbridge built in 1887 to celebrate Victoria's fifty years as queen. From here the canal heads towards Ham Mill Bridge and Lock which, like many along the way, are named after a local mill. This one is of note because here, in 1933, the last coal barge to be moved on the T&S was unloaded. A short distance further on is Bagpath Bridge. Just beyond the Hawker Siddeley works, you should peer into the distance to the left. The Phoenix Ironworks which once occupied this site produced machinery and lawn mowers but now the building produces books, as it is home to Alan Sutton Publishing.

From Bagpath, the canal curves gently south-east towards Hope Mill Lock. The lock has long since disappeared under concrete, and the road and buildings on top of it create just one of the major obstacles confronting the trust's restoration plans. The buildings to the right of the fenced footpath were formerly those of the Abdela & Mitchell boat building yard. A&M built a range of vessels varying from small sailing boats to large steam launches. The ex-mill buildings on the left were used as the offices and engine factory. A&M continued building here until the Second World War, even though the canal had closed in 1933. The part-built boats were sent out by road. Continue along the fenced path. The line of the canal shortly re-appears and 400 yd on is Gough's Orchard Lock, next to the once-extensive Brimscombe Mills. This extant portion of the line is short-lived and you soon arrive at a road. From here into Brimscombe the canal has been buried in concrete and brick. Walk along the road to a junction and turn right up Brimscombe Hill past The Ship (right) and alongside Port Mill (left). Turn left, following the advice given by the signpost, for Port Foundry. Here is a better view of the fine Port Mill buildings. The Bensons International car park to the left of the path was built on top of Brimscombe Port, once the headquarters of the T&S Company. It was here that cargo was transferred from trow to barge. Although yielding a healthy sum in port dues, this activity was a major obstacle to the smooth flow of traffic. The port consisted of a basin some 700 ft long and 250 ft wide which could hold a hundred boats. In the central area was an island on which goods liable to be pinched were safely stored. On the northern, left-hand, side was the principal wharf and the main T&S building, a huge three-storey house incorporating a store and office as well as homes for key staff. Also on that side was a transit shed, a forge and a boat-weighing machine. The building behind you (on the other side of the Frome) is one of two original structures that still remain. This was the salt warehouse. The stone building just inside the car park was another salt store. It is clearly impossible for the canal to be restored here and it is hoped that the River Frome can be used as a bypass.

To continue, follow the path which runs along the left-hand bank of the Frome. At Port Foundry the path leaves the river by bearing left to reach the canal at Bourne Lock. This was a hybrid lock, able to take both barges and trows. It allowed the trows access to the company's Bourne boat-yard which was just beyond. Above the lock, the railway swings back across the canal on an arched viaduct. The site of the boat-yard was across the canal to the left. The yard built barges and trows and maintained the company's working fleet of over fifty vessels. Boats were built in dry dock or on the bank and were launched sideways. The site has now disappeared under some post Second World War factories.

The canal now curves gently to Bourne Bridge where, on a site used by a body repair shop, Brimscombe Gasworks once received its coal supplies from the canal. After Beale's Lock and Bridge the towpath goes on to St Mary's Lock. In both cases the narrowness of the locks compared with those already past is noticeable. Just before St Mary's Lock is one of the few places where there is a connection between the canal and the Frome via a culvert under the towpath. Further on, just before the canal again passes under the railway, keep an eye open to the right of the towpath for a Great Western Railway boundary post. After the railway there is another, again on the edge of the towpath. The canal, meanwhile, goes on to Iles Mill Lock. Just after this the line is again filled in and four garages sit on top of what was once Ballinger's Lock. Within a short distance the canal reaches Chalford Wharf and roundhouse. These stone roundhouses are a unique feature of the T&S. There were five altogether: at Chalford, Coates, Cerney Wick, Marston Meysey and Inglesham. They were built as lengthsman's cottages in 1790–1. Each has three storeys: a ground floor stable, a first floor living room and an upstairs bedroom. The Chalford roundhouse, which was once a T&S museum, is now a private house. Why they were

The roundhouse at Chalford Wharf

built in this style isn't known, although there must be a good view up and down the line from the windows. Next to the roundhouse stands the former Company's Arms Inn, part of which dates back to the fifteenth century. Also on the wharf in front of the roundhouse is some old sluice gear from Sevill's Mill.

As Chapel Lock was infilled in 1964, the canal leaves Chalford via a culvert. The path leads out to cross a minor road. Go straight on behind a bus shelter and past Chalford trading estate. The canal now winds past the last group of mill buildings and under the A419. Here, bear left to cross the road (and the canal) to take a footpath that goes left opposite Springfield House Hotel. Now, with the canal to the right, walk on to pass Bell Lock and, later, Red Lion Lock, both of which were named after local pubs. The lock bridge for Red Lion Lock, about a hundred yards up from the pub, bears the inscription 'CLOWS ENGr 1785', commemorating the engineer Josiah Clowes and the date when the first commercial load left Wallbridge. Around the next bend is Golden Valley Lock. The Cotswold-stone house just beyond was formerly a millhouse, became the Clothiers Arms pub, then the Valley Inn and is now a private house. From here the canal leaves the village of Chalford to pass the large Victorian waterworks which received the coal

Bell Lock and the Bell and Red Lion inns, photographed in 1910 for a Frith & Co. postcard

Gloucester Record Office, The Boat Museum Archive

that powered its steam pumps by canal. The pound after Baker's Mill Lower Lock was one of those that leaked badly and which was lined with concrete in the 1890s. The current low water level reveals this rather nicely. Below Baker's Mill Upper Lock is Twissell's Mill, a cloth mill that later ground corn. The millpond and some of the surrounding land was bought by the T&S and converted into a reservoir. The new 'lake' is 900 ft long, 94 ft wide and contains three million gallons of water. It feeds the canal through a small culvert under the towpath just above Upper Lock.

After the reservoir there are two locks at Puck Mill. The pound between them was infamous for its leaks and water was said to flow out as quickly as it went in. Even in 1907, Gloucestershire County Council was spending vast amounts of cash relining this stretch. Temple Thurston comments on the lack of water during his visit in 1910: '. . . the draught of water is bad; in some places we just floated, and no more. . . . There were times when I thought the tow-line must give way, the strain upon it was so great.' At Puck Mill Upper Lock, the towpath changes sides. For a section above the mill site, the canal is constricted by the walls of Puck Mill Wharf. From 1927 to 1933 this was the eastern-most end of the canal. Everything from here up was officially abandoned in 1927 and the rest followed just six years later.

Whitehall Upper Lock, a surprisingly long distance from the Lower Lock, is the bottom of a flight of seven in just 1/2 mile up to the summit level at Daneway. In an attempt to limit the amount of water used to run these locks, major alterations were made in the 1820s and '30s. By then, 70 ft long narrowboats were replacing the 90 ft Thames barges. The T&S was there-fore able to shorten the locks by 20 ft and reduce downstream water flow by about 20 per cent. At the same time, the top locks were all fitted with side-pounds, limiting water loss still further by retaining some of the water evac-uated when the lock was emptied. The remains of some of the side-pounds can still be found amid the undergrowth.

The next lock is Bathurst Meadow Lock. The towpath changes sides and, within 300 yd, Sickeridge Wood Lower, Middle and Upper locks follow. At Daneway, the Basin Lock stands alongside the wharf and basin. Above the lock the canal opens into a winding hole with the wharf entrance to the right. The basin was used to unload coal, stone and timber, as well as being a place for boats to moor before venturing into Sapperton Tunnel. The con-crete-lined basin also served as a reservoir for the locks down the valley. The small wharfinger's cottage has been modernized and is now a private house, while the wharf itself has been flattened. The canal reached Daneway in 1786, three years before the tunnel was finished. It thus became the termi-nus for the line and, with a wharf, warehouse and coal-yard, was relatively busy with goods being unloaded for Cirencester and beyond. Immediately after the bridge is Daneway Lock and the canal summit, a total rise of 241 ft through twenty-eight locks. Again the towpath changes sides and the lock itself has been infilled to form a car park for the Daneway Inn. The inn was built in 1784 as accommodation for the navvies working on the tunnel. It was sold in 1807 to become the Bricklayer's Arms and has been a pub ever since.

From the road, follow the footpath sign to Sapperton over a stile, into the field below the car park and back to the towpath. The line passes over the culverted River Frome, past a derelict watchman's cottage and into Sapperton Tunnel. The tunnel was by far the most important engineering project on the T&S. Started under contract to the incompetent and often drunk Charles Jones, the work had to be finished by others, most notably John Nock and Ralph Sheppard. In the late 1700s tunnelling on this scale was an unknown quantity. Even now there are only two longer canal tunnels in the country. At its deepest point the tunnel is 200 ft below the surface and, in all, it is 3,817 yd long. As was normal for tunnels built at this time, there is no towpath. Boats were 'legged' or pushed through by men who lay on the roof or prow of the vessel and who then walked against the tunnel walls. This process could take five hours eastbound, against the flow from the pumping station, and three hours westbound. As it wasn't possible for boats to pass each other in the tunnel, only three or four passages per day were allowed and then at specific times. While the legging was in progress, the horses or donkeys followed a path over the hill. The tunnel itself has been blocked since the First World War, the crumbling layers of fuller's earth and inferior oolite proving too unstable despite the brick and stone lining. Its restoration, still a key hope of the canal trust, will be technically possible but inordinately expensive. The portal here originally bore some fancy Gothic battlements and fine tall, pointed finials.

To return to Sapperton, go back to what remains of the lengthsman's cottage and go up the hill. This moderately clear path (aim slightly right of the church) leads over two stiles to the road and Sapperton school for either the car or the bus back to Stroud.

Further Explorations

Although the T&S has been closed since 1933, there is still quite a lot to be seen. Sadly, not all of it is open to the public, although some lengths of the towpath are rights of way. If you wish to see the entire line, arm yourself with an Ordnance Survey map and Handford and Viner's book and you should be able to make your way without trespass.

One easily accessible section is that leading to the eastern portal of Sapperton Tunnel. This walk of about 1 1/2 miles starts in Coates, near the village hall. Coates can be found off the A419 to the south-east of

The Tunnel House at Coates near the eastern end of the Sapperton Tunnel

Sapperton at SO 978007 on OS Landranger 163. From the village hall walk back towards the road with the bus shelter and turn right. Go along this lane past another bus-stop. When the road bends sharply right, carry straight on along a fenced path signposted to the church. Go through the churchyard. At the lane turn left to go through a farmyard. Walk along the left edge of the field to another field. Here the clear path goes across the middle to a gap in the wall. Continue across the middle of the next field to a stile. Cross the railway to a stile and bear left along the lane towards the Tunnel House. To the left here is the eastern portal of Sapperton Tunnel and the T&S.

The Tunnel House, originally called the New Inn, was built by Lord Bathurst to house the men working on the tunnel excavation. The original building was burnt down in 1952 but has been successfully restored if on a slightly smaller scale. To reach the canal, cross over the top of the portal and turn left down the slope to the right-hand bank. The eastern portal is of a completely different design to the western. This side is much more classical in style, built of stone with columns and nooks for statuettes. Figures of Father Thames and Madam Sabrina were intended to stand on the portal but somehow never made it. Similarly, an inscription was never added to the space allowed for it; perhaps finding the right candidate for such an honour alluded them. What you presently see is in fact all thanks to some fine restoration work that was completed in 1977.

The waterway stretching out from the tunnel through the cutting was once known as 'King's Reach', following a visit to the eastern portal by George III on 19 July 1788. The king also had a quick look at the western portal and the canal at Wallbridge. Whether he walked along the towpath isn't recorded but that is your route back to Coates. The section of canal through the cutting was lined with concrete in the early years of this century in an attempt to stop the leaks. The path leads round to Tarlton Bridge and on to Coates roundhouse. This, like the one at Chalford, was built as a lengthsman's cottage in 1790. This particular roundhouse once had a concave, lead-lined roof that was purportedly used to catch water to top up the canal. This house also has a kitchen extension that was built at the behest of the incumbent's wife in the late nineteenth century. The ground floor, which was normally a stable, was converted into living quarters at the same time. Sadly, the Coates house is now pretty dilapidated.

The path continues under the railway line and along a progressively more overgrown waterway. Eventually the line dries up completely. At the next accommodation bridge, Trewsbury Bridge, go up the side and turn left. The track leads through a small farmyard and out along the right-hand edge of a field. As the wall bends left, go over a wall stile into a field. Continue now with a wall close left to another stile and a road. Turn left to return to the centre of Coates.

Further Information

The Cotswold Canal Trust aims to preserve what's left of the two Stroud waterways and to restore them for navigation if at all possible.

The Cotswold Canal Trust,
FREEPOST (GL 65),
PO Box 71,
Stroud,
Gloucestershire GL6 7BR.

Those wishing to know more about the T&S should read:
Household, H., *The Thames & Severn Canal*. Alan Sutton Publishing, 1983.

12
THE WEY & ARUN JUNCTION CANAL
Alfold and Loxwood

Introduction

Although the Wey & Arun Junction Canal is one of the country's lost water-ways – in places it is no more than a dry ditch – it has as enthusiastic a group of supporters as any in the country. Many stretches have been cleared and dredged, new locks and bridges have been built and people talk about the canal as if all the hard work had an inevitable conclusion. And, to coin a phrase, why not? It is possible, just possible, that someday boats may once again navigate along an inland route from London to the south coast.

The River Wey rises near Selborne in Hampshire and passes via Guildford to the Thames at Weybridge. The River Arun rises some 30 miles to the east amid St Leonard's Forest. It flows to the channel at Littlehampton by Pulborough and Arundel. At their nearest points the two rivers are only 10 miles apart and have tributaries that are just 2 miles from each other. The Wey & Arun Junction Canal leaves the Wey Navigation at Stonebridge Wharf, Shalford and ascends via Bramley and Run Common by seven locks to a 5 mile summit level near Cranleigh. Here the canal crosses the Surrey/Sussex watershed 163 ft above sea level. The line begins its descent near Alfold where it winds through Sidney Wood to go over a tributary of the Arun by the Drungewick Aqueduct. From here the line runs south via Malham and Rowner to New Bridge near Pulborough where the canal joins the Arun Navigation.

This walk is more akin to a nature ramble than a canal walk but is none the poorer. Here can be seen the to-be-restored and the restored in close proximity and a fascinating view of the problems involved in resurrecting a waterway that has been derelict for a hundred years.

History

There were plans to make the River Wey navigable as long ago as 1621, and in 1651 Guildford Corporation obtained an Act which enabled it to upgrade the river from the Thames at Weybridge. The Wey Navigation, completed in 1653, was 15 miles long with twelve locks. It was one of the first rivers in England to be canalized and was used primarily to ship Surrey grain into London and to carry coal from the Thames wharves. Although fraught with financial shenanigans, the navigation was highly successful and made Guildford a major inland port. So much so that the line was extended by 4½ miles in 1760 via four locks to Godalming.

The River Arun has been navigable from Pulborough to the sea since the Norman Conquest. In 1544 Henry Fitzalan oversaw a series of improvements to the main channel to North Stoke (some 2½ miles north of Arundel), including the construction of twenty-nine gated weirs. By 1623 proposals were made to extend the navigable waters along the line of the river to New Bridge near Billingshurst. The line to the sea was improved in 1732 when a new channel and harbour were built at Littlehampton. In 1785 the Arun Navigation Company was established and it raised £10,000 to improve the navigation with two cuts, the Arun Canal between Coldwaltham

Even in the early 1900s the Wey & Arun Junction Canal was partly derelict. Here an early canal restoration enthusiast admires the southernmost lock on the line, Rowner Lock
The Boat Museum Archive

and Hardham, and a stretch between Pallingham and New Bridge. The improved line was opened in 1790, by which time the company was already issuing dividends.

A bill to link the rivers Wey and Arun was brought to the House of Lords in 1641. This was the first serious attempt in Britain to link one river with another by making an artificial cut. The bill proposed that the new waterway would take a line along both rivers with a 2 mile long junction canal from a tributary of the Wey at Cranleigh to a minor branch of the Arun at Dunsfold. The main purpose of the new line was to avoid the often impassable roads between the south coast and London. At a time when Parliament was busy with its own problems, the bill was lost. Much later, the success of the Arun Navigation prompted that company to consider extending its line north of the already canalized section at Orfold and New Bridge, and in 1791 a proposal was made for a line north-west via Wisborough Green and Kirdford to Northchapel. At an independent meeting chaired by the Duke of Norfolk in Horsham on 9 July 1792, it was decided to continue the waterway to Weald Cross, Slinfold and then to Farthing Bridge near the centre of Horsham. This scheme was surveyed by John Rennie who estimated the cost at £18,133. The plan came to nought when negotiations with the Arun fell through in 1794. Another scheme, proposed in 1798 by William Marshall, suggested an extension through Horsham to Betchworth and Dorking. This met a similar fate.

By the turn of the nineteenth century a waterway was open from the sea to Arundel for vessels up to 200 tons and for barges as far as New Bridge along the Arun, and to Midhurst along the Rother Navigation which joined the Arun near Stopham. The line did not have a towpath and barges were sailed or punted upriver. The journey from Littlehampton to New Bridge took 2$^{1}/_{2}$ days and barges were constantly being held up by floods in winter and droughts in summer. The company itself was heavily burdened with debt and was unable to pay staff, even though dividends of 2–4 per cent were still being disbursed. The situation was resolved by Lord Egremont, who bought the company and became chairman in 1796. Egremont was keen to extend the line north, believing that there were advantages for the local population should a route to London be opened. Throughout these exchanges, the Wey Navigation was doing well. In 1801 the line carried 63,000 tons of cargo, an increase that was partly due to the opening of the Basingstoke.

In 1803 John Rennie surveyed and proposed a scheme for a canal to run from the Croydon Canal via Redhill, Crawley and Chichester to Portsmouth. The cost was estimated at £720,649 for a broad canal. Although the idea was presented to Parliament, heavy opposition and a lack of financial support led to the shelving of the idea. Rennie had a second attempt in 1810 with his Grand Southern Canal that ran from the Medway to Portsmouth via Tonbridge, Crawley and Pulborough. Although costing a similar amount, subscriptions were more forthcoming and the project only foundered because of a lack of confidence over traffic levels along such a circuitous route. Another plan, proposed in 1807, was the Portsmouth, Southampton & London Junction Canal, which would have run from the Basingstoke at Aldershot via Farnham and Winchester, then along the Itchen Navigation to Southampton. The cost was put at £200,000, which included a 2 mile tunnel between Alresford and Alton. Although subscriptions were raised easily, the scheme collapsed on adverse cost estimates, water supply problems and controversy over the business forecasts.

After all these comings and goings, it was left to Lord Egremont to further the Wey & Arun Junction Canal Bill. The scheme was made public at a meeting at the White Hart Inn in Guildford on 1 June 1811. The proposal was for a line of 17 miles to join the Wey and Arun Navigations from New Bridge, Wisborough Green to Stonebridge, Shalford. The new line would complete a 90 mile route from London to the English Channel at Littlehampton. Egremont had secured the support of most of the principal landowners involved, including the Duke of Norfolk, the Earl of Onslow, Lord Grantley and Lord King of Ockham. Subsequent meetings, both in Guildford and in Godalming, voiced support for the idea and a committee was set up to further the project with a petition to parliament. Josias Jessop was engaged to undertake the survey. In August 1811 the committee decided to go ahead with the project at an estimated cost of £71,217, rounded up in the name of prudence to £90,000. In proposing the canal, Lord Egremont suggested that some 1,200,000 tons of goods were shipped annually from London to Portsmouth and that it was reasonable to expect the canal to take just one-twelfth of this. The trade would primarily be in coal, Portland stone, groceries, chalk, lime, timber and manure.

By October 1811 some £51,000 had been raised, of which Egremont had promised £20,000. The issue was fully subscribed by 16 November 1811, with 132 shareholders. By May 1812 Jessop had undertaken a detailed survey that included some diversions at the behest of landowners. The survey included the diversion of the River Arun between Malham and Rowner where the canal would have otherwise crossed the river twice in half a mile, necessitating two aqueducts. The revised estimate was £86,132. Finally, after negotiations with landowners, the Act was passed on 1 April 1813. It enabled the company to raise £90,500, with an additional £9,500 if required. The Act authorized maximum tolls at 4d. per ton per mile for most cargoes, with 2d. per ton per mile for manure. Lord Egremont became the company's first chairman, Jessop was retained as engineer and May Upton was made resident engineer and clerk of works. The construction work was contracted out to a builder from Alfold, Zachariah Keppel. The canal was to be 30 ft wide and consist of twenty-three locks, each measuring

75 ft by 12 ft, able to accommodate barges carrying 50 tons. Part of the water supply to the line was to come from a new reservoir built in the grounds of Vachery House, Cranleigh.

Almost as soon as the work began, landowners held out for independent arbitration on the value of the lands being bought by the company and the estimates of any damages caused during the construction phase. As the commissioners were themselves mostly landowners, some element of bias might be supposed and the company found itself spending far more than planned for both land purchase and compensation. Overall, this took over a quarter of the final cost of the W&A.

The first sod was cut at Shalford in July 1813 and work started at the southern end in the following May. The undertaking seems to have gone well up to the point where Zachariah Keppel became bankrupt after suffering some kind of cash-flow crisis. The winter of 1814 was also a very wet one, virtually stopping the work that was now under the supervision of May Upton. A further problem occurred at the Alfold cutting, which was found to pass through beds of sand and thus needed additional lining. The first tolls were taken at Bramley on 18 December 1815 but by this time funds were running short. With extra funds from two 'final' calls on shares, the work was finished in August 1816. The completed line was 18½ miles long and the company had spent £102,626. The official opening on 29 September was marked with celebration that began at the Compasses Inn at Alfold and continued with a procession of barges up the line to Guildford. The celebration must have continued in Guildford where the price of coal fell by 20 per cent.

From the outset traffic along the canal was desultory. The average receipts for the first seven years of trading were £1,275 p.a., with less than 10,000 tons of cargo: just one-twelfth of the one-twelfth that Lord Egremont had expected. This figure appears even smaller when it is seen that the W&A carried less than half of that carried on the Arun Navigation. The disadvantages in having to tranship goods at Littlehampton, only to suffer the problems of going along the often impassable stretches of the upper Arun, made the sea passage to London look very attractive. The inland route proved more expensive with similar journey times. With the end of the war with France, the safety of the coastal route had also much improved, thereby encouraging coasters to make the trip. The lack of canalside industry cannot have helped matters. Even the W&A's initial success in delivering coal to Guildford was lost when the Wey Navigation reduced its tolls to make supplies from the Thames cheaper. As a consequence the W&A was unable to pay dividends for any of its first five years of operation. Despite these drawbacks and the generally low level of business, the canal carried seaweed to local farms, grain to watermills, and coal, groceries and merchandise inland from the coast. The boats returned with farm produce, timber, bark, flour and various rural goods. Chalk, clay, sand and gravel were also moved along the line. With the prospect of the Portsmouth & Arundel Canal completing the route from London to Portsmouth (the Act was passed in 1817), a small dividend was paid in 1821, but as this extension of the Arun Navigation did not open until 1828 this act of generosity using borrowed funds was not repeated. Instead the company set about encouraging trade through toll reductions and by signing agreements with the Wey and Arun navigations and the P&A in order to reduce the cost of carrying cargo from the coast to the Thames. This action, however, failed to produce the hoped for increase in business and in 1823 more toll reductions had to be made. Part of the problem was in navigating the Arun in periods of flood or drought. Even in good conditions the shallowness of the river meant that barges could not exceed 30 tons. After some pressure, the Arun company made improvements to its waterway so that by 1825 barges carrying loads of 40 tons could pass from Littlehampton to Guildford.

The opening of the P&A provided some improvement in the toll revenue. Income increased by 60 per cent to £1,989 in 1824. By 1826 receipts were £2,355 and a dividend of 1 per cent was announced. However, with the failure of the P&A (see below), toll receipts promptly fell by 20 per cent. There was some recovery during the 1830s, which saw good levels of trade throughout the Thames to Portsmouth lines. The Wey, for example, carried 86,000 tons of cargo in 1838 and the Arun paid a 12 per cent dividend. The W&A could only pay 1 per cent but maintained tonnage at around 20,000 p.a. (receipts peaked at £2,525 in 1839–40). By this time a £100 share was selling for £25 but at least the drop in value had been halted. The canal was still paying off its debt throughout the 1830s and it wasn't until 1842 that it was freed from this burden. Among the business on the canal at this time was coal from South Wales, imported via Arundel and delivered to Cranleigh and Bramley. The return cargo was primarily timber and stone. There were also fly-boats from Portsmouth to London.

By the 1840s the improved condition of the highways in Surrey and Sussex meant that road transport was beginning to have an impact on toll receipts on the W&A. Trade was beginning to drop rapidly: down 38 per cent between 1839 and 1840 and 1842–3, and over 65 per cent by 1851–2. The first railway in the region was the London to Brighton line, opened in September 1841. A line to Guildford and Godalming (opened in 1844 and 1849 respectively) reduced the receipts on the Wey Navigation by over 50 per cent. On 8 June 1846 a railway line from Shoreham to Chichester was opened, leading almost immediately to the total demise of the eastern half of the P&A. In 1850 the main cargoes carried on the W&A were coal (50 per cent) and timber (25 per cent) but by this time the total traffic was down to 15,121 tons p.a., with toll receipts of £1,036.

With the advent of a direct railway from London to Portsmouth, it was widely rumoured that the W&A would sell out to a railway company. Although this proved to be groundless, by the time the through line was built in 1859, traffic along the W&A was only maintained by reducing toll rates. The building of the Mid-Sussex Railway (Horsham to Midhurst) helped keep the wolf from the door for a while as some of the required materials were shipped by both the W&A and the Arun Navigation. But with completion of the Horsham to Guildford Direct Railway (which became the London, Brighton & South Coast Railway) on 2 October 1865, the W&A was forever bypassed. Indeed, it is this line that runs parallel with the W&A from Shalford to Cranleigh. So it was that in 1865 the last dividend of the W&A was paid. By 1866–7 the amount of cargo carried had nearly halved (in just two years) to 8,750 tons p.a. Cargo that went virtually straight away to the railway included coal deliveries to Cranleigh and the shipment of charcoal from Run Common.

At the W&A Annual General Meeting in May 1866, it was resolved that the company had no option but to dispose of its property. It owned over 200 acres of land, some cottages and other buildings. The intention was to wind up the company. However, at a meeting in October 1866, the motion to do so was rejected, primarily because a number of bargeowners had been buying up shares and were thus able to rally support to keep the waterway open. In response, a scheme was launched in which the Arun Navigation agreed to subsidize the line at £120 p.a. and to take responsibility for day-to-day management. Even this failed. On 11 January 1867 a poorly attended meeting decided to reject the proposal and resolved to go into voluntary liquidation. The closure of the canal was enabled by Parliament on 31 July 1868. With various rescue schemes coming to nought, the chattels were auctioned on 30 August 1870, although there was no bid for the canal line itself. This last hope having passed, the canal was officially closed to traffic on 22 July 1871. In the subsequent years it was run by the liquidator, with small portions gradually being sold off to local landowners. By 1901 all but eight of the 200 acres of the line had been sold and the company was eventually dissolved in 1910.

Meanwhile, the Arun was undergoing similar traumas. In 1882 toll receipts dipped below £100. Despite voluntary contributions from the proprietors, on 1 January 1888 closure notices were posted. The line was finally wound up on 23 September 1896. The fortunes of the Wey also declined towards the end of the century but, during the 1920s the Wey was carrying 55,000 tons p.a. and remained busy until the Second World War. Even in 1956 traffic was still using the navigation, with 16,105 tons being carried. The continuing decline eventually prompted the owner, Harry Stevens, to donate the line to the National Trust in 1963. Commercial traffic on the Wey ceased in 1969.

Even though the canal restoration movement grew following the establishment of the Inland Waterways Association, attention has never been overly focused on the W&A. Widely regarded as too far gone, the more immediate projects of the Kennet & Avon and the Basingstoke canals have been far more attractive. Despite this, the W&A has always had a band of enthusiasts. This initially small group, started in 1970 with the formation of the W&A Canal Society with, appropriately, Lord Egremont as its president. The Canal Trust (as it became in 1973) has raised funds, rebuilt locks and discussed the possibility of re-opening the line with the forty-two landowners involved. The northern end near Bramley in particular is a problem. The area is now heavily built-up and a new line will have to be sought. The hope of restoration is getting brighter but it will make the resurrection of the Basingstoke seem simple.

The Walk

Start and finish:	Alfold (OS ref: TV 038340)
Distance:	8^1/$_2$ miles/13^1/$_2$ km
Maps:	OS Landranger 186 (Aldershot & Guildford) and 187 (Dorking, Reigate & Crawley)
Car park:	Near The Crown in the centre of Alfold
Public transport:	None convenient

This circular walk goes firstly along a stretch of the abandoned canal and then a section under restoration. The return route is away from the canal and through the leafy Sussex and Surrey countryside. Casual towpathers should note that sections along the abandoned canal can be very muddy.

Alfold is a village on the B2133 that runs south from the main A281 Horsham to Guildford road. There is a limited space for car parking around the village green next to The Crown pub and St Nicholas' church. From the green, walk away from the B2133 towards the church. Bear right along Rosemary Lane to pass the Old Rectory and Linden Farm. This pleasant country road passes through some arable and then some wooded areas to reach Velhurst. A little further on, the lane reaches a fine house with some stabling. Here is a notice announcing Sidney Wood Farm and a small bridge (although it is called High Bridge) that once passed over the Wey & Arun.

By looking first to the right, the course of the canal can be seen disappearing into the middle distance. This line can be followed north by turning right just before the house and going into the Sidney Wood. However, the walk turns left just after the bridge to follow a bridleway sign which points along the right hand bank of the completely overgrown canal bed. There were once a total of six locks between here and the Onslow Arms at Loxwood (the Sidney Wood section as a whole had ten in just over 2 miles

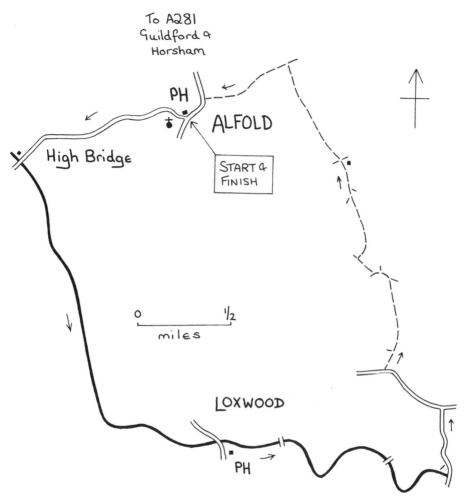

The Wey & Arun Junction Canal

for a descent of almost 90 ft). Sadly, most of these have now disappeared into the undergrowth. This, it seems, is partly a result of a certain degree of brick plundering just after the First World War and partly because what was left was blown up by the army for demolition practice during the Second World War. There was one lock, No. XII, not far in from High Bridge, although I wasn't able to find any trace of it. For avid lock-searchers, a sure sign of the positioning of any lock is a point where the canal channel becomes unusually deep and then the towpath passes down a moderately steep slope to meet it. But with the canal having closed in 1871 and the army having successfully completed their exercises, little or nothing remains

The Onslow Arms, a former favourite stopping place for passing boaters, at Loxwood

at some points and each individual lock needs some finding.

It's not all bad news though! The canal bed is now a fine area for wildlife. There are abundant rabbit holes and a wide variety of song birds. At one point, in early June, the sides of the canal bed were covered in early purple orchid. And if you loiter for a moment, the peace and quiet engulf you, a rare commodity indeed in southern England. In the days when the canal was operational, the loneliness of the wood wasn't missed by a number of locals who saw their opportunity. Passing barges were often subject to casual theft. One Alfold schoolmaster, for example, had a penchant for stealing duty-paid spirits as they made their way along the canal!

Using the criteria described earlier, Lock X is detectable as the towpath slopes down to a spot where there is a break in the canal wall to a dried-up lake bed. This lock, in the middle of Gennet's Wood, was severely damaged in 1841 by Charles Baverstock, who ran his barge into its gates. The resulting rebuilding and closure of the canal cost the company about £225. Baverstock was charged with negligence and fined £2 14s. After passing through a gate, a footbridge goes over the canal near Sydney Farm. Lock IX was, apparently, just before the bridge. Continue along to a signpost where another footpath crosses the canal. The footpath continues along the right bank, through a series of gates to a point where a dirt road crosses the canal to a house at Gennet's Bridge. Lock VIII was just before the bridge.

The act of passing down the Sidney Wood Locks was described in J.B. Dashwood's cruising guide of the 1850s as the 'great grind of the trip'. The book records that as each pound had to be left empty, downstream passage could be remarkably slow. The quickest method of passing through was to have two winches and to send a companion running on ahead to get the next lock ready for entry.

An old lock gate lies rotting in the mud at Brewhurst Lock near Loxwood

The metal quoins of the next lock (VII Southland) are still extant and thus give the first certain sighting of one of the Sidney Wood six. The lock gates worked against the quoins which were made of forged steel in 1814. The path now runs by an open field to the right, through which a footpath (from Ifold) comes to join the towpath. The canal bends left to reach a brick bridge, Devil's Hole Bridge. The stone walls of Devil's Hole Lock (VI) are, perhaps remarkably, in reasonable condition and can be seen from the bridge. The bridge itself was rebuilt recently by a combination of the West Sussex County Council and Loxwood Parish Council Youth Opportunities Project. Shortly after this, the path broadens to a dirt track which leads to a road, the B2133, and Loxwood. To the left is the site of the former Loxwood Wharf. Barges brought seaweed and lime, coal and gravel here and took away farm produce, timber and grain. To the right is the Onslow Arms. The pub was very popular with boatmen as it was the only stopping place between Bramley and Pulborough. The Earl of Onslow was one of the promoters of the original W&A bill. The pub that bears his name is recommended by CAMRA in the *Good Beer Guide*. The path continues on across the road and alongside a restored section of the canal bed. Within a short distance, this reaches Brewhurst Lock (V), derelict but definitely recognizable. The line of the canal from here on has received significant amounts of attention from the Wey & Arun Trust. Indeed, this section of the line has been leased by the trust for restoration purposes. It's worth noting as you pass Brewhurst Mill that in the original plans, the line of the canal would have turned north here via central Alfold to a point about a mile north of Alfold Crossways and would not have followed the route through Sidney Wood. The route was diverted to avoid Alfold village and various game reserves. It added 1½ miles to the length and £15,000 to the cost.

The activity of the canal trust is well demonstrated at the next lock, Baldwin's Knob Lock (IV), named after a local hillock. Here extensive renovations have been undertaken. The forged steel quoins are the originals but the masonry has been considerably reworked, as have the gates and culverts. Continue round the contours and along a wooded section that goes past a beautifully restored bridge and on into open country. The canal here was full and the whole scene idyllic. The sadness comes when one realizes how isolated it is from the rest of the waterway. The extant canal comes to a stop in the middle of a field. The original line went slightly right of centre and over the stream via the Drungewick Aqueduct. The aqueduct, which was demolished in 1957, was not one of the canal system's most ornate, being a rather heavy, even clumsy-looking, brick structure with three diminutive arches. Some canal trust volunteers at Baldwin's Knob Lock informed me that the aqueduct will be replaced by a steel channelled version faced with either stone or brick.

The towpath right of way comes to an end before the stream. To return to Alfold you should now take a route which leaves the canal. If you prefer, you can return to Loxwood and turn right to walk along the road. Alternatively, you could just walk back the way you came. The return route that I took keeps away from roads where possible but has no canal interest.

Bear left through a gate and turn left along a lane. At a T-junction, turn left. As the road bends left, there is a small side lane to the right. Go along this and turn right at the end along a bridleway. Continue past a sign for Pallinghurst Estate and along a path to the left of a gate. Go over a cross track and on through woods with fields to the left. At a T-junction, turn left to pass footpaths to right and left. The bridleway bends right and passes another footpath off to the left. Cross two dirt road junctions to reach a spot where there is a rash of footpath signs. Here go straight on through a gate to join the Sussex Border Path. This goes past a pond into an open area. Bear left through a gate and along a path with open fields to the left. After passing an attractive farmhouse, you will reach a wide dirt lane. Turn left. Follow the course of this track to reach a road and then turn left into Alfold.

Further Explorations

A walk from Guildford to Amberley along the Wey, the W&A and the Arun has been devised and is known as the Wey–South Path. As large sections of the waterway are not open to the public, the walk sometimes leaves the canal line but stays as close as possible throughout. The official route is 36 miles long. If you then follow the right of way along the River Arun to Littlehampton you will add an extra 13 miles. It could perhaps be managed over three full days. A description of the Wey–South Path can be found in Aeneas Mackintosh's booklet of the same name, available from the W&A Canal Trust (address below).

Not included in the Wey–South Path but well worth attention is the Chichester Canal. The Portsmouth & Arundel Canal Act, of which the Chichester was a constituent part, was passed in July 1817. The line was surveyed by Netlam and Francis Giles under the supervision of John Rennie. As with the W&A, the principal shareholder was Lord Egremont. The canal was in two sections: Chichester and Portsea. The Chichester line, a barge canal of 12 miles, ran from the River Arun at Ford through Yapton and Barnham to Hunston (where a branch of $1^1/_4$ miles went into Chichester), then to Birdham and Chichester Harbour. Barges for Portsmouth were then taken 13 miles around Thorney and Hayling Island to the Portsea Canal which took boats $2^1/_2$ miles into Portsmouth. An additional cut across the neck of Portsea Island to give access to Portsmouth Harbour (the Cosham cut) was added later. Following the passage of an amendment Act of 1819, the line from Birdham to Chichester was built as a ship canal, able to take vessels carrying up to 100 tons.

Despite initial enthusiasm for what was the final part of the London to Portsmouth canal route, investment was hard to find and only the combined action of Lord Egremont's wallet and Government Exchequer Bills enabled the project to go ahead. The Chichester Canal (Chichester to Birdham) officially opened on 9 April 1822, the Portsea on 19 September 1822. The whole line, including the Ford to Hunston section, was finally opened for traffic on 26 May 1823. It had cost £170,000. A barge could now make the trip between London and Portsmouth, a distance of 116 miles, in four days. The P&A was never a success. The most prosperous year was 1824, when 3,650 tons of cargo were carried against the predicted 55,000 tons. Expected government contracts for London to Portsmouth navy traffic didn't materialize. The reason was simple. Outside of wartime, the sea route was easier and quicker.

In 1825 trade halved and by 1826 it virtually disappeared with the collapse of the Portsmouth Barge Company. In the same year Lord Egremont severed his connection. An Act of 1828 enabled the raising of a further £50,000, which was used to complete the Cosham cut and to make good parts of the line that were already in disrepair. In addition, there were toll reductions, promotions of passenger traffic and a new carrying business. With this, traffic in 1833 rose to 2,500 tons, comprising mostly foodstuffs for the navy. Disappointingly, by 1838 the level was down to 750 tons. In September 1840 the last barge to sail from Chichester to London carried 6 tons of groceries. By the time the Shoreham to Chichester railway opened on 8 June 1846, the eastern end of the Chichester Canal from Hunston to Ford was already derelict and by 1853 it fell into disuse. By 1867 the line from Hunston to the Arun was reported as being 'quite dry' and various lengths sold. The P&A was eventually wound up on 3 November 1896. Various parts were then sold, although the line between Chichester and Birdham was presented to Chichester Corporation. With the turn of the twentieth century, traffic along the last remaining section dropped and was gone by 1906. In 1924 two bridges were culverted and on 6 June 1928 the corporation abandoned the line. A short section near the sea was re-opened in 1932 for yacht mooring but the rest was allowed to deteriorate. It was sold to West Sussex County Council in 1957, leased to anglers in 1972 and taken over by the Chichester Canal Society in 1984. With this, new hope has arisen, with plans to restore the canal to a navigable state.

Although the section from Hunston to Ford has long been closed, there is a good walk between Chichester and the yacht basin at Birdham. The walk starts at Chichester BR station. Go over the level-crossing along Stockbridge Road to the Richmond Arms. Here turn right along Canal Wharf to reach Southgate Basin, the terminus for the Chichester arm of the P&A. A public footpath sign points round the back of the pub and along the right-hand bank of the canal. The walk starts noisily as you go past Padwick Bridge and under the Chichester bypass. But gradually you enter the peaceful arable land to the south of the city. About a mile after the bypass, the canal reaches Hunston, where the towpath is forced left over a footbridge. Turn right at the end of the bridge to walk along the road for a short distance and then back onto the left-hand bank of the canal. This sudden 90 degree right turn appears peculiar but marks the spot where the Chichester arm joined the P&A proper. The line of the P&A ran east–west here (with the Chichester arm going roughly north). The original course east went across the road, through the farm buildings and on to Yapton and Ford, where it joined the River Arun.

Continue along the canal to Selsey Tramway Bridge. This old railway ran from 1897 to 1935 and carried both goods and passengers between Chichester and Selsey, which is a little over 6 miles to the south of here. The towpath continues to a road and Crosbie Bridge which was culverted in 1924/5. At the next road (Cutfield Bridge), the towpath changes banks. Just before entering the Chichester Yacht Basin, look left to the canal and the weir that marks the remains of the first sea lock, Manhood End Lock. Having pondered the origins of this name, continue along the road adjacent

A ketch negotiates The Hundred of Manhood & Selsey Tramway's lifting bridge at Hunston on the Chichester Canal in 1897

The Boat Museum Archive

to the canal basin. Here you will find a chandlery and a café-cum-bar which serves sandwiches and drinks. If you continue further along the remaining length of canal, you will pass a swing bridge named after Lord Egremont and then the final sea lock, Salter's Lock.

To return to Chichester, it is possible to devise a route which takes a signposted footpath across the entrance to the yacht basin and out alongside woodland and fields to New Barn. Bear left along a metalled lane to a road, turn right and then left along another lane to Apuldram. The road bends right past a manor house and then left. At the next left bend, there should be a public right of way across the field back into Chichester. When I was here, however, it was clear that walkers were not welcome and I was forced to continue along the road and on via a hazardous crossing of the bypass back to the railway station. This route is possible but frankly you would be better off relaxing in the bar for a bit longer and returning along the canal.

Further Information

The focus for all the activity along the W&A is of course the Canal Trust. It has undertaken clearance and dredging work, rebuilt locks and reconstructed bridges. The aim is not to rebuild the canal itself but to convince government at all levels of the feasibility of the project. It can be contacted via:

Mr J.R. Wood
24 Griffiths Avenue,
Lancing,
West Sussex BN15 0HW.

The Trust publishes a newsletter and a quarterly magazine. It also sells the Wey–South Path book.

The Chichester Canal has its own supporters who can be reached at:

The Chichester Canal Society,
Jaspers,
Coney Road,
East Wittering,
West Sussex PO20 8DA.

For detailed information on the history of all the waterways involved, the best reference is:

Vine, P.A.L., *London's Lost Route to the Sea*. David & Charles, 4th edition, 1986.

The Victorian guide to the canal is available in reprint:

Dashwood, J.B., *The Thames to the Solent by Canal and Sea*. Shepperton Swan, Shepperton, 1868, republished 1980.

APPENDICES

A: General Reading

This book can, of course, only provide you with a brief glimpse of the waterway network. Other authors are more qualified than me to fill the gaps and the following reading matter may help those who wish to know more.

Magazines

There are two monthly canal magazines that are available in most newsagents: *Canal & Riverboat* and *Waterways World*. Both have canal walks columns.

Books

There is a wide range of canal books available, varying between guides for specific waterways to learned historical texts. There should be something for everyone's level of interest, taste and ability to pay.

All the books listed here are available in paperback unless marked with an asterisk.

For a good introduction to canals that won't stretch the intellect, or the pocket, too far:
Smith, P.L., *Discovering Canals in Britain*. Shire Books, 1984.
Burton, A. and Platt, D., *Canal*. David & Charles, 1980.
Hadfield, C., *Waterways sights to see*. David & Charles, 1976.*
Rolt, L.T.C., *Narrowboat*. Methuen, 1944.
This can be taken a few steps further with the more learned:

Hadfield, C., *British Canals*. David & Charles, 1984; new edition, Alan Sutton, to be published 1993.

At least three companies publish boating guides:
Nicholson's Guides to the Waterways. Three volumes.
Pearson's Canal & River Companions. Eight volumes (so far).
Waterways World. Eight volumes (so far).

Readers seeking further walking books should look no further than:
Quinlan, Ray, *Canal Walks: Midlands*. Alan Sutton Publishing, 1992.
Quinlan, Ray, *Canal Walks: North*. Alan Sutton Publishing, to be published 1993.

B: Useful Addresses

British Waterways

BW is the guardian of the vast majority of the canal network and deserves our support. There are offices all over the country but the customer services department can be found at:

British Waterways,
Greycaine Road,
Watford,
WD2 4JR.
Tel: 0923-226422

Inland Waterways Association

The IWA was the first, and is still the premier, society that campaigns for Britain's waterways. It publishes a members' magazine, *Waterways*, and provides various services. There are numerous local groups which each hold meetings, outings, rallies, etc. Head office is at:

Inland Waterways Association,
114 Regent's Park Road,
London,
NW1 8UQ.
Tel: 071-5862556

Towpath Action Group

The Towpath Action Group campaigns for access to and maintenance of the towpaths of Britain and publishes a regular newsletter.

Towpath Action Group,
23 Hague Bar Road,
New Mills,
Stockport,
SK12 3AT.

C: Museums

A number of canal museums are springing up all over the country. The following are within reach of the area covered within this book and are wholly devoted to canals or have sections of interest:

THE NATIONAL WATERWAYS MUSEUM
Llanthony Warehouse,
Gloucester Docks,
Gloucester,
GL1 2EH.
Tel: 0452-307009

THE CANAL MUSEUM
Stoke Bruerne,
Towcester,
Northamptonshire,
NN12 7SE.
Tel: 0604-862229

THE NATIONAL MARITIME MUSEUM
Exeter Quay and Docks,
Haven Road,
Exeter,
EX2 8DT.
Tel: 0392-58075

INDEX